KEPLER

KEPLER

MAX CASPAR

TRANSLATED AND EDITED BY
C. DORIS HELLMAN

With a New Introduction and References by
Owen Gingerich

Bibliographical Citations by
Owen Gingerich and Alain Segonds

DOVER PUBLICATIONS, INC.
NEW YORK

Copyright

Published in Canada by General Publishing Company, Ltd., 30 Lesmill Road, Don Mills, Toronto, Ontario.

Published in the United Kingdom by Constable and Company, Ltd., 3 The Lanchesters, 162–164 Fulham Palace Road, London W6 9ER.

Bibliographical Note

This Dover edition, first published in 1993, is a corrected and expanded edition of the work originally published in 1959 by Abelard-Schuman Limited, London. New to this edition are the Introduction to the Dover Edition, Bibliographical References (which replace the section of that name in the 1959 edition), Bibliographical Citations and Index of Subjects and Places.

Library of Congress Cataloging-in-Publication Data

Caspar, Max, 1880–1956.
    [Johannes Kepler. English]
    Kepler / by Max Caspar ; translated and edited by C. Doris Hellman; with a new introduction and references by Owen Gingerich and bibliographical citations by Owen Gingerich and Alain Segonds.
        p.   cm.
    Originally published: London ; New York: Abelard-Schuman, c1959. (Life of science library ; 36)
    Includes bibliographical references and index.
    ISBN 0-486-67605-6 (pbk.)
    1. Kepler, Johannes, 1571–1630. 2. Astronomers—Germany—Biography. I. Hellman, Clarisse Doris, 1910– II. Title.
QB36.K4C33   1993
520'.92—dc20
[B]   93-14292
                                           CIP

Manufactured in the United States of America
Dover Publications, Inc., 31 East 2nd Street, Mineola, N.Y. 11501

INTRODUCTION
TO THE DOVER EDITION

IN 1959 the distinguished novelist Arthur Koestler surprised his readers with *The Sleepwalkers*, a departure from his accustomed literary genre. In a nonfictional book that explored the psychology of creativity and discovery, Koestler composed as his centerpiece an engaging, and sometimes exasperating, biography of Johannes Kepler. Koestler had long held an interest in scientific matters, having worked before World War II as science editor for a newspaper chain. When he learned about Kepler's fascinating personality and remarkable achievements, Koestler became increasingly vexed that the public knew so little of the German astronomer's life and works. While "Kepler and Galileo were the two giants on whose shoulders Newton had stood," why was "the name of one of the giants familiar to every schoolboy, but the second known to only a small number of intellectuals?" Koestler deliberately set out to rectify this imbalance with a polemical work designed to supplement the meagre information then available in English about Kepler. The appearance of Koestler's well-written account sparked considerable interest in the sixteenth-century astronomer-cosmologist.

Before the year was out, however, Koestler's readers were able to discover the source of much of his information: Max Caspar's *Kepler* was finally published in English translation. Caspar was eminently qualified to write the standard biography. Like Kepler himself, Caspar was born in southern Germany (in 1880), had been trained in both theology and mathematics, and had studied at the University of Tübingen. In 1923 he published a German translation of Kepler's *Mysterium cosmographicum*, and he followed this in 1929 with a German version of the important *Astronomia nova*, in which Kepler had enunciated his first two laws of planetary motion. He became a co-worker in Walther von Dyck's project to prepare a new collected edition of Kepler's works, and, as a preliminary, the two published a German edition of some of Kepler's letters. After von Dyck's death, Caspar continued amassing an archive of 15,000 photocopies of all known Keplerian manuscripts, and in 1937 he brought out the initial volume of the *Johannes Kepler Gesammelte Werke*. By the time he produced the first edition of his major biography of Kepler (in 1948), he had completed three more volumes of the *Gesammelte Werke* as well as a translation of

3

Harmonice mundi, and by then five more volumes of Keplerian correspondence were in the works.

It is interesting to compare Koestler's shorter biography with Caspar's, which has become the standard account of Kepler. With the sure hand of an accomplished novelist, Koestler rearranged and juxtaposed his materials, telling a compelling story but one strongly colored by his own particular predilections. Caspar's *Kepler*, too, has its drama: there are pages that amaze and passages that can bring the reader to tears. But his biases are more modest, and generally not enhanced by selectivity. Caspar's book is a straightforward march through the thickets of Kepleriana. Each of Kepler's books is described in perceptive, if not technical, detail. The religious turmoil of the age and Kepler's own anguish are graphically conveyed. Laid out before the reader is an extensive recounting of Katharina Kepler's witchcraft trial, which exhibits the whole range of human fear, greed, and stupidity.

For good reason, then, Caspar's *Kepler*, in any language, is the closest we have to a definitive account. Thus it is all the more strange that this treatment, so rich in extensive quotations, was prepared with virtually no footnotes or citations. Its translator, the late C. Doris Hellman, remarked on this deficiency, but elected to indicate only a few of the sources. For several decades I have been marking the references in the margins of my copy, and in preparation for this Dover edition, Alain Segonds and I have systematically completed this documentation. We managed to find essentially all of the appropriate sources—nearly 1200—and these have been listed by page and line in the Bibliographical Citations supplement at the end of the book. I have also included as Bibliographical References some of the major items in the growing literature about Kepler (this replaces the small section in the earlier editions), and, at the end, I have added a new Index of Subjects and Places.

The work of Segonds and myself was ably assisted by colleagues who found several of the more elusive citations, or drew our attention to relevant references. We particularly thank Volker Bialas, Mary Ellen Bowden, William Donahue, Judith Field, Friedrich Seck, and most especially James Voelkel, as well as the late Eric Aiton who made the new translation of Kepler's *Harmony of the World* available to us.

Harvard-Smithsonian Center for Astrophysics,   *Owen Gingerich*
September, 1992

Editor's Foreword

When Max Caspar died on September 1, 1956, the world lost the recognized dean of Kepler scholars. Nearly two-thirds of Professor Caspar's seventy-six years were devoted to assembling, cataloguing, describing, analyzing and editing the works of Johannes Kepler. Thirteen volumes of the monumental edition of Kepler's works, begun by the late Walther von Dyck and continued by Professor Caspar, have now been published and it is hoped that the other volumes will soon follow. This edition makes Kepler's works and Professor Caspar's valuable comments available to scholars the world over.

The definitive biography of Johannes Kepler which we now have before us first appeared in German in 1948 (reprinted in 1950 and in 1958) and is a by-product of Professor Caspar's immense erudition. It is unfortunate that he did not live to see its translation into English. Until 1948 the best biography of Kepler was that published in 1871 in Latin in the Frisch edition of Kepler's Opera Omnia, and which Professor Caspar's scholarship has done so much to supplement. Published in German, the new biography had but a limited audience. There are many readers of English for whom it was inaccessible and therefore it seemed eminently desirable to translate Professor Caspar's work. In doing so I have tried to retain the flavor of his language and I have not shortened the passages about Kepler's mother's witch trial and about the religious conflicts as I had originally intended to do. In fact, I have not altered anything but have put all my comments into the footnotes.

Professor Caspar, addressing himself to the layman, thought that footnotes would get in the reader's way, but I feel that the American or British reader would welcome the addition of certain bibliographical references as well as citations of recent articles. The added notes are clearly indicated as such and I am solely responsible for them.

Any faults in translation or interpretation are entirely my own. But much of what is good in the translation is due to a host of scholars on both sides of the Atlantic to whom I express my sincere thanks. My indebtedness to Professor Julius S. Held of Barnard College, Columbia University, must be singled out. His tireless advice, based on vast learning and a remarkable understanding of the nuances of both languages, has saved me from many serious errors. My husband, Morton Pepper, read the entire manuscript and I am, as always, grateful to him for his help.

C. Doris Hellman

TABLE OF CONTENTS

Table of Contents

III. IMPERIAL MATHEMATICIAN
IN PRAGUE 1600–1612

8

Table of Contents

IV. DISTRICT MATHEMATICIAN
IN LINZ 1612–1626

V. LAST YEARS IN ULM AND SAGAN AND
DEATH IN REGENSBURG 1626–1630

9

Table of Contents

REVIEW AND EVALUATION

I may say with truth that whenever I consider in my thoughts the beautiful order, how one thing issues out of and is derived from another, then it is as though I had read a divine text, written into the world itself, not with letters but rather with essential objects, saying: Man, stretch thy reason hither, so that thou mayest comprehend these things.

KEPLER,
in his calendar for the year 1604

PREFACE

WHENEVER Kepler's name is pronounced, many people prick up their ears. The world of scholars praises him for having opened new paths for astronomy by his discovery of the planet laws. The specialists are not alone, indeed, not even foremost, among those concerned with him. The circle of his admirers reaches far beyond them. It can be said that among the men whose genius enriched and deepened human knowledge by creative achievements in the area of exact science there is hardly one who enjoys the sympathy of as many as does Kepler, despite the facts that his principal field of activity is unfamiliar to most and that the result of his labors is difficult to understand and appreciate. It is the halo of his personality which draws many under his spell, the nobility of his character which makes friends for him, the vicissitudes of his life which arouse sympathy, and the secret of his union with nature that attracts all those who seek something in the universe beyond, and different from, that which rigorous science offers. In their hearts they all quietly bear veneration and love for this exceptional man. For no one who has once entered the magic sphere that surrounds him can ever escape from it.

In these circumstances, it is surprising that to this day we possess no biography of Kepler which, in extent and contents, meets the requirements which the scholar as well as the layman in astronomy has a right to expect. What has been published recently is slight and inadequate, forming a survey which merely hints at the great wealth of his life and work. A few earlier biographies, while wider in scope, are obsolete or incomplete, and occasionally one-sided in their approach. I refrain from enumerating these works. In the last volume of his valuable edition of Kepler's collected works (1858–1871), Christian Frisch offered the most comprehensive vita so far. Yet this is not a consecutive systematic biography, but rather a collection of material, assembled with great industry and arranged by individual years of Kepler's life. It proves to be a most useful mine of information for a biographer. The earliest presentation of Kepler's life is by Michael Gottlieb Hansch, the Leipzig scholar, in the introduction to his 1718 edition of a part of Kepler's correspondence. It forms, thenceforth, the basis of the later biographies. However, since then, research, particularly by Christian Frisch, but also by others after him, among whom Walther von Dyck may be singled out, has brought to light considerable new material which is of great importance for our knowledge of Kepler's life and works. Anyone, who does not assimilate the result of this research, easily falls into the danger of perpetuating and further spreading the errors which once crept in. The literature about Kepler contains false or at least slanted statements, which one author took from another because he neglected to go back to the sources. Furthermore—and this appears self-evident—in order to portray and evaluate not only Kepler's life but also his intellectual contribution, it is necessary to have studied at least his principal works, difficult as they are. Yet, it is easy to see that this condition has not been fulfilled by all authors.

No reproach is intended in these critical remarks. The extraordinary vastness of the material in the works and letters of Kepler and in the literature about him readily explains the type of fault described. It is necessary to be able to devote the labor of many years to the study of this material. I consider it an especial piece of good fortune that it was granted to me to be able to concern myself with the life and work of the great man for many decades, and for the past several years to be active, as editor, in the new edition of his collected works still in process. Therefore, I hope that the readers of this book will admit my qualifications to spread before them a survey of Kepler's life and work. What I offer to them has grown out of the studies which I have undertaken as my lifework. The opportunity to write the book presented itself to me when, after the (German) collapse, a pause occurred in the continuation of the edition of the complete works.

He who sets about describing Kepler's life from the sources, is surprised at the wealth of detail known to us. About four hundred of his letters are preserved. In his communicative manner, he often expressed himself in these about the events and moods, sorrows and joys of his life, side by side with scholarly discussions. In addition, the approximately seven hundred extant letters written to him throw valuable light on his interests and connections. Several hundred other documents, relating to his life, shed light on his financial situation, his professional activity, his relationship to the princes and authorities on whom he was dependent, as well as on some family events. Added to these are also the numerous more intimate memoranda which he made for himself and the many personal remarks which he allowed to slip into his published works. Since photographic copies of all these documents, in thousands of sheets, as well as Kepler's own works and, in a large measure, the literature about him, are at my disposal, I have been able to compose this book, although at the time the public libraries were not accessible to me.

It is on these documents above all, together with the works published by Kepler himself, that the present book is based. Nothing essential has been taken over from other writings without having been re-examined from the sources. For the literature used, I may refer to the Kepler bibliography which I edited in 1936, where the whole Kepler literature is listed. Although the book claims to satisfy the requirements of scholarly thoroughness, I have omitted to hang around it a cloak of learned apparatus in order to make it seem more important. The material for this is at hand, but since the book was not written for scholarly purposes alone, I wished to give it a form, in which the attention of the reader would not be continually interrupted by notes with critical discussions, individual analyses and references to literature. Similarly, I have omitted the references for the many quotations in which I let Kepler speak. They are, however, reproduced with the greatest possible accuracy, so that they can be trusted.

The wealth of the material at hand, satisfying though it is, nevertheless presents certain difficulties. It demands painstaking sifting and exceedingly judicious planning in organization and formulation. Without abandoning the main lines, I have enlivened the picture by including details. Also, it has been necessary to delineate as completely as possible those circumstances of Kepler's life which seemed to me important. Particular value has been placed on the description of the contemporary background and the interrelation of Kepler's personal life with political events and situations. Only thus does the figure of this man emerge clearly, a man who was able to master, in exemplary fashion, the adverse fortunes which he encountered.

But the description of the intellectual climate in which Kepler lived is of the greatest importance. It becomes evident that the origin and publication of his works are bound up and intertwined in an often remarkable manner with the outer developments in his very eventful life. This observation has induced me to combine the discussion of his works with the presentation of his life. I preferred to exhibit the whole figure as a living organism, with the interaction of its parts, rather than to cut out the individual limbs in anatomical manner and to prepare and present them separately as is often done. It is necessary to discover what Kepler thought and worked out in the changing phases of his life, what harassed him or made him happy, what advanced him and what hindered him, in what he succeeded or what escaped him. Life and work form a unity with him which must not be torn up, an organic creation which must not be plucked to pieces.

Special attention has been paid to the analysis of Kepler's two principal works, the *New Astronomy* and the *World Harmony*, in which he offered his planet laws. This seemed so much the more necessary since the material in the available biographies is in every way insufficient. Besides, it was my endeavor to weigh out correctly the description in its separate parts, by not stressing or setting aside what this one or that one held as important or unimportant; I rather gave the individual divisions the meaning due them in accordance with Kepler's own thought. Thus it was necessary to discuss his astrological activities again and again and to treat them with the emphasis which they occupied in his thinking. He who believes it possible to pass over them, with a few kind apologies, distorts the picture. Here it is a question not of our opinions on the subject, but his. In accordance with this principle, his religious convictions, as well as the complex of the questions on harmony, required exhaustive treatment.

I am completely aware that something of the spirit of the author enters into every biography. Each writer presents his hero as he sees him in the mirror of his own ego. One must be alert to the danger that lies therein, and suppress one's own person in order not to introduce false traits into the picture. When, however, one occupies oneself over a period of many years, as I have had the good fortune to do, with the man whom one wishes to serve, a certain conformity in character and thought appears which helps one to see correctly. In this sense, I should like to hope that the love which I bear Kepler has not prevented me from drawing a true picture of him.

In the composition of this book, Miss Martha List, co-worker in the edition of the collected works of Kepler, has given me the utmost support. With her excellent knowledge of the manuscript sources, she rendered me valuable service in the choice, sifting and assembling of the material. I benefited as much from her active concern for the whole as from her excellent hints on details, when in continuous daily conversation we reviewed the sections of the book. For this I should like to express to her my most heartfelt thanks.

Many thanks are also due to Dr. Fritz Rossmann, who took the trouble to read one proof.[1]

München-Solln, July, 1947   *Max Caspar*

---

[1] ED. NOTE. One proof of the German edition.

INTRODUCTION

1. *Philosophic and scientific thought in the Renaissance*

IT was in a spiritually and politically rent and riven period of German history that Johannes Kepler carried out his lifework. The year 1600 divides the time of his earthly sojourn almost in half. One needs only to think about the fact that his life extended for twelve years into the tragic Thirty Years' War, in order to understand that for him, as for everyone who played any part on the stage of the great world, it must have been a life full of unrest and worry. The war, which did not occur by mere chance, had cast its ominous shadows for a long time. Even if the decisions of the leading statesmen largely determined the course of the historical development and even if everything could have turned out differently, had one or another of them felt or thought differently, nevertheless, they all were influenced by the predominating contemporary views and tendencies. They thought and acted in accordance with the concepts of the world and life accepted by that epoch.

These tendencies and concepts, as well as the course of political events, must be known, at least in their general outline, in order to understand and appreciate Kepler's life and work, the deep tragedy of his personal fate and the shining triumph of his intellectual activity. He, the unpolitical one, had to learn that political events intervened in his life more than he would have wished. He had close relations with some of the main participants and saw himself lifted by fate into positions which were lapped by the waves of political events. In a famous university whose influence shone far, he absorbed what the intellect of his time had to offer. By the alertness of his mind he perceived the forces which formed the spirit of the age and found the direction in which he had to guide these forces to discover new land. And since his inner life, in the last analysis, fed on the sources which religion unlocks, he found himself in the midst of the denominational fights which above all characterized his times. The great war was, surely, in its first phase a war of belief, grown out of the unbearable tensions which had formed among the various sects.

In our description of Kepler's life we shall come across details about the course of political events and about a good many tendencies which were then at work. Nevertheless, to prepare for a better understanding we might make a few introductory remarks about the intellectual situation toward the close of the sixteenth century, at least in so far as it deals with the provinces which have importance for Kepler's life and work.

For approximately the two hundred preceding years, a profound change in philosophical and scientific thought had taken place. Scholasticism, which reached a peak in the magnificent system of St. Thomas Aquinas, had seen its main task in developing, systematizing and penetrating intellectually the lofty truths of Christian teaching, so far as this is possible for the human mind. It had admirably solved this problem for its own time, and not only for its own time; but in its subsequent development it degenerated more and more into subtle speculation, which could no longer satisfy open-minded and independent thinkers. These felt themselves entwined and caught in a system of conceptual construction which chained their intellect. The authority of Aristotle, who since the peak of scholasticism had reigned over both the philosophical and the physical domain, had in a manner been intensified, so that it was believed that finding and demonstrating the truth called for and demanded supporting a thesis with citations from the philosopher. As time went on, this difficulty became unbearable and it was necessary to find a way out. In this situation the ever restless and inquiring spirit turned toward the contemplation of nature and man's place in it. An empire full of riddles and secrets opened invitingly before man—a new world, a cosmos of astonishing beauty, an apparent perplexity of hidden connections and associations, behind which he sensed and felt a grandiose order. Not that men had previously been completely neglectful of nature and blind to her strength and stature or that they now wished to loosen the tie to God and the supernatural. But formerly the desire had been to comprehend nature from within, or, if you like, from above as a whole, always mindful of the other-worldly destiny of man. Now men's eyes were turned to the fullness of the facts which were nevertheless considered as a work of the omnipotent and infinitely good Creator. If man had previously looked down, as it were, from the other world upon the earth and the whole material world, he now placed himself inside of things and looked from these up to the heaven. The difficult point in the thinking resulted from the supernatural in nature. Together with the disclosure of God by the word came the disclosure of God in the work; together with the book of Holy Scriptures came the book of nature, the explanation of which would be considered man's great duty. It was the concern of the theologian to interpret God's word. But the investigation of God's work was for the thinkers, who ardently depended on the phenomena of nature. A secularization of learning and philosophy began to appear and the new aim helped in the gradual liberation of man from the authority of the Church, by which the spiritual life had hitherto been held together.

However, this new pursuit was not yet science in the sense in which we now understand it. The unutterable patience and toil required to discover the secrets of nature by experiment and observation were still unknown. The concept of laws of nature that establish causal relationships between phenomena and put them in formulas was not yet held. Men had not yet learned the inductive method, by which consequences are drawn from an hypothesis, consequences which must be tested by experience in order to prove the correctness or at least the probability of the hypothesis. How, then, was it possible to get correct answers from nature before one had learned how to put the correct questions to her? So, in the first instance, the concern was not actually with science but rather with natural philosophy. Men wished to learn immediately what held the world together in its very depth. They sensed the order and called it harmony. They speculated about the Earth Soul and the World Soul, about sympathy and antipathy between objects, about the elements and vital spirits, about the macrocosm and microcosm. They did not think so much in terms of causes as of ends. They asked themselves how knowledge of nature is possible and in what it consists. Their minds were affected by the magic of Platonic and Neo-Platonic philosophy. For many, Plato and Plotinus replaced Aristotle. They became enthusiastic about the thought that God made the world as beautiful as possible, and they admired in the Platonic ideas the thoughts of God which came to light in the perceptible phenomena.

So it is a very variegated picture in its multiplicity of efforts and directions which is offered to us by the scholarly thought of that period of time which we are accustomed to call the Renaissance, as can be seen from these short references. We would greatly overstep the borders of this introduction, were we to enumerate the names and accomplishments of the men who individually molded, or carved, the spiritual form of their age. To name Nicholas of Cusa and Paracelsus rouses a wealth of thoughts not easily caught in a few words and recalls a period in which each thinker builds his own world for himself, each roams and feasts in his fantasies and judgments, or in what he regards them to be, each tries to seize the truth by another corner. Old and new intermingle. One swears by Plato, another by Aristotle, a third looks for a synthesis of the two. Scholasticism still holds its place for a long time, and its abstractions perform excellent services. Alchemists and astrologers dig for new treasures of knowledge. We shall see how these various tendencies cross each other in Kepler's thoughts. He was possessed and enchanted by the idea of harmony; he erected an astrological system for himself on the basis of his psychology; he held the thought that a World Soul embraces and professes Plato's idealistic theory of knowledge. But he shows himself also schooled in the spirit of scholasticism, advocates its theory of perception, uses its fundamental notions in the conception of organic growth and at every opportunity pushes his views on the track of Aristotle's doctrine of material and form. Yet he determinedly turned against Aristotle's physics and in this field proceeded on new and promising paths.

2. Revival of astronomical research; Copernicus

Astronomy drew the first and most important advantage from this turn to nature. The impulses came from various directions. In the world of stars, aesthetic metaphysical thinking was faced with a realm of nature which, because of its impressive beauty was designated, with a special emphasis, as cosmos. Even in antiquity, thinkers ardently desired to fathom its secrets. Now that such aesthetic-metaphysical considerations were revived, minds were challenged when they realized how the continuous flux of earthly phenomena, with their growing and withering, their birth and death, was confronted with unshakable stability and permanence in the skies, and how endless variety here below was confronted by eternal harmony and simplicity in the firmament. Did not harmony, which in the rest of nature was hidden under an almost impenetrable veil, seem to shimmer through here? Did not, perhaps, precisely that which should be meant by harmony, disclose itself here, a structure of significant numerical proportions? And was this radiant world, so far beyond the reach of human beings, not a reflection of Divinity itself, of the fountainhead of harmony, so that man may feel himself closest to it through the contemplation of the stars?

Yet, from other sides also, came stimulations. Practical exigencies asserted themselves. It had long been apparent that an improvement of the calendar was necessary. The men who at that time set out on venturesome journeys to discover new lands both asked for and needed help from the astronomers for geographical topography. And not the least significant in stimulating the struggle for closer investigation of the motions of the planets was the belief in the influence of the heavenly phenomena on earthly occurrences. The desire to lift, in any way whatsoever, the curtain which concealed the future, always formed a strong impulse for the exertions of man caught up in "world-fear."

Hitherto the heavens had been thought of as composed of crystal spheres, hollow globes touching one another, which carried the fixed stars and the separate planets. Aristotle had contrived a system consisting of a large number of such spheres in order to explain the motions of the celestial bodies, especially those of the planets with their irregularities. Beyond the sphere of the fixed stars was assumed the empyrean, which in the Christian Middle Ages, as well as with Dante, represented the place of the departed. It was believed that from there the parts of the universe descended in rank step by step down to the earth, which held the lowest place. The individual spheres were supposed to be turned by angels or other spirit-beings. Then was recalled the great accomplishment of Claudius Ptolemy of Alexandria, who in the second century after Christ had worked out an admirable system for the calculation of the motions of the heavens without using such spheres. Among the numerous Greek manuscripts which had come into the western world, especially to Italy, after the conquest of Constantinople by the Turks, there was found one of Ptolemy's main works which is usually entitled *Almagest* and which had hitherto been known only through a Latin translation from the Arabic.[1] The preoccupation with this work gave considerable ascendency to the awakening interest in the study of the heavens. Men did not stop at assimilating its contents. Attempts were also made, by means of observations with simple instruments, to bring the numerical data of Ptolemy into closer accord with the actual phenomena and to perfect the necessary computations. In the forefront of the men taking part in the reawakening of astronomical studies stood Georg Peuerbach from Upper Austria and his pupil, Johannes Müller, called Regiomontanus after his birthplace, Königsberg in Franconia. In spite of the fact that both men died young, through their indefatigable activity, they exercised an extremely effective and far-reaching influence in Germany and Italy. The invention of the art of printing was highly conducive to their undertaking.

But a greater man was still to come, who not only repaired the old view of the world but also opened the gate to a new one. It was Copernicus who brought the change and thereby erected a milestone in the development of Western thought. He was chosen to shake the world. Like the seafarers of his time, he left the path in which the thinking of his time moved. He shifted the tiller with dauntless grasp

[1] ED. NOTE. For information concerning translations of the *Almagest* from the Greek (*Syntaxis*) in 1160 and later, see C. H. Haskins, *Studies in the History of Mediaeval Science*, 2nd ed., Cambridge: Harvard University Press, 1927, pp. 104 ff., pp. 157 ff.

and, led by his genius, followed a new direction to a promised new land. For several decades he wrote and polished his masterpiece: *De Revolutionibus Orbium Coelestium*, on the revolutions of the heavenly spheres. It was published the year of his death, 1543. As everyone knows, in it he placed the sun in the center of the universe and let the earth, as one of the planets, revolve around the sun and turn on its own axis. He could show that with this assumption the motions of the celestial bodies permitted of the simplest explanation. And since nature loves simplicity, he held fast to this assumption despite all objections which he himself had to make in accordance with the thought of that time. Could he foresee what revolutionary consequences for further development would result from his conception?

Like everything new, provided that it is really great and pregnant, the work of Copernicus was widely challenged. It was discussed everywhere, and people joked about the foolishness of the new contentions. Here and there a man of sense, who occupied himself more seriously with the new teaching, lifted his voice in its favor. For the most part, however, the recognition concerned, not that in which we today perceive the kernel of the Copernican theory, but rather the new type of astronomical calculation which the master had introduced. On the whole, the excitement was not very great and the interest sustained itself only in learned circles. The objections which were raised came from various sides. The theologians above all categorically rejected the theory of the motion of the earth, because they considered it contrary to Holy Writ. Luther expressed adverse opinions about Copernicus, and Melanchthon even felt that the interference of the supreme power of the state against the innovation was indicated. The Catholics held back—after all Copernicus had dedicated his work to Pope Paul III. The conflict with the Catholic Church was not kindled until much later. The physicists pointed to the flight of birds, the motion of the clouds, the perpendicularly falling stone and similar things in order to disprove the rotation of the earth. The thought, that everything which lies within the reach of the earth's attraction takes part in its rotation, lay entirely outside their comprehension. Besides, they were entirely entangled in the Aristotelian theory of heaviness and lightness. But even the astronomers could not make friends with the new theory. It did not simplify their task, of calculating in advance the positions of the planets, which they considered the summit of astronomical investigation. Thus they could not decide to abandon their trusted views and methods of calculation for the sake of a theory which contradicted the appearances of the senses, placed high demands on the imagination, and excited the opposition of the theologians and physicists. Besides these objections, a great number of other arguments against Copernicus were put forward. These prove how basically different is our present view from the paths in which thinkers at that time moved. The opposition to the new theory is more easily understood if it is realized that Copernicus could not introduce a real proof for his opinions. There first had to be someone who had the strength to push through the whole thicket, who was able to refute or shove aside the objections, who grasped and recognized the substance and the potentiality for development in the Copernican theory—that here it was a matter of more than a new method of calculation, namely the setting of a new goal for the study of the heavens, a new fashioning of the picture of the universe. We shall see how Kepler felt called to this task and one of the principal aims of this biography will be to show how he met this summons. We shall become acquainted with the triumph accorded his efforts to extract from the Copernican conception what was latently embedded in it.

3. *Religious conflict in the sixteenth century*

But now we must turn to the circumstances out of which grew the tragedy of his personal life. They are concerned with the denominational conditions and the state of church politics that had developed from and since the Reformation. The movement led by Luther had stirred up the people more deeply and more radically than the changes in scientific thought, where it was a question of a development which took place among the intellectual leaders and whose influence sifted down but slowly and gradually. It is as though a person of ripening age alters his point of view without realizing it and finds himself pushed into a new province of intellectual activity. It is impossible to say, there, on this day, the new burst forth. The Reformation, however, was a storm arising suddenly, a revolution stirring all ranks, high and low, intellectual and non-intellectual, into powerful surging. It was not a matter of sun, moon and stars, nor of the priority of Aristotle or Plato. The call which sounded hit the very heart of the people. They feared for their salvation, their deepest and last concern, and in the knowledge of their sins longed for deliverance and wrestled for their justification before God. It was more than dissatisfaction with the abuses reigning in the Church which aroused so strong a reverberation of the new proclamation in the widest circles. Had not a deep religious sense been imbedded in the people, the change in beliefs could not have spread to that extent. Luther placed justification in belief alone, without the works. By repudiating the sacramental priesthood and eliminating the priest as dispenser of divine mercy, he placed man directly face to face with God, before whom, according to his conscience, he had to answer for his moral actions. Luther repudiated the clerical office of teaching and announced the freedom to expound the Scriptures. He shattered the hierarchical order and gathered the faithful into an invisible congregation. With all the theses he placed himself in sharp opposition to the old Church which, despite all tensions and wars, had up to then preserved the unity of Christianity. The storm, which he initiated, carried him further than he had originally wished. He found that only too soon very mundane interests mingled with the striving for religious revival. The word of the freedom of the Christian person rang temptingly in the ears of the many discontented and led to conclusions of which its announcer had not thought. Still more disastrous evils were the greed for power and the avarice of the princes, who forthwith perceived what advantages the new state of affairs brought them. The dissension became a lengthy one, to the detriment of Germany and to the suffering of all who recognize and worship Christ as the Saviour of the world.

Our introduction is not the place to set forth the dramatic course of events full of discord, the arguments and negotiations, the contradictions in theory, the political chess moves, the deeper background of the happenings during those decades, and even much less to deliver a judgment on the Reformation movement. The events are known and the judgment may be left to the reader. However, it is necessary to discuss several points of dogma and church policy with which Kepler was concerned so that we may understand and evaluate his adventures. They were conditioned and determined by the confused era as well as by his inner attitude toward the questions of faith, which in connection with the circumstances of the times showed him his difficult path.

The Augsburg Confession stands first among the symbolic books on which the Reformers anchored their teaching in opposition to the teaching of the Catholic Church. After the religious schism reached full expression at the Reichstag in Speyer, it was the task of the Reichstag, which started at Augsburg in 1530, to bring the disunited together. In order to create a base for the negotiations, the Protestant princes submitted that confessional formulation in which the fundamental points of Lutheran doctrine were established. Melanchthon, who had drawn it up or at least edited it, in accordance with his mild and more conciliatory attitude, chose a form of presentation which kept the points of disagreement in the background, and he preferred phrases which could more readily be brought into harmony with the Catholic dogma. But the harsh contrasts, which, after all, existed, could not be disposed of by such a course and came clearly to the fore with the ensuing negotiations. The desired unity could not be attained. We shall see how Kepler, whose manner was akin to that of Melanchthon, always professed himself faithful to the Augsburg Confession.

The development of the dogma did not cease with the writing of the Confession. In addition to the opposition of the Catholic Church, another contradiction soon showed itself and still further stirred up Church affairs in Germany, subsequently leading to sharp cleavages and conflicts. At about the time of Luther in Germany, Ulrich Zwingli in Switzerland took up the fight against the old Church and attacked its dogma and discipline most violently. While both reformers went the same way in most of the essential points and were united in their opposition to the Catholic Church, they differed sharply in their teaching about the Lord's Supper. Although unity between them could not be achieved, their antagonism did not halt the continuation of the work of the Reformation in Germany. The situation changed, however, when several years later Calvin raised his tyrannical church army in Geneva and, as a third reform leader, spread his teaching. His doctrine of communion also differed from Luther's. The dispute about the sacraments burned most fiercely. The Calvinistic doctrine found entry into Germany. In the year 1562, the Prince Palatine, Frederick III, introduced it as the foremost in his country. In the following decades other princes of the realm joined. Even Melanchthon inclined toward the Calvinistic doctrine of communion. Through his authority, especially after Luther's death, it was further spread, particularly in Saxony. The intense anger of the old Lutherans turned against the "Cryptocalvinists" or "Philippists," as the followers of Melanchthon were called. Today it is difficult to picture the passion and acrimony with which the adversaries declaimed against one another. Those who agreed with the Augsburg Confession had equal hate for the Calvinists and the followers of the Pope. In 1576–1577, in order to erect a dam against the hated Calvinist teaching, Jacob Andreae, the Tübingen theologian, and a few other men of similar ideas, composed a new work about the sacrament, the so-called Formula of Concord, in which the Lutheran doctrine was fixed in all its sharpness. However, the dissension did not end here, because not all the princes devoted to the Reformation accepted the formula. Therefore, the recognition of the Formula of Concord was more relentlessly promoted in all the countries which followed the Augsburg Confession, including, above all, Kepler's native land, the duchy of Württemberg.

The chief stumbling block was the interpretation of the Lord's Supper. The Catholic Church, following the biblical words instituting the sacrament, insisted, according to the doctrine of transubstantiation, that during Mass the substance of the bread is changed into the substance of the body of Christ. However, Luther, who repudiated Mass and rejected transubstantiation, still held to the real presence of the body and blood of Christ in the communion celebration. In place of transubstantiation he accepted consubstantiation, that is the sacramental permeation of the remaining bread substance by the substance of the body of Christ. In order to found this doctrine in opposition to the objections of the reformed theologians, he established the dogma that Christ, by virtue of the hypostatical union, that is, the union of the human and the divine nature in one person, is also omnipresent in body. This remarkable doctrine of ubiquity, untenable on the grounds of the traditional Christology and given up again later by the Lutheran theologians, formed a cornerstone of the Formula of Concord. Calvin rejected it. According to him, the believer in the sacrament indeed partakes of the body and blood of Christ, however, in such a manner that simultaneously with the partaking by mouth of the material substances, which in all respects remain what they are and only denote the body and blood of Christ, the spirit is offered a strength which flows out of Christ's body, present only in heaven. Pursuant to his fearful theory of predestination, according to which, part of mankind, without any consideration of works, is preordained by God to everlasting damnation, Calvin inserted into his theory of the Eucharist the addition that only the chosen, by partaking of the communion, shared in the partaking of the body of Christ.

It was these very disputes which lay as an oppressive burden on the whole life of Kepler.

In respect to church politics, the Religious Peace of Augsburg of 1555 assumed a central position in the history of the Reformation of the sixteenth century. It was no longer a question of reconciling the alienated parties. The Protestants had already gained such strength that all one discussed was a peace in order to restore a bearable situation for the followers of different faiths who lived together. In accordance with the decrees of the Reichstag, the Estates of the Reich were given the choice between the Catholic and the Augsburg creeds. But still more—the decision of the Estates of the Reich was to be valid for the whole extent of their dominion. With that the principle "cuius regio, eius religio" (to whom the land, to him the religion) became law. By this, to our present way of thinking completely monstrous legal position, the sovereign was made lord over the innermost domain of the human heart. Freedom of conscience was at an end. The sovereign ordered, and the subject had to believe what pleased the master. A subject unwilling to submit could emigrate. That right was explicitly reserved for him. Imagine into what conflicts of conscience fell even the best who took their religion seriously. They were confronted with the choice, either of leaving home and property or abandoning what was most sacred to them. It is to be noted that the choice of creed did not apply to Calvinism. In the imperial cities, both the Catholic and the Augsburg creeds were allowed to remain if, in the past, they had been practiced side by side. In the following years, the Protestants realized the greatest benefit from the new regulation. The Catholic Church remained, to begin with, in the defensive position into which it had been pushed some time before. Only toward the end of the century, just at the time when Kepler entered public life, did the Church, with the help of the Jesuits in the so-called Counter Reformation, set to work to win back its lost positions.

Such, then, is the period we enter into when we prepare to review Kepler's life, from its very beginning. An immense number of princes and other Estates of the Reich loudly asserted their rights. Some were Catholic, others Augsburgian, the third group Calvinist. Each church claimed to be in possession of the faith which alone can save. To the already existing political antagonisms were added the more dangerous and more serious religious ones. What was left now of the freedom of conscience which Luther had proclaimed? What of the idea of an invisible community of believers which was present in his mind? The claim to authoritative leadership on the part of the Church, against which he had so passionately fought in the old Church, had again risen up in his own ranks. An oath on the book of Confession was demanded and carried through in the Protestant lands with the same severity with which the old Church proceeded in matters of faith. The place of the bishops was taken by the rulers whose power was thereby considerably increased. On every side the positions stiffened. The Jesuits were at work to set up again, in its old magnitude, the Catholic Church, purged, renewed and strengthened by the Council of Trent. Everywhere there existed tensions, antagonisms, frictions, explosive situations. In the face of the increased power of the princes, the power of the emperor was lessened and endangered. The centrifugal forces were stronger than the power of order. In the east the Turks at the same time persisted in ever new attacks against the borders of the Reich. In the west France waited impatiently for an opportunity to make use, for her own ends, of the weakness of the imperial rule. What could still come out of this? A time fraught with disaster, a time in which one would gladly flee to the stars in order to find home and security there.

I

CHILDHOOD AND YOUTH

1571–1594

1. Birth and ancestry

THESE were the times into which was born the first child of
Heinrich Kepler and his wife Katharina, née Guldenmann. His
birth took place in the little Swabian imperial city of Weil, today
called Weil der Stadt, on Thursday, December 27, 1571, at two-thirty
in the afternoon. The infant was baptized Johannes after the saint of
the day, the Apostle John.[1]

The Kepler family,[2] from which the child was descended, had been
located in Weil der Stadt for about fifty years. About 1520 the great-
grandfather of Johannes, named Sebald, emigrated from his native city,
Nuremberg, and settled in Weil der Stadt. He belonged to the craftsman's class and was a furrier by trade. The family which he founded in
his new dwelling place was very numerous. His sons soon gained respect
by their ability. Several were council members and the second of
them, named Sebald like his father, became mayor and administrator
of trusts in the city. His marriage to Katharina Müller from neighboring Marbach was richly blessed with children. The fourth child,
Heinrich, was the father of our Johannes. Like his wife he was twenty-
four years old at the birth of their first child. Johannes' mother was the
daughter of Melchior Guldenmann, the innkeeper and village mayor

[1] AU. NOTE. It is superfluous to enter into the controversy whether Kepler was born in
Weil der Stadt or, as has been asserted, in the nearby village of Magstadt or in the neighboring town of Leonberg. Kepler himself so clearly and certainly testified to the fact that
Weil der Stadt was his birthplace and the grounds which are brought forward for the
other places are so easily disproved that there is no basis for doubt.

[2] AU. NOTE. The question of whether the spelling of the name "Kepler" or "Keppler"
is correct is meaningless. Both spellings are equally justified. In that period such differences
in spelling a name were disregarded, and our astronomer himself is not consistent in this,
since he sometimes wrote his name "Kepler," sometimes "Keppler" or even "Khepler"
and "Kheppler." In the frequently used Latinized form he always wrote "Keplerus." The
statement occasionally made that he always used the written form "Kepler" in Latin texts
and, on the contrary, in the German always "Keppler" is completely false; many German
documents signed "Kepler" can be found. Others also wrote his name "Köpler."

in nearby Eltingen. The family tree can be traced still further back. The father of that Sebald Kepler who moved to Weil der Stadt was the master bookbinder in Nuremberg, Sebald Kepner. Thus, not "Kepler," is his name spelled in Johannes Kepler's hand, in a document of a later date on which the above genealogical particulars rest. This was an arbitrary oral modification of the old family name Kepler, perhaps in assimilation of the name Kepner, frequently seen in the fifteenth century in Nuremberg.

While the immediate ancestors were craftsmen, if we go still further back in the family history, we get a different picture. Sebald Kepner or Kepler, the Nuremberg master bookbinder, was of noble descent, but driven by need he put aside the nobility and entered the craftsman's class in Nuremberg; perhaps the change of the name Kepner to Kepler is connected with this change in social position. According to a tradition worthy of belief, this Sebald was a son of Kaspar von Kepler, who toward the end of the fifteenth century appeared in Worms as court post-equerry. And this Kaspar von Kepler was in turn a son of Friedrich Kepler, the warrior who was knighted by Emperor Sigismund on the Tiber bridge in Rome at Whitsuntide, May 31, 1433. Not only did Johannes Kepler explicitly certify the report of this rise in rank later when, without boasting, he cited it in opposition to a Venetian nobleman, but it is much more amply corroborated by the patent of nobility of the year 1433, still extant in the Vienna peerage archives, according to which the brothers Konrad and Friedrich Kepler, because of their military services in the army of the emperor, had been distinguished in the way described. In this patent of nobility the coat of arms of the Kepler house also experienced a corresponding embellishment. The shield of the coat of arms is divided into a yellow upper and a blue lower field. In the upper field can be found the half figure of a red-robed angel with golden wings, who places his hand on the division line. On the helmet with red-yellow cover stands a blue-bordered yellow peaked cap, which is adorned on top by a yellow-blue-red padded roll, and from which issues a black heron's feather tuft on which is scattered golden tinsel. Upon request, this coat of arms was confirmed to the grandfather Sebald and his brothers in the year 1564 by the emperor. Johannes Kepler was in the habit of sealing with this coat of arms. It cannot be stated where that knightly ancestor, Friedrich, had his home and property. From the remark of our Kepler that Friedrich was knighted "with other Swabian knights" by Emperor Sigismund, it might be concluded that the home was in Swabia. Yet such a far-reaching conclusion should scarcely be drawn from this remark. From the explanation in the letter of nobility, that the emperor wished to treat with special distinction those men "whose forefathers had always shown themselves useful to the Holy Empire," it follows that their distant ancestors were confirmed as valiant bondsmen, as, indeed, old documents report various heroic deeds by bearers of the name "Keppler" or "Kappler" without its being established that these men belonged to our Kepler family. Nor is it certain that Friedrich Keppler, the thirteenth century noble who originated from Salzburg, was a member of the family. A document in the Vienna archives of nobility reports that Friedrich had proved himself in wartime and peacetime by his bravery and faithfulness; it is noteworthy that this nobleman likewise carried an angel on a shield as a family coat of arms. That soldier's blood flowed in the veins of Kepler is also confirmed by the tale that both the great-grandfather, Sebald, and subsequently the grandfather, Sebald, were supposed to have received privileges for military laurels earned under the banner of Charles V and his successors. We do not know what moved the great-grandfather, Sebald, to transfer to little Weil, leaving Nuremberg, where applied art and industry had created a brilliant tradition and promised a capable man rich opportunities. Did he visit Weil der Stadt on his travels and get stuck, or had he perhaps been induced by relatives in that town to settle there? At any rate, it is noteworthy that evidently toward the end of the fifteenth century bearers of the name Kepler were already living in Weil as can be proved by the registers of the University of Tübingen. Further information about this, as well as some other details worth knowing concerning the Kepler family history, so far as they deal with Weil der Stadt, cannot be established since the relevant archives are no longer extant. They "vanished in dark smoke" at the end of the Thirty Years' War when the French, in October, 1648, just at the time when the Westphalian peace was signed, besieged the city, setting it afire. A large proportion of the buildings was reduced to ruins and ashes and both the parish registers and most of the archives went up in flames.[1]

---

[1] Au. Note. The information about Kepler's ancestors so far as he himself does not attest to it directly, is based on the handwritten document entitled "Genealogia Keppleriana" that his grandson Johann Jakob Bartsch left and which already has been exhausted by M. G. Hansch for his vita. Although Bartsch first came into the world when his grandfather already had closed his eyes forever, the information given there may be trusted, even though certain scruples cannot be entirely suppressed. The contents of this piece of writing certainly go back indirectly to Kepler, who had shown himself a faithful guardian of the family tradition. Later researches could add nothing essential. The credibility of the information, which also contains details about the military accomplishments of the ancestors, will, after all, not be harmed by the fact that Kepler, in the above cited place, erroneously gives Heinrich instead of Friedrich as the name of the knighted ancestor.

2. *Weil der Stadt*

Small and crowded together though Weil der Stadt was, its inhabitants, nevertheless, were always inspired with self-esteem and pride in their independence, as privileged citizens of a free imperial city. Founded by the Hohenstaufens, this little city received imperial freedom toward the end of the thirteenth century after the interregnum under Rudolph I. An idea of the city's earlier appearance in Kepler's time can well be formed from the picture that the city presents today. The little streets, the spacious market place, which is surrounded by high gabled houses, the towers and gates of the city wall, which is still preserved to a considerable extent, are seen, as formerly, grouped in friendly fashion. In a rolling landscape on the edge of the Black Forest, surrounded by gardens and meadows, fields and woods, the place was built up against a gentle slope which falls away toward the broad valley of the little river Würm. The crown of the picture and its loveliest jewel is the three-steepled Gothic church, situated on high and visible from afar, towering above the tangle of roof-tops like a magnificent minster. It gathers the houses around it and takes them under its care, as a brood-hen does her chicks, an eloquent witness to the pious soul of the ancestors, who well knew what to make the focal point of their existence. With Swabian diligence, the inhabitants attempted to keep their city in neat order and, in a democratic spirit, tried to protect its rights. Mostly peasants and craftsmen, among whom the leather dressers and weavers stood out, they had to direct their cares and hopes to the exigencies of life. They let the path of the sun, the moon and the stars go just as it went, and advanced science lay beyond their intellectual horizon, although some clever heads went forth from the city. Since the community in that earlier time consisted of only about two hundred burghers and their families, the free imperial city of Weil had no weighty voice in the state proceedings in the Holy Roman Empire. When once in a century an emperor came to visit, it was an event that was carefully recorded in the annals. What otherwise aroused men's minds were squabbles over tolls and hunting rights with the neighboring duke of Württemberg, whose land surrounded the urban district. Military events, to be sure, also repeatedly aroused the townspeople from their quiet. That they fought well when they fought for their freedom is proved by the part which they took on the side of the confederation of towns in the unhappy battle in neighboring Döffingen in the year 1388. In this battle against the duke of Württemberg sixty of their citizens were killed.

In Weil der Stadt the Reformation led to long continued tensions and disputes. Although the evangelical doctrine found followers among the citizens very soon after Luther's appearance, it did not succeed in winning over the majority. The parish church always remained in the hands of the Catholics. At the time of Kepler's birth, there was still no Protestant preacher in the city. A few years later, the followers of the new creed, with the support of the Württemberg duke, struggled in vain with the city councilor for the toleration of the evangelical faith, for the granting of a special church or chapel and for the employment of a clergyman of their own. The councilor considered it a special concession when he allowed the Protestant townsmen to receive sermon and sacrament abroad and permitted a preacher of their creed to come and administer communion when they were in danger of death. It was, indeed, a victory for the Protestants that a few years thereafter evangelical baptism was permitted in the city. The Kepler family, especially Johannes' grandfather, Sebald, were among the most prominent and most active advocates of the Lutheran teaching. That the latter, as leader of his coreligionists, retained the office of mayor, in spite of the Catholic majority, speaks for his ability and the high regard that he was able to earn among his fellow citizens. In the same period, members of the Fickler family were prominent among the promoters of the Catholic interests. This was particularly true of Johann Baptist Fickler, prothonotary of the archbishopric of Salzburg, who was active in the Counter Reformation as an influential opponent of Protestantism. Despite the difference in faith, the Kepler and Fickler families were related by marriage. For this reason, many years later, Kepler's son Ludwig succeeded to the scholarship that a relative of the Fickler family had founded in Tübingen. The circumstances described above explain why no one knows where Kepler was christened, whether in the parish church by a Catholic clergyman, or, what is more probable, by a Protestant preacher in one of the neighboring villages, most likely Magstadt.

Grandfather Sebald's house was situated, tradition has it, near a corner of the market place, in a short alley leading to the church, so that a glimpse of the market fountain with the statue of Emperor Charles V and the mighty west tower of the house of God was afforded. This family house fell a victim to the city fire of 1648. There is reason, however, to believe that it was rebuilt in its earlier form. It is correctly considered the birthplace of our Johannes, since his father, Heinrich, continued to live in it after his marriage on May 15, 1571. Although it appears small from the outside, inside it is sufficiently roomy to hold a large family. The mayor, Sebald, apparently attained greater wealth only in later years, mainly by inheritance.

3. Kepler's family

When he was about twenty-five years old, Johannes Kepler made
notes about the characteristics of his grandparents and parents, as well
as about some accidents and misfortunes, such as life brings, so that we
gain a picture of their character and the life in the house in which he
passed his first years. He did this in an appendix to the birth-horoscope
of these ancestors, because at that time he was much occupied with
astrology and believed that personality was influenced by the position
of the planets at the time of birth. He described his Grandfather Sebald
as proud and arrogant in manner, hot-tempered, impetuous, stubborn
and sensual; his face was red and rather fleshy; his beard lent him an
air of importance. He was able, without being especially eloquent, to
give good and wise instructions and to enforce their observance.
Kepler's grandmother was, according to his description, very restless,
clever, inclined to lie, but zealous in religious matters, slender, of fiery
nature, lively, ever on the move, jealous, spiteful, resentful. Kepler
says that Saturn in trigon to Mars[1] in the seventh house, had made his
father, Heinrich, an immoral, rough and quarrelsome soldier. The
mother, too, does not come off so well; she was small, thin, dark-
complexioned, garrulous, quarrelsome and generally unpleasant. This
is no brilliant ancestral gallery which Kepler parades before us, and his
character portraits are all the more surprising as it is known that
reverence for his relatives was a prominent trait in his nature. It must
be remembered here that he made these notes only for himself to
demonstrate the agreement of the characters with the heavenly
constellations. In this way it could easily happen that he searched
the heavens mainly for an explanation of the bad qualities as a
kind of apology and thus let the good ones recede into the background.

At any rate, it is clear that life was not exactly peaceful and harmonious in the Kepler house, where several of Heinrich's younger brothers
and sisters also dwelt. It would have needed no further explanation on
the part of Kepler to make us understand that the marriage of his
parents was an unhappy one. The father treated the mother harshly
and rudely and the mother opposed his unloving behavior with pout-
ing stubbornness. By fighting and arguing they made their lives miserable. Even little Johannes, their first-born, did not bring the father
and mother closer to one another. As a seven-months' baby, he was of
delicate constitution. He was not spoiled by any parental love. He

[1] ED. NOTE. Saturn and Mars were separated by two signs of the Zodiac.

surely received his characteristics more from the maternal side, as is not seldom the case with men of genius. So also as a man he was of small and graceful build, dark-eyed and black-haired. Nor did he have anything soldierly about him like his father. The mother appears to have been a strange woman. Her manner is not completely characterized by the few epithets given above. We shall get to know her better in the ugly witch trial in which she was involved in her old age. In this trial it was alleged that a kinswoman, who later died at the stake as a sorceress, had reared Frau Kepler. The latter was obviously very active and restless, interested in everything possible, reflective, but also alert and talkative. She gathered herbs and prepared salves, inspired by belief in magical forces and relationships, as if she had seen through the things of nature. After her first child, she bore six more, only three of whom, however, grew up, all different in type. While the genius of our Johannes won undying fame for the name of the family, his brother Heinrich, who was two years younger, was a veritable good-for-nothing. He was an epileptic and a trial to his mother; he was roughly beaten, bitten by animals, brought home bruises and wounds, was nearly drowned, nearly frozen to death, and nearly died of illness. When he was fourteen years old, he was apprenticed to a cloth-shearer, then to a baker, was further beaten, and ran away to Austria when his father threatened to sell him. He served some soldiers in Hungary in the fight against the Turks, earned a meager livelihood in Vienna by singing and bread-baking, became the servant of a nobleman, was discharged, robbed, wounded and begged his way home. Soon he went forth again to Strasburg, Mainz, and Belgium, became drummer for a regiment, and at Cologne was plundered by the robber-band "cock's feather." Later he was imperial guardsman in Prague, and returned home a poor and badly beaten-up man and was a drag on his mother until he died at the age of forty-two. The gentle daughter, Margarete, is a friendly contrast to this son. She, of the whole family, was closest to her great brother. She married a clergyman, which suited her well. The youngest son, Christopher, was honorable, correct, mindful of his reputation; he became a respectable craftsman, a pewterer. The military blood of the Kepler family ran in him greatly diluted, in so far as, on the side, he took pride in being active in the militia as a drill master for the duke of Württemberg. We shall hear more about him later.

The father, Heinrich, finally could no longer stand staying at home. The heavy atmosphere which prevailed there and the restless blood in his veins drove him forth. Whether he had learned anything at all in his youth we do not know. There is no mention of it anywhere. Probably he helped with the management of his father's estates. His taste aspired to a different activity. When the call for recruits sounded in 1574, he started for the Netherlands, where the terror regime of the Duke of Alva had led to insurrection and rebellion. That was the atmosphere which pleased Heinrich. He wanted to earn his spurs in this tumult of war. He left wife and children at home. In the following year his wife, Katharina, who got along badly with her mother-in-law and felt oppressed by her, followed after her husband. Little Johannes was abandoned to the care of his grandparents, who had little love for him and treated him roughly. While his parents were away he was so sick with smallpox that he nearly died. After his parents' return in 1576, his father renounced the right of citizenship in Weil der Stadt and, with his family, moved to the nearby town of Leonberg, which belonged to the duchy of Württemberg. He had bought himself a house there and wanted to start life anew. But in the very next year we again find him in the Belgian military service. Luck seems to have been against him, for he nearly ended his life on the gallows. After his return he lost his fortune by acting as a bondsman, sold his house, quit Leonberg and in 1580 rented the then much frequented Gasthaus zur Sonne in Ellmendingen, a small town in Baden near Pforzheim. But here also he did not remain long. As early as 1583 we see him again in Leonberg, where he acquired some real estate. Five years later he abandoned his family forever. He is supposed to have taken part in a naval war of the Neapolitans as a captain and on his way home to have died in the vicinity of Augsburg. His family did not see him again.

Children believe the course of the world must be as it appears to them when they begin to think and they accept the storms as they happen. However, the quiet, sensitive youth, Johannes, must have had much trouble getting rid of all the wounding impressions which he received. The world order which presented itself to him was difficult for his childish thought to comprehend and the bad pictures which were fixed in his soul were not easy to erase. His devout attitude shone forth very early. In his trials he sought the help of God, whom he revered as an omnipotent, all-governing and all-embracing power.

4. First schooling

There was something else, too, which diverted him from his inner suffering, aroused his self-assurance and offered nourishment to his mind. That was school. And it was his good fortune that, in Württemberg, instruction was well developed. Not only were there German schools everywhere, in which one learned, in a haphazard fashion, to read, write and do arithmetic. In addition, after the introduction of the Reformation, the Württemberg dukes had arranged for the establishment in all little towns of Latin schools as well, to take over the tasks of the earlier monastery schools and to educate a sound new generation for the clergy and for service in civil administration. In Leonberg there was such a school, divided into three classes. In view of the importance of Latin at that time as the language of scholars and the agent of a higher culture, instruction in it was pushed with the greatest zeal, and therefore the pupils were required to learn to read, write and speak Latin fluently. This teaching began in the very first year of school attendance. First, reading and writing were taught. The second year was devoted to grammar drill and the third year to the reading of classical texts, comedies by Terence in particular, because special advancement in oral expression was expected from this. The school regulations demanded most strictly that the boys speak Latin to each other. Little value was placed on the cultivation of the German language, because it was believed that through the writing of Latin the writing of German would also be "grasped." The inevitable result was that men who knew how to set down the loveliest sentences in the Latin language, which forced them to think clearly and logically, frequently expressed themselves in a stilted, twisted, chopped-up, often almost incomprehensible manner when they wrote German.

It was in such a school that Kepler now laid the foundation for the stylistic perfection with which he later was able to express his thoughts in the Latin language. It appears that his parents first sent him to the German school. They cannot be expected to have understood the aims of the Latin school. However, the teachers of the German school gladly sent their talented pupils over to the Latin school, in order to smooth their path toward a better future. Consequently Kepler, who gave early evidence of a keen intellect, soon came to the institution which should lead to higher goals. He entered the first class at the age of seven. But it was five years before he completed the three classes of his school. This was not because his attainments were insufficient but because he had to interrupt school attendance for months, even years, due to his parents' move to Ellmendingen and to their limited understanding and precarious position. The boy was used for hard agricultural labor and had to help himself during these interruptions as best he could.

Kepler distinctly remembered two events from his boyhood which pointed toward his later calling. In the year 1577 his mother led him up a slope and showed him the great comet then in the sky. And in

1580 his father took him out at night under the open sky to observe a lunar eclipse. Both heavenly phenomena made a vivid impression on his receptive mind, so that later he could still recall details.

5. *The seminary*

What was to become of the boy? His constitution was too weak for heavy agricultural labor. His exceptional talent pointed toward a loftier objective. The advice of the teachers, the pious disposition of the child, and also financial considerations may have induced the parents to destine him for the clergy, a choice to which Johannes surely consented heartily. The path to this goal was marked out and smooth. A student who had completed the Latin school and had passed a competitive examination, supervised by the state, was sent to one of the seminaries, which prepared its pupils for further study at the University of Tübingen, where in turn they were admitted to a college for their theological studies. This has been the path trod by thousands of hopeful youths in Württemberg up to our day. Among them not a few subsequently made their names famous throughout the world. This path Johannes Kepler also took.

A large number of such seminaries was established in little Swabia by the intelligent foresight of its dukes and their counsellors. They were set up in monasteries which, like the famous Abbey Hirsau, at one time had stirred with life but after the introduction of the Reformation had been abolished. They were divided into lower and upper seminaries. Whereas the first ones, the "grammar-cloister schools," continued and completed the instruction begun in the Latin schools, the higher seminaries prepared the pupils directly for university study. House and school regulations were strict. In the summer at four o'clock, in the winter at five, the day's work began with psalm singing. Every hour had its prescribed work. There was no time off. A uniform costume which consisted of a sleeveless knee-length coat distinguished the scholars at the monasteries and supported the feeling of solidarity. The director of a seminary was designated as Abbot, still recalling the Catholic past. The instruction was given by tutors, often young theologians who had just ended their studies in Tübingen. Latin dominated and formed the everyday speech of the scholars. To this was now added instruction in Greek. By the perusal of the old classics, above all Cicero, Vergil, Xenophon and Demosthenes, the young heads were supposed to mold their world of ideas. According to the cultural schemes of the trivium and quadrivium, rhetoric, dialectics, and music would be taught next, then in the higher seminaries, the elements of spherics and arithmetic. Bible reading, to which zealous attention was given, was supposed to fill head and heart with the Christian heritage. Maintenance and instruction in the seminaries were free.

On October 16, 1584, the thirteen-year-old candidate Kepler placed his foot on the lowest rung of the ladder, which he was to mount: after passing the state examinations, he entered the convent-school Adelberg, which had been established in a monastery of the Premonstratensian order located near Mt. Hohenstaufen. Two years later he climbed further up and on November 26, 1586, moved into the higher seminary, which had found a home in the Cistercian monastery, Maulbronn, which is famed for its artistic construction and historical significance.

It was an unusual boy who entered the community of the convent school, unusual not so much in his achievements, although in every way he gained the praise of his teachers and punctiliously and conscientiously discharged what was required of him. He was distinguished from the average student by his introspective nature which drove him to almost torturesome self-observation, by the nature and content of his intellectual activity which found pleasure in special exercises, by the religious anxiety with which he fulfilled the demands of his conscience, by early participation in the confessional conflict of his time which disturbed him, and by the great sensitivity with which he reacted to the conflicts of community life. Such a nature must have had difficulty in asserting and maintaining itself against the robust manners of those who, without proper qualifications, frequently in similar communities, want to dominate and who take pleasure in oppressing and tormenting others, especially when young, inexperienced educators do not understand how to master the gross manners of the youthful crowd.

Kepler later made note of the results of his introspection and individual incidents from his boyhood and youth, which give us a glimpse into his inner life and his position in the boarding school. Reciting names and stating reasons, he reports about quarrel and dissension as well as friendship and alliance with comrades. Not seldom were either individuals or the majority against him. Rivalry in competing for higher places played an important part there. At times he had to defend himself against the disparagement of his father or drive off obtrusive friendship. Lack of self-control when speaking, sauciness and sharp criticism provoked the opposition of the others. He aroused indignation and rebellion among his comrades when under moral pressure from above he became an informer. However, he tried to set the matter right again and discharge his conscience by saying a good word for the wrong-doers. He put much stock in the praise of his superiors; he could not stand it when they were not pleased with him. He suffered no less when he noticed that envious talk about him was circulating among his comrades. To practice the virtue of thankfulness, and also to show it outwardly, was easy for him. Always his efforts were toward moderation, "because he weighed carefully the causes of things." He made good use of time. He was always busy, but did not stick to one thing, because new thoughts and goals always pressed upon him. He wrote what occurred to him on little slips, which he carefully kept. He never disposed of any books he could obtain, believing they might at some time or other be useful to him. He considered himself created to while away the time with difficult things, before which others shrank. At an early age he occupied himself with the various meters. Soon he attempted original poetry. He wanted to write comedies. Later he wrote lyric poems in imitation of the ancient poetical forms. He had an especial partiality for riddles; he liked to play with anagrams and daring allegories. In his compositions he pleased himself with para-doxical assertions, such as—the promotion of knowledge is a sign of the decline of Germany—or—one ought to learn French before Greek (this assertion, too, he considered paradoxical). He always digressed from the draft in copying his compositions. He exercised his memory by learning the longest psalms by heart. He also attempted to memorize all the examples in Crusius' grammar.

Since boyhood, Kepler's religious disposition was pronounced. It assumed exaggerated forms. He tells us that once when he fell asleep without evening prayer he made it up the next morning. It pained him that, because of the worldliness of his life, the gift of prophecy was denied him. If he had committed an error, then he set himself a penance to atone for his fault. This penance consisted of repeating certain sermons. When at the age of ten he was able to read the biblical histories, he took Jacob and Rebecca as a model, should he ever marry, and resolved to follow the precepts of the Mosaic law. The proclaimers of the Word did not control the flame in his forma-tive soul, so susceptible to religious instruction. On the contrary, their harsh confessional controversy was tinder which filled his soul with suffocating smoke. In Leonberg, when he was only twelve years old, so he tells us, he was seized with a great tormenting unrest over the disunity among the churches, because of a sermon by a young deacon declaiming at length against the Calvinists. Often, thereafter, a preacher who argued with his opponents about the meaning of the scriptures did not satisfy Kepler. He consulted these passages in the text himself and gained the impression that the interpretation of the opponent, which he had learned from the account of the preacher, had its good points. In Adelberg, the young teachers, who also managed the job of preaching, were very busy refuting the reformed doctrine of the Eucharist. Thereafter, as a result of their exhortations to beware of the Calvinist distortions and to steer clear of them, he often meditated in solitude—trying to decide exactly wherein the argument consisted and what was the manner of participation in the body of Christ. He concluded that the right manner is precisely the one which he soon thereafter heard condemned from the pulpit. In addition to the doctrines of communion and ubiquity, the boy racked his brains over that of predestination which gave him serious qualms. In his very first year in Adelberg, he had a treatise sent to him from Tübingen, because of which, in one of the scholastic disputations, a comrade questioned him in school jargon: "Bacchant [freshman] hast thou also doubts about predestination?" Kepler could not accept the belief that God would simply damn the heathens who do not believe in Christ. Even that early his peace-loving nature cared more for that which unifies than that which separates in religious matters. As he advised peace between the Lutherans and the Calvinists, so he also meted out justice to the followers of the Pope and recommended this in his conversations. We see from all this how, even in his youth, he laid the foundation for an attitude which was to have such tragic consequences for him in later life.

6. The Stift in Tübingen

In September, 1588, Kepler passed the baccalaureat examination in Tübingen. However, after this first step into the promised land, he had to return to Maulbronn for another year, to complete his studies there as "veteran." On September 17, 1589, the gates of the university in the city on the Neckar finally opened for him. He had attained the goal toward which, in the long years of preparation, he had so yearningly stared in his hankering for knowledge. His heart must have beaten faster when, stepping out of the magnificent woods of the Schönbuch, he espied the fortified castle of Hohentübingen as he looked upon the enchanting landscape of the Neckar valley and entered the narrow streets of the city which climbed up from the river to the fortress.

No one was better provided for in Tübingen than a theologian. When he arrived he knew where to turn. A study, a set table, a bed—

all was ready for him. He needed to bring along only heart and soul for his profession, a good training and the belief that here flowed the source of wisdom. The seminary, called Stift, which since 1547 was accommodated in the former Augustinian monastery, accepted the candidates who, thirsty for knowledge, crowded together from all parts of Swabia. Here, on the foundation of Duke Christopher's church regulations, arose a training school in which were mirrored the philosophical and theological fights of the following centuries with their victories and defeats, the ups and downs of the evolution of intellectual life, and the manifold tendencies of the various epochs. Not a few men, who once had acquired their scholarly armor here, later appeared as leaders in the arena of intellectual life. Through all time's changes, the basic form of this intellectual workshop has proved its worth and created an educational type which, while tinged by Swabian characteristics, yet must be considered representative of a general, open-minded and noble humanity. The joy in speculation and disputation, the tendency toward brooding and meditation, the craving for distances which can never be reached, that plunging into depths which can never be fathomed are just as apparent here as are a strong sense of reality, an inclination to criticism and contradiction, an openness to new ideas and, last but not least, delight in humor and satire. Only smaller souls, in whom the drive for education led to pedantic surface knowledge of many things, neatly arranged, in the manner of pharmacists, the many little doses of their material in the various compartments of their brains. And if the claim always to be right had sometimes settled deep down in the heads of those lacking in the necessary self-criticism, the cause can be found in exaggerated self-assurance due to the knowledge of belonging to an important community. A further cause is the pleasurable habit of disputing, by which one is forced to defend one's view by any arguments whatever.

As in the lower seminary, life in the Stift was strictly regulated. Although in consideration of the more advanced age of the students restrictions were looser, there was still no question of what is called academic freedom. The rigid discipline kept the candidates in theology from the disorderly goings-on to which at that time large groups of the student community abandoned themselves. The course of instruction was regulated in such a manner that the new arrivals had to hear lectures in the faculty of arts for two years before they could begin the study of theology. Ethics, dialectics, rhetoric, Greek, Hebrew, astronomy and physics were treated in those lectures. The students' progress was carefully supervised; grades were given every quarter. The studies in the arts faculty would be terminated by the master's examination. Three years were applied to education in the theological disciplines. After completing this training the scholars were constrained to remain for life in the ducal service and needed the explicit consent of the prince who had borne the cost of their education if they were to accept a position in another country.

The founder of the seminary, Duke Ulrich, had prescribed that the scholars should be "children of poor, pious people, with an industrious, Christian and God-fearing character." Although Kepler's father did not exactly fill this requirement completely as regards piousness, he himself conformed so much the more to the restrictions set. His parents were not blessed with wealth. Since, however, instruction and maintenance were free here, and since the scholarship students received, in addition, 6 gulden annually for their other expenses, and since besides Grandfather Guldenmann, by letter, placed the yield of a meadow at the disposal of his daughter's son "for better and more dignified upbringing," the studies of their son did not cost the parents much. The young student's circumstances became even better when, in the second year of his residence at the university, he obtained a stipend amounting to 20 gulden annually, for which the magistrate of his native city had to propose fit candidates.

Kepler was in his element in the new surroundings in which he was installed. He used the opportunity for broad education to the best of his ability and soon earned the reputation, with his teachers and fellow students, of being a diligent, sedate and pious young man. Later he could say of himself, that his life had been free of conspicuous short-comings, with the exception of such as had arisen out of fits of anger or high-spirited and thoughtless pranks. Here also disputes with fellow students were not lacking, yet he was no spoil-sport. He co-operated in the public theatrical presentation, which the students gave each year at the time of Shrovetide and in which biblical or classical themes were produced. He himself tells that in February, 1591, he played the part of Mariamne in such a production, when a tragedy about John the Baptist was given. Since the female roles all had to be taken by the students, he had been chosen for that part because of his graceful and slight figure. The production, which was held in the market place in spite of the inclement time of year, was bad for his health. As a result of the excitement of these days he was seized with a feverish illness. His delicate constitution was not infrequently subject to such seizures. Headaches, intermittent fever and violent rashes hindered him repeatedly in his studies, just as, in the years of his youth, he had also frequently had similar complaints. On August 10, 1591, he passed the master's examination in second place among fourteen candidates.

(Hippolyt Brenz, a professor's son, a grandson of the reformer Brenz, attained the first place.) The young master especially attracted the attention of his professors. When, soon after his examination, he applied for renewal of the scholarship granted him in the previous year, the senate supported his request with the remarkable words: "Because the above-mentioned Kepler has such a superior and magnificent mind that something special may be expected of him, we wish, on our part, to continue to that Kepler his stipend, as he requests, also because of his special learning and ability." His teachers were not to be disappointed in their expectations.

7. Studies and teachers at the university

Unfortunately, what Kepler relates about his university studies, about his teachers whose names are all known to us, about the stimulation which he received from them, about the sources on which his zeal for learning fed, and about the subject matter to which he applied himself, is incomplete. To explore his strongly marked personality and magnificent life work, as well as to illuminate the cross currents of intellectual history, it would be of interest to know more than he himself tells us. For example, all he says of his study of philosophy is that, of the writings of Aristotle, he had read particularly the *Analytica posteriora* and the *Physica*, pushing aside the *Ethics* and *Topics*. We see, however, that from the very beginning his whole thinking was stamped in accordance with Platonic and Neo-Platonic speculation. From this, just as from the system of ideas which tradition connects with the name of Pythagoras, he received the strongest impetus for his work. We do not know on which sources in particular he drew. These speculations of antiquity were, to be sure, so firmly anchored in the world of thought of his time, that his familiarity with them is easily explained. Although he says nothing specific about it, Kepler may have received stimulation and instruction about these questions which so fascinated him from Vitus Müller, the professor of philosophy. It is further established that he had known and read various writings by Nicholas of Cusa, whose geometrical mysticism agreed so closely with his own thinking that in his very first work a few years later he began with considerations which he had taken over from that thinker. The great esteem in which he held him is shown by the fact that he does not hesitate to confer on him the epithet "divus," divine.

The young student's interest, however, did not stop at philosophical speculation. His intellect grappled with the most diverse questions. So he tells of the impression made on him by Julius Caesar Scaliger's *Exercitationes exotericae*, a book which at that time was passed from hand to hand among the students and was eagerly read. This work, so he informs us, awakened in him all possible thoughts about all possible questions, about heaven, souls, spirits, the elements, the nature of fire, the origin of springs, the tides, the shape of the earth and surrounding seas, and so forth. Anyone who knows Kepler's work finds expressed everywhere thoughts which go back to this early youthful preoccupation. Along the same lines lie his studies of Aristotle's *Meteorologica*, on the four books of which he conducted disputations. Into one field far removed from this, but to which he later dedicated much time and energy, namely chronology, he entered with inquiries about the Roman calendar, the weeks of the year according to the prophet Daniel, as well as the history of the Assyrian empire. The professor who taught the Greek language was particularly well disposed toward the eager student. This teacher was Martin Crusius, the famous Hellenist, who had such command of the Greek language that he was able to take down the sermons in the Stiftskirche in this language. Later he exchanged letters with Kepler and tried, unsuccessfully, to persuade him to collaborate in his commentary on Homer, by asking him to interpret the astronomical and astrological allusions in the poet's work. Kepler makes an interesting observation comparing his type of intellectual activity with that of Crusius. Common to both was attention to minute detail. However, whereas Crusius surpassed him in industry, his own power of judgment was greater. Crusius worked by gathering, he by separating; the former was a hoe, he a wedge. Additional poetical attempts testify further to the intellectual agility of the eager student, and he was pleased to be able to present his friends with printed copies of a few very artistically constructed poems written for special occasions.

All these activities and efforts still do not herald the summons which the theological candidate Johannes Kepler was to receive and which led him to the highest achievements in the sphere of astronomy. He remained unconscious of this calling as long as he was at the university. But inclination and talent marked out the path which he would have to follow, and without knowing the goal, which was hidden from him, he here laid the first foundations for the high achievement, which later lifted him so far above his time. An experienced teacher was able to awaken the slumbering forces, to guide the first steps and to plant in the prepared soil that seed which in due time was to grow and develop magnificently. He was Magister Michael Maestlin, the professor of mathematics and astronomy. He was a good twenty years older than his famous pupil, was born in Göppingen, had been deacon in the Swabian town of Backnang and for a few years professor of mathematics at Heidelberg before receiving the chair at the university in his native land in 1583. His predecessor had been the well-known astronomer Philip Apian, who had been dismissed from his post because of his refusal to subscribe to the Formula of Concord, and who still lived in Tübingen when Kepler began his studies there. Maestlin was one of the most capable astronomers of the time and enjoyed great esteem in the learned world. As was customary at that time, his instruction in geometry was based on the *Elements* of Euclid, to which was probably added an occasional glance at Archimedes and Apollonius. In addition he introduced his audience to the elements of trigonometry. For his course of astronomical lectures he brought out a text book, *Epitome Astronomiae*, which appeared for the first time in 1582 and was reprinted several times in the following decades.[1] Maestlin soon saw that there was something special about his pupil, who showed a great preference for mathematics and proved his aptitude by often devising propositions and constructions which he only afterwards discovered were already known.

Through Maestlin, Kepler also came to know Copernicus, the man whose prophet he was to become. To be sure, in his public lectures, as well as in all the editions of the *Epitome*, the astronomy professor sided completely with the system described in Ptolemy's *Almagest*, because the Copernican theory was strictly prohibited among his theological colleagues, due to its supposed contrariness to Holy Writ.[2] He did not want to risk his secure professorship and could not speak out of turn without endangering the peace and order in a college welded together by numerous bonds of kinship and intermarriage, where the theological faculty wielded the baton. Accordingly, only with careful restraint and in an intimate circle, did he tell what Copernicus had taught about the structure of the universe. Yet in the youthful enthusiastic head of his pupil the spark ignited. Maestlin's considerations and repressions were alien to the young and unencumbered Kepler who, open and dauntless, entered into disputations in favor of the new astronomical theory. Several years later he wrote about the stimulation he received which was so exceedingly momentous for his work. He also described the impression which it aroused. These are his words: "Already in

[1] ED. NOTE. For a listing of the various editions of the *Epitome*, see C. Doris Hellman. *The comet of 1577: its place in the history of astronomy*. New York: Columbia University Press, 1944, p. 143.

[2] ED. NOTE. See Hellman, *op cit.*, p. 138, n. 40, for a discussion of the additions made to the appendix to the *Epitome* in the later editions, in which the Copernican theory was presented.

Tübingen when I followed attentively the instruction of the famous Magister Michael Maestlin, I perceived how clumsy in many respects is the hitherto customary notion of the structure of the universe. Hence I was so very delighted by Copernicus, whom my teacher very often mentioned in his lectures, that I not only repeatedly advocated his views in the disputations of the candidates, but also made a careful disputation about the thesis that the first motion (the revolution of the heaven of the fixed stars) results from the rotation of the earth. I already set to work also to ascribe to the earth on physical, or, if one prefers, metaphysical, grounds the motion of the sun, as Copernicus does on mathematical grounds. For this purpose I have by degrees—partly out of Maestlin's lecture, partly out of myself—collected all the mathematical advantages which Copernicus has over Ptolemy." The impetuous youth could not foresee whither these first tentative beginnings were leading him and what enormous difficulties he would have to overcome before he reached the goal. In any case, he did not obtain Copernicus' work itself to read. When he made his first trials, he did not even know the *Narratio prima*, the first report by Joachim Rheticus, in which the latter, a few years before the appearance of the *Revolutiones* had communicated to the world the new theory of the Frauenburg canon.[1]

For many years after Kepler's departure from the university, he and Maestlin kept up an active correspondence. We shall see how the older man proved a loyal helper and adviser to the younger and facilitated and furthered his entrance into the scholarly world. Later, however, he became retiring, and Kepler had to use all his art of persuasion to get him to answer his letters. Toward his former teacher, Kepler remained reverent and loyal for his whole life, even when he had long surpassed Maestlin and had earned a great name for himself. The manner in which he repeatedly gave public expression to his gratitude, devotion and veneration contrasts clearly with the sullen humor into which the aging Maestlin withdrew more and more until, surviving his famous pupil, he passed away at a patriarchal age.

As early as his student days, Kepler worked happily and thoroughly on astronomical questions. This is shown by the fact that in those days he wrote a disputation concerning the heavenly phenomena and how they appear from the moon. It contains the first germ of a work which

[1] AU. NOTE. The copy of the *Revolutiones* which Kepler later had in his possession and in which, in 1598, he entered one of the poems he had written about the new discovery, is now in the library of the University of Leipzig.

ED. NOTE. Georg Joachim, better known as Rheticus after Rhaetia, the Roman name for the Tyrol where he was born, visited Copernicus, the canon of Frauenburg, in 1539 and in 1540 published a brief description of the new theory of the universe. Copernicus' own great work *De Revolutionibus Orbium Coelestium* did not appear until 1543, when its author was on his deathbed.

we shall come to know as the last of the books he published. His friend Christoph Besold, who was a few years younger, and subsequently became a distinguished professor of jurisprudence at the University of Tübingen, took from it a number of theses which he wished to uphold in a disputation under the chairmanship of Vitus Müller.

In addition to astronomy, Kepler also busied himself with astrology. This not only fitted in with the bent of his time, but also was in complete accord with his own form of thinking. He was important among his fellow students as a master in casting horoscopes. The deeper and purer conception, which he advocated for this field and subsequently developed (we shall become more closely acquainted with it later), had already been pronounced in its general form by Melanchthon in the preface which is printed in the later editions of Georg Peuerbach's *Theoricae planetarum*. There is no doubt but that Kepler already knew this widely circulated work.

But Kepler had not gone to Tübingen to become a philosopher, a mathematician, or an astronomer. Everything which he absorbed in the faculty of arts was only supposed to serve as preparation for the theological studies which opened the gate to the desired church office. What can he tell us about this? How did the incipient churchman, whose religious sensibility and scruples we know, see his way, when he entered the province of powerful men who interpreted Holy Writ according to rigid rule, and when he sat at the feet of warlike theologians who resisted each Calvinist tendency with the same passion as everything which came from the Roman Church? He tells us nothing about Jacob Heerbrand, the successor in the position of chancellor to Jacob Andreae, the old rough fighter, who had known Luther and Melanchthon personally and as a towering pillar had supported the edifice of the first reformers. Likewise, we learn nothing about Johann Georg Sigwart, who let loose with sharp pen against the Calvinists. Although he deplored the lack of clarity of Stephen Gerlach's[1] lectures, Kepler had a closer relationship to him. Nor did Gerlach offer Kepler a solution to his old theological doubts, which oppressed him and which revolved about the doctrines of predestination, communion and the ubiquity of the body of Christ. Yes, the weight of the objections, which Kepler raised against the announced doctrines, grew and so oppressed him, as he said, that he invariably pushed aside the whole complicated question and had to sweep it completely out of his heart, when he attended Holy Communion. The Bible commentary by the Wittenberg professor, Aegidius Hunnius,

---

[1] Gerlach once wanted to get the leaders of the Greek Church in Constantinople to unite with the Lutheran Church. (Transferred from text.)

which Kepler cherished for its clarity, helped him over several difficulties, especially when he came upon Luther's *De servo arbitrio*, on the
lack of freedom of the will. In sharp controversy with Erasmus of
Rotterdam, the reformer had developed his well-known determinism
and taught that man is evil by nature, that it is God who produces
everything in us, the good and the bad, and that mankind with simple
passive necessity is at the mercy of God's actions. Kepler tells us that,
under the influence of Hunnius' writings, he again "became adapted
to a healthy condition." His principal scruple, even Hunnius, who
adhered to the orthodox Lutheran conviction, could not dissipate.
The theological squabbling, to which he was witness, finally repelled
him to the extent that, as he says, little by little he grew to hate the
whole controversy. The dogmatic attitude, at which he arrived toward
the end of his student days, he later clarified with the words: "I had
gradually learned to comprehend that, in respect to the articles about
the person of Christ, the Jesuits and the Calvinists agree and that they
refer similarly to the Church Fathers and their followers and to their
scholastic interpreters. The agreement, so it appeared to me, corresponded thus to early Christianity while our aforementioned conflict
was something new developing with reference to the conception of
the Eucharist and not directed from the beginning against the papists.
Accordingly, I had pangs of conscience about chiming in with the
frequent condemnation of the Calvinists, and even with reference to
the interpretation of the Eucharist. For I told myself if injustice is done
them in regard to the one article about the person of Christ, so will
undoubtedly injustice also be done them in regard to the other article
of the Holy Communion."

As faithful mentor, indeed as warm-hearted friend, Matthias Hafenreffer, the youngest of the professors of theology, met the struggling
disciple of theology. In contrast to his colleagues, Hafenreffer, only ten
years older than Kepler, had a mild, conciliatory nature, with which he
captured the hearts of many of his listeners. He had especially cherished
the young Kepler because of the sincerity of his character and his
distinguished intellectual gifts, and the latter reciprocated his teacher's
love by wholehearted veneration. The mutual attachment lasted far
beyond the university period up until the death of Hafenreffer. Of
course, the confidence between the two did not go far enough for
Kepler to confess his secret spiritual troubles and his doubts in matters
of faith to his older friend. In spite of his peaceful way of thinking,
Hafenreffer was, after all, subject to the dominant conviction of the
Faculty, so that Kepler knew from the very beginning what answer
to expect. Thus, likewise, later, in Kepler's conflict with the church

authority, Hafenreffer, as we shall see, supported the point of view of his Faculty. As its spokesman, although this was painful to both of them, he had to communicate to his former student the Württemberg church authority's verdict about him. Similarly, even though, as Kepler believed at least in private, Hafenreffer himself secretly adhered to the Copernican theory, it was he who, for the time being, restrained Kepler from openly supporting its consistency with Holy Scripture.

8. Summons to Graz

Kepler's theological studies were to end during the year 1594. But before this came about, in the first months of that same year, there occurred a decisive turn in his life. Death had summoned Georg Stadius, the mathematics teacher at the Protestant seminary in Graz. The Styrian representatives then asked the senate of the University of Tübingen to recommend a successor. Why did these Graz rulers turn to far distant Tübingen? The reason lay in the importance of that university as one of the chief centers of reformation life and activity. Numerous preachers and teachers had already gone forth from there to the Austrian lands in order to publicize and spread the new teaching. For example, the then superintendent in Graz, Wilhelm Zimmermann, one of the inspectors of the convent school, had come from the Tübingen seminary. The choice of the senate fell on the candidate, Kepler. He was astonished when the call came to him. Should he accept it? Various considerations caused him to ponder; he could not accept so quickly. He had already imagined himself in priest's robes at the pulpit, and after the success he had attained up to then in his studies, he could count on an honorable career in the service of the Church. He said that up to this time, by the grace of God, he had come to esteem and cherish the study of sacred theology so highly that, whatever were to happen to him, he would never think of interrupting it provided God continued to preserve his healthy mind and his freedom. And should he now, so near the end of his studies, stop and accept a position to teach at a boys' school, a post which, according to him, was almost despised in the opinion of his time when compared to a position in the Church? Clearly, the summons showed recognition of his mathematical accomplishments to date. He considered himself still insufficiently educated for a teacher of mathematics, although he believed he certainly possessed the talent for this profession. He easily understood the geometry and astronomy prescribed by school regulations. But those were only obligatory studies, nothing which had shown a very special

leaning toward astronomy. Yet, on the other hand, he felt the appeal
to his moral discipline from which it was difficult for him to escape.
He had frequently noticed how fellow students, about to be sent into
a foreign country, that is beyond the Württemberg boundaries, in
their love for home used all possible excuses to avoid going. So, being,
as he described himself, "tougher than I actually was," he had previously resolved that, should he receive a call, he would follow
willingly wherever it should lead. In order to come to a decision, he
asked his grandfathers and mother for advice. Of course, they would
rather have seen him enter the pulpit soon while they themselves
shone in his reflected glory. Yet they wanted to leave the decision to
the theological faculty, which hitherto had shown so much good will
toward their little offspring. Could he not, perhaps, while holding his
teaching position in Graz, obtain through Pastor Zimmermann the
opportunity for practical exercise in church service and further educate
himself by private theological studies, so that later he might transfer to
the church service? This path would be preferable because, in age and
outer appearance, he did not yet quite fit the pulpit. So he accepted,
while explicitly reserving the right to return and enter the clerical
profession. How important was this decision, not only for his personal
fate but also for the whole history of astronomy! Looking back later
when, through the discovery of his planet laws, he had become aware of
his ability, he recognized the voice of God in the call which had come
to him. It is God who by a combination of circumstances secretly
guides man to the various arts and sciences and endows him with the
sure consciousness that he is not only a part of the creation but also
partakes in the divine providence.

It is often said that Kepler was pushed off to Graz by his Tübingen
teachers because he had laid himself open to suspicion by reason of his
straying theological opinions. This allegation is incorrect. Its propa-
gators rely on the fact that Kepler himself once said he had been pushed
away (*extrusus*) from Tübingen. By that, however, he only wanted to
say that it had needed pressure on the part of his teachers to move him
to the acceptance of the position which did not appear very agreeable
to him. Nothing is said in this about the reasons which guided the
theologians in Tübingen. It can be argued against those who try to
press the meaning of that word [*extrusus*], that Kepler says in two places,
lying far apart in time, that it had been a lucky accident (*commode
accidit*) that he had been summoned to Graz. Certainly, this expression
says nothing about the motives of the theologians. We only recognize
from it that Kepler considered the turn of events a blessing, a piece of
luck for his further intellectual development. But a positive refutation

of the claim, that Kepler was pushed out of Tübingen because of his theological opinions, is found in his explicit declaration that, realizing his youth, he had kept to himself his heterodox conceptions and had not disclosed them to the servants of the Church. The Tübingen professors may well have shaken their heads, when they heard the young zealot so enthusiastically supporting Copernicus. Yes, they may also have gotten wind of his doubts. But it is to appraise the Tübingen professors as very poor pedagogues if one credits them with so little understanding of the outbreaks of youthful temperament that they let a candidate, who far excelled the others in character and accomplishment, go away because in youthful exuberance he expressed opinions which were questionable in their eyes. No—Kepler was sent to Graz because, on the basis of his mathematical and astronomical knowledge, he was by far the most suitable candidate for the teaching position there, the only one worthy of consideration and likely to bring honor to Tübingen University. He was completely at peace when he left Tübingen. The state of confidence between him and his teachers remained unchanged for the next years, and a conflict did not arise until some years later, when he came forward with his theological scruples. It is true, certainly, that Kepler, with his candor and integrity, would assuredly soon have met great difficulties if he had completed his theological course and had taken a church position.

After the duke had given permission for Kepler's departure and even received him, the new mathematics professor took leave of his beloved university on March 13, 1594, and began the long journey to Graz. Because his cash funds were meager, he borrowed 50 gulden from Professor Gerlach, the superintendent of the school. Could Kepler divine that he would never work in his native land and would be permitted to see it again only as a visitor?

II

DISTRICT MATHEMATICIAN AND
TEACHER IN GRAZ

1594–1600

---

1. Church politics in Graz

KEPLER arrived in Graz on April 11, 1594. Since Württemberg stubbornly clung to the Julian calendar, he lost ten days when he entered Austria, or rather prior to this in Bavaria, where the new calendar was in use. This was his first big journey into the world. As he approached the lovely city on the Mur and glimpsed the towering castle-crowned hill, he may have thought of the lovely Neckar city in which he had studied and which likewise nestles around a hill crowned by a prince's castle. In the softness of the landscape, too, he could find something akin to the Neckar valley which he had left behind. The face of the city and the behavior of its inhabitants exhibited a more southerly character. These seemed friendly in comparison to the high-gabled houses of his native land and the slow manner of his countrymen.

However, the intellectual climate in his new dwelling place was very different from that to which he was accustomed. In Württemberg, duke and people were entirely and unqualifiedly dedicated to Luther's teaching, so that this land, with its intellectual center at Tübingen, formed a high citadel of Protestant belief in the empire, and the confessional tensions were discharged by academic utterances. This was not so in Graz, although in Styria, also, the noble lords in their numerous castles and strongholds as well as the townsmen had, for a long while, been adherents to the new doctrine. However, the rulers at the head of the land not only clung personally to Catholicism but even considered it a duty owed to their conscience to suppress the new doctrine and to lead the inhabitants of the land back to the old belief. They had not forgotten the legal maxims which had been set up by the Religious Peace of Augsburg in 1555, according to which the princes could choose between the Catholic and the Augsburgian creed

53

for their whole domain. This necessarily meant that the religious tensions in Graz were perceptible not only in polemical treatises and disputations but also directly in the life of each individual, threatening his religious security. It can easily be seen that Kepler was about to enter on a critical time because of his profoundly religious attitude. Spiritual battles, because they involve the highest sphere, are passionately fought and in Kepler's time doubtful means were often used.

After the death of Emperor Ferdinand I, in 1564, the Austrian lands were divided among his sons. Archduke Charles assumed the rule in inner-Austria, that is in Styria, Carinthia, and Carniola. The archduke, in the so-called Pacification of Bruck in 1578, had promised the free exercise of religion to the Lutheran Estates and their relatives in their castles and in the cities of Graz, Judenburg, Laibach, and Klagenfurt. However, in the years immediately following, efforts were begun to cancel the concessions granted to the Protestants, and thereafter no year slipped by without this effort leading to angry and embarrassing controversies. After the death of Charles in 1590, his widow, Archduchess Maria, of the House of Wittelsbach, showing even greater zeal, continued the endeavor to regain the land for the Catholic faith. And in the very first year which Kepler spent in Graz, negotiations were in progress for the assumption of the rule by Charles' minor son, Ferdinand. At that time he was still in Ingolstadt, where, along with Duke Maximilian of Bavaria, he was being brought up and where he was studying under the guidance of the Jesuits. In a few years, as we shall see, he was going to complete that effort.

To support his reform plans, Archduke Charles had summoned Jesuits to the city and in 1573 had erected a large building for their college. They not only dedicated themselves with zeal to ministerial activities but also promptly started a Latin school, through which, by their well-known pedagogical skill, they were able to influence the young people. In 1576, they erected a commons next to this building, partly for noble, partly for poorer youths who wished to dedicate themselves to the clergy and finally, in 1586, with Papal endorsement, they founded a university with faculties of philosophy and theology. It is evident that through these organizations they gained a strong position in the intellectual life of the city and the country.

2. *The Stiftsschule*

Opposed to these Jesuit schools, there was the Protestant Stiftsschule to which Kepler had been summoned. It had been opened on July 1,

1574, and thereafter had developed into the principal seat of the Protestant party of the city, due to the zeal of numerous clergy and teachers, who worked there. It was founded by the nobility for their sons, but was also used by the burghers for their children. In the narrow confines of the old part of the town, wedged between houses, can still be seen the impressive buildings which enclose a square courtyard surrounded by arcades and galleries and which afforded space not only for the pupils and the boarding school but also for the dwellings of some of the teachers. Kepler, too, found his first shelter here, in rooms which had been left vacant by the death of his predecessor.

The Protestant organizations had regulated the school with discretion and care. They had summoned David Chytraeus to set up the curriculum. He was a very well-known and esteemed liberal churchman and pedagogue who, like Kepler, originally had come from Swabia but at that time was active in Rostock as professor of theology. In Wittenberg he had lived with Melanchthon, been his docile pupil and later had been frequently called upon to regulate Protestant church matters and to execute tasks connected with church politics. The constitution of the Stiftsschule was comparable to that found in other similar schools of the time. A committee of supervisors was appointed from among the councilors. Under them were several inspectors, who were specially qualified clergymen and members of the community. The rector managed the instruction proper. He himself gave instruction and by frequently visiting lectures by other teachers was supposed to obtain insight into their performance. The teaching staff usually consisted of four preachers and twelve to fourteen teachers. At the time when Kepler took his position the following belonged to the committee of the representatives: Messrs. Balthasar Wagen von Wagensberg, Matthes Ammann von Ammansegg, Gregor von Galler, Wilhelm von Galler. The inspectors were the chief pastor, Wilhelm Zimmermann, the court lawyer (*Schrannenadvokat*) Adam Venediger, the court clerk (*Schrannenschreiber*) Hans Adam Gabelkofer and the district secretary Stephan Speidel. Johann Papius officiated as rector, but a few months later, to Kepler's sorrow, he was summoned to Tübingen as professor of medicine.

The school itself was arranged in two divisions. First a boy's school was composed of three "Dekurien" [decurions or groups of ten] in which instruction was given according to a curriculum introduced elsewhere by Melanchthon. An upper school contained four classes. The highest was called "publica classis." Its teachers were called professors. This class was again split in three. In one division were the incipient theologians. In a second the juridical subjects as well as history were taught. In the philosophical division were taught logic, meta-physics, rhetoric and classical reading, as well as mathematics with which astronomy was also connected.

That was the Graz circle into which the young Master entered as teacher of the last-named subjects. His humble annual salary amounted to 150 gulden; his predecessor had received 200. The representatives granted him 60 gulden for the expenses of his journey. He arrived as a novice who must first earn his spurs and establish a position in that circle. The inspectors reported to the representatives after his arrival: "We have talked with him as much as necessary and have formed the opinion that we sincerely hope he will be able to succeed the late Master Stadius nobly. Yet we wish to try him out for one or two months before we hire him on a permanent basis."

With his young, fresh vigor he soon found his way about in the new circumstances, but at first could not feel at home. His thoughts often lingered in his native land, where he had been spurred on by inter-course with friends of similar interests and stimulated by the teachers at a highly esteemed and famous university. In Graz his contacts were narrower. Since the job of district mathematician and calendar maker was joined to his teaching position, he did come in touch with wider circles, especially among the nobility, where there was great interest in astrological prediction. He could not expect to find understanding of his scholarly endeavors here. For, as a friend of his, Koloman Zehentmair, secretary to a Baron Herberstein, remarked, the nobles were grossly ignorant in all things and had a barbarian attitude in their judgment; they hated the sciences and cared for no one less than for savants and outstanding scientists. Kepler's sanguine temperament, his amiable behavior and the wealth of his thoughts won him sympathy and regard, so that many gladly associated with him. To be sure, his unrestrained frankness of speech, in which he exposed the weaknesses of others, sometimes embarrassed him, as when he told Pastor Zimmer-mann to his face that it was the mother's fault that their ill-bred son had been expelled from Tübingen, because she had spoiled him. On the whole, at first he felt almost as though exiled, so that after a year he contemplated returning to Tübingen.

The subject which Kepler taught in the school was not popular with the sons of the nobles and burghers. So it came about that he had only a few listeners the first year and none at all the second. The inspectors were, however, sufficiently broad-minded not to blame this on the teacher "because the study of mathematics is not everyone's meat." To compensate for this, Kepler, with the approbation of the rector, was ordered to teach arithmetic, Vergil and rhetoric in six lessons in the upper classes, "to which he obediently complies, until an opportunity will arise to make use of his knowledge of mathematics." Later, it appears, he was drafted to teach other subjects. At any rate, it was attested in the letter of recommendation that was given him at the end of his teaching activity in Graz that he had "besides his regular prescribed mathematics teaching, also taught history and ethics faithfully and with outstanding skill." Kepler got along so well with the first rector, Papius, that the two exchanged letters for many long years thereafter. However, there were differences immediately with his successor, Johannes Regius, since the latter reproached the mathematics teacher with not revering him sufficiently as his superior and with disregarding his directions. Kepler reports that consequently the rector was unbelievably hostile to him. For all that, the testimony of the inspectors, in a report on Kepler to the representatives in the second year of his teaching, sounds very favorable. He had, in their words, "distinguished himself first as a speaker [*perorando*], then as a teacher [*docendo*] and finally also as a debator [*disputando*], in such a manner that we cannot judge otherwise than that he was, considering his youth, a well trained and modest master and professor, and one who fitted well into this worthy district school."

It seldom happens that a great scholar, rich in ideas, or a creative genius is at the same time a good teacher. This applies also to Kepler. If he found few listeners, then the fault was certainly in part his. He expected too much of his pupils and assumed they would have the same intellectual flexibility and receptiveness, the same enthusiasm for his subject and the same devotion to the knowledge of truth by which he himself was animated. In a penetrating self-characterization which he wrote in 1597, he cited qualities which also throw light on his teaching. He speaks there about his overpowering "cupiditas speculandi,"[1] of his philosophical urge which throws itself at everything and always tackles new things, which rushes on and robs him of leisure to think a thought through to the end. He would always think of something to say before he had a chance to reflect on how good it might be. So he always talked too quickly. He would always think of new words, new subjects, new modes of expression and new arguments while talking and writing. Sometimes in the middle of a lecture he would think that it might be better to change its purpose or even to suppress what he was saying. Imagination and memory were remarkable, in the case of a thought sequence, in which one thing followed from another, whereas otherwise he could not easily remember what he heard and read. That accounts for the many parentheses in his

---

[1] ED. NOTE. Eagerness to speculate.

speech. Since his memory brought back all related subjects at the same time and as everything occurred to him at once, he also wanted to say everything at once. Therefore his manner of speech was wearying or at any rate confused and hard to understand. Besides, he would not be prevented from his passionate drive for learning by his professional activity. Indeed, neglecting his very praiseworthy profession, he wandered off wherever his mind drove him, so that he would not have escaped reproof, had he not been able to satisfy with his knowledge all the demands of his position on the spur of the moment. If he was mindful of his position, it was still only with these reservations. For never did he lack something on which he could exercise his drive for knowledge, his fiery zeal, and his desire to grasp especially difficult things. Since thousands of things occurred to him at once and since they did not follow an orderly time sequence, a certain degree of carelessness in their explanation was less a hindrance than punctilious care. A teacher of this kind is only suitable for outstanding pupils, and these are scarce. Kepler himself was certainly the one who derived the greatest profit from his activity as a teacher, since it variously stimulated him in his field and forced him to put his thoughts into words.

3. *The district mathematician's first calendars*

Only a few months after his arrival the young district mathematician issued his first calendar—for 1595. Five more followed during his stay in Graz. Unfortunately, of these, only two copies, for the years 1598 and 1599, are extant. All the rest are lost. In those days of belief in the influence of stars, calendars played a different role from nowadays. High and low were permeated with the belief that future fate could be predicted from the path of the heavenly bodies. So the calendar men, of whom there were very many, were expected to furnish information about weather and the harvest prospects, about war and danger of pestilence, about religious and political events. People wanted to know on what day to sow and harvest, when to bleed, when to expect hail and storm, cold and heat, illness and hunger. This is not the place to go into greater detail concerning Kepler's attitude regarding astrology. That will be done later. Only this much should be said, that it is true he thoroughly spurned the customary rules and prophecies as horrible superstition, as "necromantic monkeyshines," but on the other hand held just as positively to the conviction of an influence of the stars on earthly events and human fate, a conviction that one cannot imagine as absent from his view of nature. What he understood to be his position as calendar maker becomes clear from his own words: "He whose job it is to write prognostications must before all, with high courage, despise two contrasting but popular points of view and guard against two emotions which arise from a low and despicable attitude, namely ambition and fear. That is to say, since the curiosity of the masses is great, it denotes intellectual dishonesty if, to please these crowds and purely out of vain search for fame, one reports things which nature does not contain or instead announces real wonders of nature without entering into their weightier causes. On the other hand there are those who maintain that it does not behoove an earnest man and philosopher in a domain defiled each year by so many silly and empty prognostications to sacrifice the fame of his talents and his honor in nourishing the overcuriosity of the masses and the superstitions of the fatheads. I must acknowledge that this reproach has a certain foundation and that it is sufficient to frighten away an honorable man from such writing if he does not possess more earnest reasons. But if he has reasons at hand for his plans, which will be approved by intelligent men, he must really be a coward if he lets himself be frightened away from his purposes by these foreign outer hindrances, to be bothered with idle talk and worried about unwarranted opprobrium. For even though many of the rules in this Arabic art amount to nothing, still all that which therein is contained of the secrets of nature is no nothing and may therefore not be discarded with nothingness; rather one must separate the precious stones from the dung, one must glorify the honor of God, by taking for one's purpose the contemplation of nature, must lift up others by one's own example and exert oneself to move into bright daylight from the darkness of the unknown all that which at some time may be exceedingly useful to the human race." So from the very beginning, as calendar man, Kepler strongly advised against depending on astrological prognostications especially for political decisions. Therefore, at the close of his *Practica* for the year 1598, he says to the fighting men: "Heaven cannot do much harm to the stronger of two enemies, nor help the weaker much. He who strengthens himself with good advice, with soldiers, with weapons, with courage, also brings heaven on his side, and if the stars are unfavorable to him, he conquers them and all misfortune." He expresses his moral intent as follows: "We make use of the disordered and corruptible desires of the masses, in order to instill in them (as cure) proper warnings, disguised as prognostications, warnings which contribute to the removal of illness and which we can scarcely introduce in another manner." So we see Kepler continually shifting his point of view in his calendar writing. He prophesies because he is not averse to playing with the rules of astrology, but he immediately adds that one should not depend on prophecies. He preaches and scoffs. He writes calendars because he must. However, he writes them not unwillingly, because thus he has the opportunity to give his opinion to people who do not read his Latin writings and understand nothing about science. He writes them because, on the whole, he likes to write, even if he also occasionally rebels against the irksome drudgery. He writes them—and this is not the least important reason—in order to earn money. To be sure, he is never entirely happy writing calendars; he worries lest he lose his scientific reputation with the intelligent. When he presented the calendar for the year 1598 to Maestlin, he wrote in this vein: "There is much therein which must be deliberately pardoned or else it injures my reputation with you. The thing is this: I write not for the large masses, nor for learned people, but rather for noblemen and prelates, who claim a knowledge of things which they do not understand. No more than four hundred to six hundred copies will be distributed, none goes beyond the boundaries of these lands. With all prognostications I see to it that I give my above described circle of readers a happy enjoyment of the vastness of nature with sentences which happen to occur to me and appear true, in the hope that the readers might be tempted thereby to raise my salary."

Indeed, for the presentation of his first calendar, Kepler did receive an extra honorarium of 20 gulden from the representatives. In it he had hit the mark with his prophesying. He had, as is seen from his letters, predicted bitter cold and invasions of the Turks. Both happened. Many herdsmen in the mountains were rumored to have died from the cold; upon arriving home, many lost their noses when they blew them; the Turks were supposed to have ravaged the entire region below Vienna. This success turned attention to the young district mathematician and soon procured him prestige in the land, so that the lords gladly sought his astrological opinions and nativities. Kepler complied with the various requests which came to him because he was thus offered the opportunity of increasing his modest income.

4. Mysterium Cosmographicum

Yet all such cheap success could not satisfy an intellect like Kepler's. His *cupiditas speculandi* strove for higher things; it flew through the breadth of the world and fathomed the depths up to the boundaries which are set for mortals. In Tübingen, as he says, he had already grasped philosophy as a whole with prodigious avidity, as soon as he was old enough to taste its sweetness. Now his thoughts revolved primarily about the great eternal questions, which from time immemorial have been proposed to mankind by the wonder of the heavens in their mysterious beauty. It was not only the instruction compatible with his duty which offered opportunity for this. His maturing and searching intellect fixed on the field which was appropriate for him, on which he could develop his best strength and which he was summoned to develop in an impressive manner. The science of the heavens after him had an entirely different form than from before his entrance on the stage.

Copernicus' cosmography, under the spell of which Kepler had already been drawn in his student days, appeared to him with increased intensity. The longer he contemplated it and the more absorbed he became in its details, that much clearer, cleaner, more convincing did it appear to him, and the more brightly did the enthusiasm which it had already ignited in him blaze up. He understood that Copernicus had made far from the final pronouncement, that here was "a still unexhausted treasure of truly divine insight into the magnificent order of the whole world and of all bodies." The sun was made the center of the universe. It was its heart, the Queen around whom the wandering stars, six in number—Mercury, Venus, Earth, Mars, Jupiter and Saturn—moved in eternally uniform rhythm. The new doctrine offered a very special advantage over all earlier theories because it made it possible to calculate, from the observations, the relative distances of the planets from the sun. Had not the Greeks even without this knowledge speculatively divined a harmony in these distances which could now be tested by fact? Was it not necessary for structural connections and inner relationships to exist between all the numbers which the theory of Copernicus offered? Could the beautiful order be accidental? Did not the princely household of the sun desire a measured ceremonial?

Consciously or unconsciously, Kepler's thoughts were connected with everything which he had heard and read of Pythagoras and Plato, of Augustine and Nicholas of Cusa and many other great men of the past and with that which Christian teaching about God and the world and the position of men regarding both had implanted in him. The time had come when these whirling thoughts of Kepler's took on a distinct form and pressed toward a goal. By the first half of 1595 he was busy with the new questions which he saw himself forced to ask of Nature.

What is the world, he asked himself. Why are there six planets? Why are their distances from the sun exactly such and such? Why do they move more slowly the further away they are from the sun? With these bold questions about the reasons for the number, the size and the motion of the heavenly paths, the young seeker after truth approached Copernicus' picture of the world. Just as the latter had to some extent marked the borders of the universe, so now Kepler tried to prove physically and metaphysically that these bounds were the plan of the Creator, who in His wisdom and goodness could create only a most beautiful world. Nothing in the world was created by God without a plan; this was Kepler's principal axiom. His undertaking was no less than to discover this plan of creation, to think the thoughts of God over again, because he was convinced that "just like a human architect, God has approached the foundation of the world according to order and rule and so measured out everything that one might suppose that architecture did not take Nature as a model but rather that God had looked upon the manner of building of the coming[1] human." Those questions present the root of Kepler's lifelong work in astronomy just as they reveal his mode of thinking in its distinct individuality.

He sought the answers to his questions in geometry, in the structure of space. For the geometrical figures have their foundation in the divine being, and consequently it is in them that one should seek the numbers and sizes which appear in the visible world. Everything is regulated according to mass and number. The world is established in accordance with the norms of the quantities provided by geometry. That was why God also gave people a mind which is able to recognize these patterns. For "as the eye was created for color, the ear for tone, so was the intellect of humans created for the understanding not of just any thing whatsoever but of quantities. It grasps a matter so much the more correctly the closer it approaches pure quantities as its source. But the further something diverges from them, that much more do darkness and error appear. It is the nature of our intellect to bring to the study of divine matters concepts which are built upon the category of quantity; if it is deprived of these concepts, then it can define only by pure negations."

But which geometrical constructions could provide the numerical relationships he sought? In vain the restless ponderer tried all possible starting points. He wasted the whole summer in this difficult work. At last, during a teaching hour, he saw the light. "I believe it was by divine ordinance that I obtained by chance that which previously I could not reach by any pains; I believe that so much the more readily because I had always prayed to God to let my plan succeed, if Copernicus had told the truth." On July 19, 1595—he preserved his great day forever by recording the date—the thought came to him: "If, for the sizes

[1] ED. NOTE. "About to be created."

and the relations of the six heavenly paths assumed by Copernicus, five figures possessing certain distinguishing characteristics could be discovered among the remaining infinitely many, then everything would go as desired." Now, does not Euclid's geometry already teach us that there are exactly five and only five regular solids, the tetrahedron [four-sided], the cube or six-sided, the eight-sided, the twelve-sided, the twenty-sided? Do not these regular solids perhaps permit of being inserted between the planet spheres in such a way that each time the sphere of one planet will be circumscribed by one of the regular solids which in turn is inscribed in the sphere of the next lower planet? Immediately he wrote down the sentence: "The earth is the measure for all other orbits. Circumscribe a twelve-sided regular solid [dodecahedron] about it; the sphere stretched around this will be that of Mars. Let the orbit of Mars be circumscribed by a four-sided solid [tetrahedron]. The sphere which is described about this will be that of Jupiter. Let Jupiter's orbit be circumscribed by a cube. The sphere described about this will be that of Saturn. Now, place a twenty-sided figure [icosahedron] in the orbit of the earth. The sphere inscribed in this will be that of Venus. In Venus' orbit place an octahedron. The sphere inscribed in this will be that of Mercury. There you have the basis for the number of the planets."

As he later acknowledged, the enthusiastic scholar felt as though an oracle had spoken to him from heaven. After this vision, he compared the relationship of the numbers supplied by the regular bodies with those which Copernicus stated for the distances of the planets from the sun and found at least partial agreement. He was extremely excited. He believed that he had lifted the veil which hides the majesty of God and given a glimpse into its profound glory. The experience loosed a flood of tears. He was amazed that precisely he, although he was a sinful person, had received this revelation, especially as he had not intended to enter the whole matter as an astronomer, but rather had undertaken the investigation purely for his intellectual diversion. "It will never be possible for me to describe with words the enjoyment which I have drawn from my discovery. Now I no longer bemoaned the lost time; I no longer became weary at work; I shunned no calculation no matter how difficult. Days and nights I passed in calculating until I saw if the sentence formulated in words agreed with the orbits of Copernicus, or if the winds carried away my joy. In the event that I, as I believed, had correctly grasped the matter, I vowed to God the Omnipotent and All-merciful that at the first opportunity I would make public in print this wonderful example of His wisdom. Although these investigations are in no way terminated and there still are missing from my basic thoughts some conclusions whose discovery I could reserve for myself, as soon as possible others, who have a mind for it, should make with me as many discoveries as possible for the glorification of the name of God and sing unanimous praise and glory to the All-wise Creator."

The elaboration of his invention, its systematic foundation and the calculation, which he had to work out for more exact proof, caused endless trouble in the following weeks and months. Until then he had paid attention in his studies more to the big lines, to the basic thoughts. Now, however, it was a matter of executing scholarly minutiae; it was still necessary, as he himself admitted, to learn much and to fill in some gaps in his earlier astronomical and mathematical education. There followed then, as is always the case after hours of highest elation, weeks of torturing work and gnawing doubt. Letters were sent to Tübingen to his old teacher, Maestlin, asking for advice and help. Maestlin was keenly interested in the discovery of his promising pupil and gave it his complete approval, although, of course, he also matched the latter's youthful exuberance with an experienced cautiousness.

So passed the winter months of 1595-1596. In the beginning of February Kepler took a leave and journeyed to his Swabian homeland. Both his grandfathers were old and sick and desirous of once more seeing their grandson of whom they were proud. His paternal grandfather, indeed, died just at that time and the other was also close to his end. At the same time the stay at home gave the young scholar the welcome opportunity to discuss personally with Maestlin the completion and publication of the book, in which he wanted to announce his discovery. He wanted to speed the matter as much as possible, although he clearly understood that it was a matter of "not yet fully grown and half-fledged doves." The work, which he wanted to publish, would help him to improve and make secure his position in Graz, which the previous year, as already noted, he had wanted to renounce because of professional disagreements. In the meanwhile, however, his heart had caught fire in Graz; as early as the December before (he recorded this date, too) "Vulcan" had, as he noted secretively, "made the first mention of the Venus, whom he should join in wedlock."

During his stay in his homeland, he engaged in still another enterprise. It had occurred to him to make an artistic model to illustrate his plan of the construction of the universe. "A childish or fateful desire to please princes" drove him to Stuttgart to the court of the duke of Württemberg. He once had a scholarship from the latter and now he wanted to interest him in his plan and dedicate the model to him for the enrichment of his cabinet of curiosities. The duke was well disposed, after he had sent for a recommendation from Maestlin which proved very favorable. The negotiations and trials dragged on over several years. The first plan was to give the model the form of a graceful goblet. In due course, this was replaced by an ingenious planetarium, for which Kepler prepared detailed proposals with "blueprints." In the end, however, the plan was not carried out, because of the provoking failure to act on the part of the craftsmen entrusted with it and the difficulties inherent in the plan itself.

For the greater part of his visit in Württemberg, Kepler remained in Stuttgart. There he sought and found a place in the ducal castle at the so-called "Trippeltisch," at which the middle and lower ducal office-holders dined. In Tübingen he was a welcome guest; the fame of his discovery had spread and earned him esteem. Crusius, the well-known Hellenist, who had the habit of noting in a diary all the little occurrences of his daily Tübingen existence, even the seating order of the guests whom he invited, broke through his matter of fact style by the notation of the participation of Kepler at a meal with the interjection: *pulcher iuvenis.*

In August Kepler returned to Graz. He had been granted a leave of two months but had remained away nearly seven. His superiors, nevertheless, in consideration of the duke, were sufficiently generous to overlook this transgression. Moreover, Kepler had already composed a dedication to the estates of Styria for the book that was about to come into existence and which had such great significance, according to the testimony of the eminent Tübingen college professors.

After the senate of the university had given its approval, the printing began, at Gruppenbach's in Tübingen. Naturally, the senate had previously assured itself of the assenting judgment of its member, Maestlin. What Kepler had done, the latter explained, was highly ingenious, thoroughly worthy of publication and entirely new. Never had it occurred to anyone to derive *a priori* the number, arrangement, size and motion of the orbits, that is, as it were, to draw them out of the secret plans of the Creator. Now, it would no longer be necessary to derive the sizes of the orbits *a posteriori*, that is, from observations. Since those measurements are known *a priori*, it is possible to calculate the motions of the planets with far more success than hitherto. What Maestlin objected to was the unclear and at times confused presentation. He felt that Kepler had written his book as though all who came in contact with it would know the very complicated exposition of Copernicus and were completely at home in the mathematical material, as though others knew as much about it as he. This criticism provoked Kepler to make improvements then and there; he smoothed and enlarged the text in various places. To be sure, since Maestlin was on hand, he had to take charge of the main work, the supervision of the printing. This cost him much time and effort. The beginner, in his first work, had delivered much that was not ready for printing. Maestlin writes that day after day he went to the printing press, often two or three times in a day, in order to instruct the printer personally. He did not fail to hold up to his former pupil his own services on behalf of the completion of the book. In return, Kepler gave lively words of thanks. Certainly he exaggerates when he writes to his old teacher: "I have but slight basis to call it my work. In the origin of this work I was Semele, you Jupiter. Or if you would rather compare the work with Minerva than with Bacchus, then I as Jupiter carried it in my head. But if you had not performed the midwife's task as Vulcan with the axe, I should never have given birth." And when Maestlin, moreover, communicates to him that by caring for the printing he had had to postpone the composition of an opinion about the Gregorian calendar for which he was commissioned and consequently had drawn upon himself a reproach by the senate, Kepler gives him the consolation that his services in this work will bring him everlasting fame.

In the spring of 1597 Kepler received the first copies of his finished book. The title reads: *Prodromus Dissertationum Cosmographicarum continens Mysterium Cosmographicum de admirabili Proportione Orbium Coelestium deque Causis Coelorum numeri, magnitudinis, motuumque periodicorum genuinis et propriis, demonstratum per quinque regularia corpora Geometrica.* Shortened, it is rendered as *Mysterium Cosmographicum* or "The Mystery of the Universe." The little work, today rare and very valuable, then cost 10 kreuzers (less than a nickel). The author had to pledge himself to take two hundred copies from the printer, for which he had to pay 33 gulden. As evidence of his gratitude, he transferred fifty copies to Maestlin, who was to distribute them in Tübingen; in addition, he gave his teacher a gilded silver goblet, which he had received for casting nativities. From the representatives in Styria, to whom the work is dedicated, Kepler hoped, according to the custom of the time, for an appropriate "appreciation." However, he had to wait until 1600 for it; in the end, 250 gulden were presented to him and proved very useful in financing his involuntary departure from Graz.

Kepler's manner of thinking, which was entirely innate in him and formed from the most varied influences, is shown in the systematic presentation of his discovery. He treats the regular solids according to rank and class; indeed, for him they are not only representations with so many planes, edges and corners, but rather visible carriers of the relationships of measures which existed in the divine being from the beginning. He shows how close is the agreement between the distances of the planets from the sun which he has taken *a priori* and those furnished by investigation. He looks for reasons why this agreement is not complete. In every dilemma he knows how to help himself. The question, why, keeps reappearing. Why is the earth's position between Venus and Mars? In his arrangement, why is the cube in the first place from the outside in, between Saturn and Jupiter, the tetrahedron in the second, and so on? Why is the cube to be associated with Saturn? Why does the earth have a moon? Why are the eccentricities of the orbits precisely such and such a size? Five manners of approach to the examination of the world enable him to answer these and similar questions: the aesthetic which finds the principle of the beautiful primarily in symmetry; the teleological which begins with the assumption that "man is the goal of the world and of all creation"; the mystic by which he is convinced that "most causes for the things in the world can be derived from God's love for man"; the metaphysical which holds that "the mathematical things are the causes of the physical because God from the beginning of time carried within himself in simple and divine abstraction the mathematical things as prototypes of the materially planned quantities"; and finally the physical which starts from the principle that "each philosophical speculation must take its point of departure from the experiences of the senses." In his mind are crossed and intertwined teleological and physical principles, induction and deduction, unconditional veneration for the facts and a passionate drive for *a priori* thinking, theological and mathematical speculations, Platonic and Aristotelian points of view. His basically pious attitude is shown in the hymn in praise of the glory of God, with which he closes his little book.

One thought in the book is of very special importance for the further development of astronomy. With the answer to the question about the cause of the planet motions, Kepler follows an entirely new path. Here he already seeks a relationship connecting the periods of the planets with their distances from the sun. Assuredly, another quarter of a century had to elapse before he found the right law. But the born master was revealed by the fact that already in his youth he propounded this question. No less significant is the idea which guided him in his search. It is the new thought that in the sun there is situated a force which produces the planet motions, and which is so much the weaker, the further removed the planet is from the source of the force. To be sure, in his book he speaks of an "anima motrix," a moving soul; but already in a letter of this period he uses the word "vigor," force. In this idea is hidden the first germ of celestial mechanics. We shall see later how this germ developed further in Kepler's mind.

Originally our eager seeker after truth intended to point out, in an introductory chapter, the consistency of the Copernican doctrine with the Bible. At the request of the Tübingen senate, however, he had to omit this portion. The friendly letter on this subject which he received from Matthias Hafenreffer, the rector, throws light on the intellectual situation of the time: "I give you the brotherly advice by no means to support and defend that agreement openly; for many good people would take offense, and not unjustly, and your entire work could be entirely hampered or else burdened with the grievous reproach of having created disunity. For I have no doubt that, should that point of view be supported and defended, it would find adversaries, some of whom might be well equipped [with scientific armor]. If, therefore, my brotherly advice is heeded, as I certainly hope, so proceed in the presentation of such hypotheses clearly only as a mathematician, who does not have to bother himself about the question whether these theories correspond to existing things or not. For I am of the opinion that the mathematician has achieved his go when he advances hypotheses to which the phenomena correspond as closely as possible; you yourself would also withdraw, I believe, if someone could offer still better ones. It by no means follows that the reality immediately conforms to the detailed hypotheses of every master. I do not want to touch upon the unrefuted reasons which I could take from Holy Scripture. For in my opinion, here, not learned disputations, but rather brotherly advice is in place. If, as I strongly trust, you follow these and satisfy yourself with the role of pure mathematician, then I do not doubt that your thoughts will give the greatest enjoyment to a great many people, as they certainly do to me. However, should you attempt to openly defend and bring that theory into agreement with Holy Scripture, which God the Omnipotent and All-merciful may prevent, then I certainly would fear that this thing would lead to schisms and drastic measures. In that event, I should only wish that I, for my part, had never heard your thoughts which considered from the mathematical standpoint alone are excellent and distinguished. There already reigns more dispute in the church of Christ than the weak can endure." To Hafenreffer's great happiness, Kepler yielded, without, however, changing his standpoint. His answer is contained in a letter to Maestlin, where it says: "The whole of astronomy is not worth so much that one of the little ones who follow Christ should be angered. Since, however, the majority of scholars cannot rise either to the high conception of Copernicus, well now, we shall imitate the Pythagoreans also in their customs. If someone asks us for our opinion in private, then we wish to analyze our theory clearly for him. In public, though, we wish to be silent." However, several years later, when his scholarly position was secure, Kepler could no longer restrain himself. In the introduction to his *Astronomia Nova*, he developed explanatory axioms, which afterwards were universally taken over by the theologians.

Thus the work, with which Kepler introduced himself into the scholarly world, was completed. (That his name "Keplerus" was wrongly given as "Repleus" in the Frankfort Fair catalogue, which advertised the book in the spring of 1597, was, however, an annoying error.) He sent the book to various scholars asking for their opinion. What is preserved of such opinions in letters to Kepler and other documents is partly agreement, partly refutation or critical reserve. These observations reveal the deep contrast in the scientific and philosophical trends of that period which was so disturbed intellectually and politically. It has been mentioned that Maestlin, one of the ablest critics of his time, agreed completely. On the other hand, Johannes Praetorius, the professor from Altdorf, expressed himself as disagreeing entirely. He would not have anything to do with these sophistries. In his opinion, these things belonged to physics, not to astronomy which, as a practical science, could derive no profit from such speculations. The planet distances should be determined from observations; it signified nothing if afterwards they agreed with the size relationships of the regular solids. Professor Georg Limnäus in Jena arrived at an entirely opposite opinion. He is ecstatic that at last someone had again revived the time-honored Platonic art of philosophizing. The whole scholarly world should be congratulated for this work. Kepler would gladly also have learned the opinion of Galileo, who at that time lived in Padua. Galileo was seven years older and had already produced works on physics but had not yet made a name for himself as an astronomer. Kepler sent him his book. By return mail Galileo sent a few civil lines as answer, saying that in the short time he had read nothing in the book other than the preface, but that he was looking forward with pleasure to reading the rest which promised much that was fine. Kepler was not satisfied with this answer. In a warm and lively letter he entreated Galileo to stand openly with him in favor of the Copernican doctrine ("confide, Galilaee, et progredere") and urgently repeated the request for an opinion about his book. "You can believe me, I prefer a criticism even if sharp from a single intelligent man to the ill-considered approval of the great masses." Galileo remained silent. Yet a couple of years later, Kepler was informed by a friend from Italy (let it remain undecided whether rightly or wrongly) that Galileo included in his lectures thoughts from the book as his own.

Much more important for Kepler's life and works and of decisive significance for the further development of astronomy are the connections which he made with Tycho Brahe by presenting his book to him. Tycho, then fifty years old, was rightly considered the outstanding astronomer of his time. Recognizing that the discrepancies between theory and reality existing since Ptolemy and even after Copernicus could not be eliminated unless the realities were first established with greater certainty by observation, he had in decades of untiring work improved the art of astronomical observation in a manner hitherto unheard of. With his remarkable instruments and, on the broadest foundations, with the aid of numerous assistants, he had gathered an extremely valuable treasure of observations. Uraniborg, which this ingenious observer and organizer had built for himself on the Danish island Hven, was the intellectual center of astronomical research, the first and most important astronomical observatory at the beginning of modern times. After twenty active years, Tycho Brahe had quit this site of his activity as a result of disagreements and had found refuge in Germany. Just then Kepler's book, with a covering letter by the author, reached Tycho by a devious route. The latter's rich experience told him at once that there was something promising in the young investigator, and since he was accustomed to having a staff of young collaborators about him, he forthwith thought of winning Kepler for himself. He sent him a long letter, giving his opinion, carefully balanced in approbation and criticism, of the "Mystery of the Universe." He takes unusual pleasure in the book. Connecting the distances and the orbits of the planets with the symmetrical characteristics of the regular solids he calls a very clever and polished speculation. A great deal of it seems to be sufficiently correct. However, it could not easily be said whether everything could be agreed to. Certain details give him pause. Yet the zeal, the fine understanding and acumen ought to be praised. The opinion which Brahe pronounced in a letter[1] written at the same time to Maestlin is clearer and somewhat more critical: "If the improvement of astronomy should be accomplished rather

[1] ED. NOTE. *Tychonis Brahe Opera Omnia*, edidit I. L. E. Dreyer, VIII (1925), 52–5. The letter is dated April 21, 1598. See especially p. 53: "Si ab anteriore per corpora ista regularia dimensione facta, ea restituenda erit, potius quam ex accuratis observationibus a posteriori sumptis, veluti insinuas, utique nimis diu, si non prorsus in perpetuum frustra expectabimus, antequam tale quid a quoquam praestari poterit." Part of this letter, including the part quoted, is reprinted, in *Johannes Kepler Gesammelte Werke*, XIII, 204–5 (letter number 94).

a priori with the help of the relationships of those regular bodies than on the basis of observational facts gained *a posteriori*, then decidedly we will wait much too long for it if not forever and in vain until someone is able to accomplish this." Kepler was not entirely satisfied with the master's restrained caution. Yet the gap between the two men was bridged. And because Tycho Brahe invited the aspiring novice, the latter had a prospect for the near future which attracted him and which he might properly expect could be advantageous in various respects.

So, with one stroke Kepler had made a name for himself in all circles which had the knowledge of the stars at heart. The first promising step was taken. As with many men of genius, the great decisive conception of his life occurred in early manhood. It was no exaggeration when Kepler as a man of fifty, looking back, stated: "The direction of my whole life, of my studies and works, took its departure from this one booklet." "For nearly all astronomical books which I published since that time were related to some one of the main chapters in this little book, presenting themselves as its more detailed argument or perfection. And that was not because I had let myself be led by the love for my discovery (far be such a folly from me), but rather because things themselves and the dependable observations of Tycho Brahe have taught me that for the perfection of astronomy, for guaranteeing the calculation, for constructing the metaphysical part of heavenly knowledge and of heavenly physics no way can be found other than that which I outlined in this booklet either explicitly or at least sketched through timid expression of my opinions, since deeper insight was still lacking." We can understand if, in proud self-assurance contrary to his custom, he praised himself: "The success which my book has had in the following years loudly testifies that no one ever produced a first work more deserving of admiration, more auspicious and, as far as its subject is concerned, more worthy."

5. Marriage

The title describes the book as a "Prodromus," a first precursor of a succession of cosmographical treatises. Kepler had sundry further plans, however only plans, in his head. But now he had to carry out another plan which lay close to his heart, namely his marriage. It was noted above that already in December, 1595, he had been presented to a person who would be a proper match for him. He very soon caught fire. She was Barbara, the first-born daughter of the wealthy mill owner, Jobst Müller "zu Gössendorf" as he signed himself, who resided on the estate Mühleck which was situated in the township Gössendorf, about two hours south of Graz. She was in her twenty-third year, pretty and plump, as is shown by a medallion of her which today belongs to the Pulkova observatory near Leningrad. In spite of her youth, she already had been married twice and a few months before had become a widow for the second time. At the age of sixteen she was wedded to the well-to-do cabinetmaker, Wolf Lorenz in Graz, to whom she presented a little daughter, Regina. When he died after scarcely two years of matrimony, she shortly thereafter gave her hand to a man who, like Lorenz, was no longer young. He was Marx Müller, a respectable Styrian district paymaster or clerk. In spite of the elevated position which he filled by his jobs in the service of the province of Styria, this marriage was not happy; the husband was sickly; he brought with him from an earlier marriage children who were obviously ill-bred and was himself, as is reported, guilty of some irregularites in his service which became evident at his death in 1595. Jobst Müller, the lady's father, appears to have migrated in days past from the empire to Styria. He was a very active man, proud of his possessions, and very successful in increasing his money and property by all sorts of businesses and enterprises. Yet he himself was not of noble blood, no more than his first wife, Barbara's mother, who was born a Niedenaus.[1] However, with the ambition which was characteristic of him, he may have aspired to nobility. He did not bear the predicate "von Mühleck" frequently applied by biographers to Kepler's wife, Barbara. His son, Michael, first assumed this after the death of his father, Jobst Müller, in 1601; in 1623 Michael, in recognition of his and his forefathers' services to the empire and the Austrian reigning house, was raised to knighthood and the authorization was granted to him to sign himself "von und zu Mühleck" and to seal with red wax. Since Michael, however, left no masculine heirs, the title of nobility promptly expired again. These relationships, in themselves subordinate, must be clarified because, according to the universal account, they play a part in Kepler's marital affairs.

When the latter decided to sue for the hand of the rich Frau Barbara, most likely in January, 1596, two friends, the physician Dr. Johannes Oberdorfer, inspector at the Stiftsschule, and Heinrich Osius, former

---

[1] AU. NOTE. In the relevant literature the maiden name of Barbara's mother is given everywhere as Margarete von Hemetter. This information is, however, certainly wrong. It is based on a thoroughly unreliable note of a later date by Kepler's son, Ludwig (see footnote to p. 73). He confused his grandmother's maiden name with that of the wife of Barbara's brother, Michael Müller. The name Niedenaus, sometimes appearing in letters by its bearer, is attested by Kepler in his memorial writing at Barbara's death.

professor at this school and now deacon at the collegiate church, proceeded, in accordance with the prevailing custom, as "gentlemen delegates" to the father Jobst Müller in order to explore the position he and his relatives had taken toward the marriage plan and to recommend their client. Now, as is usually further related, the proud man made his consent contingent on proof of the suitor's noble descent. Therefore, Kepler is supposed to have gone to Württemberg to procure an adequate document in Stuttgart itself, the seat of the ducal government. This account is, however, surely incorrect. The reasons for Kepler's journey to Swabia, in February, 1596, have been given above. That the immediate cause was the wish of his two very old and sick grandfathers, he himself alleges in the official petition, in which he excuses himself for the infraction of the furlough, and there is no reason to doubt the correctness of this assertion. The importance, then, for him of the other errands which led him home, the preparation of his book for printing and the construction of the model, is established by the very circumstance that because of them he stayed away fully seven months when only two were granted him. The providing of that other document plays no role in this. Besides, he could not obtain one of that kind in Stuttgart; for that he would have had to turn to Vienna. And why should Jobst Müller have insisted on noble descent with his daughter's third husband after he twice previously had not only approved, but really had urged, her marriage with middle-class husbands?[1] No, his gruff opposition to Kepler's wish had its origin in the poverty of the wooer; Müller did not want to give his daughter to a husband who, in the poorly paid and little esteemed position of a school attendant with its inconsiderable income, promised a wretched

[1] AU. NOTE. The conventional presentation, which one biographer takes over from another, in the last analysis, goes back solely and alone to one note by Kepler's son, Ludwig. In Kepler's own observations about his marriage, so far as they are set down in writing, not the slightest clue is to be found. Ludwig Kepler intended to compose a biography of his father. The plan was not carried out. But one leaf is preserved, on which he had recorded a number of dates from his father's life. The following remark can be found there: "Ao. 96. In Württembergiam reversus ad inquirendam originem familiae, quia sponsae cognati nolebant admittere coniugium, nisi originis nobilitatem demonstrare posset." This "Synopsis Vitae" is, however, unreliable. In it there are various incorrect assertions; for example, place and date of birth of his father, Johannes, are equally wrongly reported. Since this synopsis in any case was assembled long after Kepler's death, assertions such as the above-mentioned going back to a time when the author, Ludwig, was not even alive are unreliable. Ludwig Kepler has here arbitrarily dreamed up something from his memory of statements by his father. During his visit to his native land, the latter may also have spoken about the family descent with his old Grandfather Sebald Kepler, who set a high value on it. In later years Johannes Kepler may have told his son what he knew about his knightly forbears. After all, he could well meet Jobst Müller with the information that his grandfather had been Burgermaster of a free imperial city, and that he did not need to be ashamed of his descent. Kepler undertook the journey to Stuttgart, not because of, but in spite of his courtship.

future. The rich man relied entirely on money and property; he had not the slightest appreciation of scholarly accomplishment. As for the rest, the proceedings of the marriage transactions are not entirely clear. While Kepler tarried in Swabia, his matchmakers continued their efforts to win over the bride's relatives. That they were successful, he learned from Professor Papius in Tübingen, who carried on an active correspondence with his old Graz friends. In Graz there had long been doubts about the mathematician's marriage; now the matter was settled; the bride was secure for him. But Kepler should hasten back to Graz. Papius also advised the suitor "To get ready to have a whole dress made at Ulm for you and your fiancee with very good silk fleece or at least the best double taffeta." However, almost a whole quarter of a year more passed after this advice before Kepler came to Graz again. As he returned home he experienced a great disappointment. No one congratulated him on his arrival as he had expected. Instead it was reported to him in secret that he had lost his bride. For half a year he had lived in hope of this marriage. The cause of this turn of events is not certain. A suitor should certainly not stay away as long as Kepler had done. He neglected to forge the iron while it was hot. Especially his Swabian countryman, the district secretary Stephan Speidel, esteemed because of his position, worked against this union with Frau Barbara. He wanted to have this good match for somebody else so as to increase his own influence; also, as Kepler himself frankly writes, Speidel wanted to see that woman better provided for. Some months passed, during which the distressed master slowly made himself familiar with the thought of a new course of life without, nevertheless, completely giving up his hope. The negotiations continued. The rector of the Stiftsschule also was active in favor of his mathematics teacher. It is apparent that such a union was not only a matter between bridegroom and bride, nor only between the relatives, but far more a concern in which the community took an active interest. Before his journey to Württemberg Kepler had given his binding word to the woman chosen by him. So he could now turn to the church authority to free him from his promise or else, as intermediary, influence the bride and her relations. The latter happened. The authority of the church government made an impression on those concerned. Besides, by this time they were afraid of the mockery of the people. In a joint charge, then, the stronghold was taken in January, 1597, and, on February 9, the solemn promise of marriage was celebrated, followed on April 27 by the wedding. This was celebrated, after the marriage ceremony in the collegiate church, in Frau Barbara's previous residence in Herr Georg Hartmann von Stubenberg's house in Stempfergasse[1] and with great splendor, according to the custom of the time. After all that went before, it is understandable that the celebration did not take place in the bride's parental home in beautiful Mühleck as would have been normal. One can imagine the sour expression of the bride's father at the festivities. From the commissioners, whom he had invited to the celebration, Kepler received, as "veneration," a silver cup worth 27 gulden. Also, at his request, his annual pay was increased from 150 to 200 gulden because he relinquished the dwelling in the school and moved to the Stempfergasse.

A week before his marriage, in a letter to Maestlin, the bridegroom analyzed his new economic situation: "My assets are such that if I were to die inside of a year, barely anyone could leave worse conditions behind after his death. I must make great outlays out of my own pocket, for here it is the custom to furnish the marriage most flashily. If, however, God grants me a longer life, then it is certain that I am tied and chained to this place whatever may become of our school. For here my bride has properties, friends, a wealthy father. It appears that I should not, after a few years, need a salary any longer, if that would suit me. Also, I could not quit the land unless a public or private misfortune arose. It would be a public one if the land was no longer safe for a Lutheran or if it was further pressed by the Turk, who is already said to stand in readiness with 600,000 men. It would be a personal misfortune if my wife should die. So a shadow hovers over me. Yet I dare not ask more of God than He in these days allots to me." It is not amiss to suppose that in the choice of his wife, consideration of her fortune was not the last thought by which Kepler was guided. Money is always important to him. He knows: "He who is in want is a slave; and scarcely anyone is that voluntarily." At any rate, these remarks show how he played with the thought of procuring an independent position for himself with his wife's fortune, a thought in which one forgets too easily that this freedom is often purchased by another more embarrassing dependency. Yet this dream remained a dream. The shadow, of which he spoke, would soon enough cast a gloom over his life. "Calamitoso coelo," beneath ominous constellations, he had celebrated his wedding, he noted for himself in his diary. The stars announced "a more agreeable than happy marriage, in which, however, there was love and dignity."

Just as he then and also later always associated character and fate with the heavens, so a few years later he expressed himself in a letter

---

[1] Au. Note. It appears doubtful whether the house today designated on a plaque as the house, Stempfergasse No. 6, is really Kepler's dwelling. At any rate, in the meantime the house has been changed by rebuilding.

about the influence of the heavenly bodies on his wife, without naming her. "If you behold a person at whose birth the good planets Jupiter and Venus were not propitiously placed, you will see that such a person can be righteous and wise and yet will lead a dismal and cheerless existence. Such a woman is known to me. She is praised in the whole city because of her virtue, modesty and humility. But yet she is simple-minded and fat. From childhood on she was sternly treated by her parents; barely grown up, she married against her will, a man of forty. After his early death, she married another of the same age, who was of livelier disposition; but he was not much of a man and passed the four years of his married life in illness. For the third time she, who previously was rich, married a poor man in a disdained position. Her fortune was withheld unjustly. She can have only one maidservant, who is deformed. In all transactions she is confused and perplexed. Also, she gives birth with difficulty. Everything else about her is similar. You can here recognize in soul, body and fate, the same character which is in fact analogous to the position of the stars. Yet it is impossible to say that this soul forged its fate, because that fate comes from outside and is something strange."

As Kepler wrote these words his view had become more critical. At first, however, he was happy about the new household and the hopes it contained. His seven-year-old stepdaughter, Regina, was also part of that which made him happy and which he loved. The thought of leaving Graz was abandoned just as was the thought of a church position. He knew where he belonged and had founded his social position more soundly through his marriage into an esteemed and established family. Through this tie his life was also chained to the grave events which were in store for the province of Styria and thereby was pushed in a direction that was decisive for his work and for the development of astronomy. In this, Kepler perceived the providential hand of God.

Great joy ruled in the house in Stempfergasse, as Frau Barbara presented a little son to her husband on February 2, 1598. He was christened Heinrich, a name common in Kepler's family. Again the stars are questioned and they promise all possible good: noble character, nimbleness of body and limb, mathematical and mechanical aptitude, fantasy, diligence, and so forth; the child would be "charming." A favorite thought of Kepler's, the conviction that the foetus was influenced by strong appetites and mental impressions of the mother, is expressed when he reports to Maestlin that the genitals of the little boy are malformed so that they look like a cooked tortoise in their receptacle. But cooked tortoises are a favorite food of his wife! The joy in the house was, however, of short duration; after sixty days the child died. "No day can soothe my wife's yearning and the word is close to my heart: O vanity of vanities, and all is vanity." The little daughter Susanna, too, who came into the world in June of the following year lived for only thirty-five days. Presentiment of death hovered over the soul of the grieving father as he carried her to her grave. "If the father would soon follow, this fate would not meet him unexpectedly. In fact, everywhere in Hungary, bloody crosses have shown themselves on the bodies of people and other similar bloody stains on the doors of the houses, the benches and the walls (what history shows as an omen of a general pestilence). As the first now in our city, so far as I know, I have glimpsed a little cross on my left foot, the color of which turned from blood red to yellow." The cause of death was the same with both children, "apostema capitis," probably meningitis.

6. The beginning of the Counter Reformation

Yet the domestic tribulations were not all that weighed upon Kepler. In the same letter in which he informed Maestlin of his little son's death, he gave the first news of the new danger which crept up. At that time, in the city, with which he had just bound himself most closely by his marriage, the tension increased from year to year. In that atmosphere, life with its joys and sorrows was enacted. Thus, not only his own existence, but also that of the whole community to which he belonged by creed was most severely threatened. The eighteen-year-old Archduke Ferdinand took the oath of allegiance of the representatives and assumed the rule on December 16, 1596, only a few months before Kepler's wedding. Then, after the prologue, of which we have spoken, began the drama which was later described by the unpleasant title, Counter Reformation. Although he played no main part therein, Kepler was nevertheless drawn into the catastrophe with which this drama ended. The detailed accounts which he gives in his letters of that time vividly supplement the documentary sources from which we gather the course of events.

In the capital of Styria two parties opposed each other with harsh hostility. The Protestant was supported by the majority of the middle-class and by the noble estates, who possessed special rights in military and financial questions. The strong prop of the Catholic party rested in the person of the sovereign and in the Jesuits. The Catholic Restoration party possessed able and learned leaders in Bishop Martin Brenner of Seckau and Bishop Georg Stobäus of Lavant and followed its far-flung goals with great hopes. The Protestants, on the contrary, did not show the same unity in the defense of their cause, no matter how zealously individuals lent themselves to the preservation of the freedom of religious practice. Hotspurs in both camps fanned the fire, causing vexatious incidents. The Protestants drew up their complaints and submitted them to the emperor, who, however, referred the petitioners to the archduke. The latter thus received the welcome opportunity to move against his opponents.

In the year 1597 his procedure still was limited to measures in particular cases. The situation, which became more acute, as a consequence in the following year led to the first great blow. From April 22 to June 28, 1598, the prince traveled to Italy, where he met with the Pope and visited the Holy Shrine Loreto. Here he, so the story goes, vowed to lead his country back to Catholicism. Various incidents on this journey, which were heard as rumors and immediately pointed out as omens, led the Protestants to expect bad things. "Everything trembles," Kepler writes, "in anticipation of the return of the prince. One says that he is at the head of Italian auxiliary troops. The city magistrate of our creed was dismissed. The task of watching the gates and arsenal was transferred to followers of the Pope. Everywhere one hears threats." Barely had the prince returned from his journey when new and agonizing incidents arose. In Protestant circles, caricatures of the Pope were distributed. The prince was angered. He sent for the chairman of the church ministry and declared: "You would spurn peace even if I would give it to you." Arrests were made. At the same time Protestant receivers of alms were maltreated and passed by in the community hospital. The Lutherans complained that they were taxed too heavily for burials. As thereupon the preachers in the collegiate church begged from the pulpit for gifts for their own hospital and for their own cemetery, there followed a prohibition by the prince. An attack by Lorenz Sonnabenter, the Catholic archpresbyter, followed this skirmishing and released the main blow. He forbade the evangelical preachers every exercise of religion, the administering of the sacrament and the consecration of marriages, using as a pretext the right which long belonged to every archpresbyter of the place, if his fees had diminished as a consequence of the exercise of that performance by other church servants. Thus the question of the recovery of clerical property and clerical right, which for a decade was handled in Graz in a theoretical manner, was transferred to the practical domain. The church officers objected emphatically. The matter gathered impetus. Appeal was made to temporal authority. The prince explained he owed his protection not only to the Protestants but also to his own coreligionists, and on September 13 published the order to the councilors to dismiss the preachers and abolish the entire collegiate chapter, church and school ministries in Graz and other cities within fourteen days. In a memoir dated September 19, the councilors begged for the repeal of the decree. The archbishop refused and decreed that the collegiate church be kept closed. On September 23 he ordered the preachers and collegiate teachers under threat of death to quit the city inside of eight days. The situation reached a crisis. Troops were mobilized and it looked as though open combat would ensue. The diet was hurriedly summoned. Because of floods not all could appear. Once again the councilors asked for repeal of the decree of banishment, which "pierces" them "to the very marrow of the bones." Yet, instead of the hoped-for relief, on September 28 a more severe edict was promulgated. The collegiate preachers, rectors, and school employees received, by virtue of the power of the sovereign, the order "that this very same day during the sunlight hours they each and all surely depart from the serene highness' own city Graz and its peace [burg-friede] and that subsequently within the previously given term of eight days they surely clear out of all his highness' land and after the end of the same prescribed eight days not be seen therein at the cost of life and limb." Nothing remained but to follow this order. So on that day the preachers and teachers, among them Kepler too, left, all on the advice and orders of the representatives, some here, others there, to Hungarian and Croatian territory, where the emperor ruled. Because they hoped for an early return, they left their wives behind. Maintenance was paid to them, also travel money. The hope for a return was futile. Kepler alone received permission to return to Graz, which he did at the end of October.

It is not clear on what grounds an exception was made with Kepler. He himself reports that he returned "on order" of servants of the prince. His friend Zehentmair, in a letter referring to a particular remark by the Captain-General Baron Herberstein, writes that from the beginning Kepler had been expressly excepted by the prince and would not have needed to leave the city. It is otherwise reported in the reference which the representatives gave their district mathematician two years later when he left the city for good. After he, too, had been expelled and dismissed as a teacher at the Stiftsschule, the representatives had "by the most obedient intercession humbly begged and procured" for him from the prince "*salvum redeundi conductum* and permission to remain in their town as a worthy district mathematician." The truth may never be known. At any rate, as a precaution, because the decree of banishment was general, Kepler had petitioned the prince to declare that his neutral job be excepted so that he would not be in danger if he tarried in the land. The petition was complied with and it was decreed: "His Highness will hereupon have granted out of particular mercy that the supplicant in disregard of the general banishment, etc., may remain here still longer. Yet he should in every way show proper discretion and avoid offense so that his Highness will not have cause to withdraw such mercy again."

How did it happen that an exception was made with Kepler? The petition of the representatives might lead one to think that there had been a desire to distinguish between the mathematics professor and the district mathematician, and that residence in Graz had been granted to the latter as the incumbent of a neutral office. Yet this reason was hardly decisive. Some biographers assume that the Jesuits, who would gladly have converted Kepler to Catholicism, had had their fingers in the pie; others say otherwise. At any rate, it can be said that if Kepler had been objectionable to the Jesuits, he, too, would have been affected. Various circumstances, however, testify that he enjoyed certain sympathies not only with the Jesuits but also at court. The prince took pleasure in his scientific discoveries, he was told. For his favor at court he credits a certain regimental counsellor, Manechio (probably identical with that Manicor who appears in various documents), with whom he was in contact. There is, however, still another connection which was very important for Kepler and required consideration. In the fall of 1597, the Bavarian chancellor Hans Georg Herwart von Hohenburg, through intercession of the Jesuit Father Grienberger of Graz, came to Kepler with a scientific question, which will be discussed later. From this first "feeler," there developed a correspondence extending over many years and bringing the two men very close together. The influential chancellor proved himself a zealous patron of the rising young astronomer and showed great sympathy for him. He also warmly supported his research. Herwart von Hohenburg was an ardent Catholic and a friend of the Jesuits. The correspondence between him and Kepler began at the very time when Duke Wilhelm der Fromme transferred the rule to his son Maximilian, a cousin of Archduke Ferdinand. Both these young princes previously studied in Ingolstadt under the care of Johann Baptist Fickler, who was a friend of the Jesuits. Fickler also came from a family in Weil der Stadt, related by marriage to Kepler's family. Since he now lived in Munich, Kepler took the opportunity in his first letter to Herwart to send his respects to Fickler and in doing so likewise recommended himself to the Jesuits. Fickler thanked Kepler immediately for the greetings. Herwart sent his letters to Kepler via the Bavarian agent at the imperial court in Prague, who directed them further to Ferdinand's secretary, Capuchin Father Peter Casal, and proposed that Kepler send his in the reverse order. All these circumstances raised Kepler from the mass of his colleagues and it is understandable that he found special consideration with the ruling Catholic party and was treated differently from the other teachers who lacked that kind of influential connection. It may be remarked in passing that a brother of Kepler's father had become Catholic and belonged to the Jesuit order. However, only very little is known about him.

Besides these advantageous circumstances, Kepler also recommended himself by his nature. At heart he was a peace-loving person. Not that he would have avoided argument nor docilely agreed with everyone. Quite the contrary. He loved a dispute and always advocated his cause vigorously. Only the means had to fit the point in question. That which was sacred in religion was to be taken up, administered and advocated by sacred means. In this domain, the most serious affair of conscience, neither outside pressure nor authoritative decrees from above should be allowed to extort a decision. Similarly, he considered it thoroughly unworthy and offensive when anyone supporting his religious convictions indulged in baiting and slandering others. He thought, spoke and acted according to the principle: *sancta sancte.* As a consequence, the outer difficulties and drawbacks in the proceedings which he witnessed did not oppress him as much as the deep need of his heart, which he felt in the face of the tyranny, intolerance, hate and volley of abuse. He would pray: "God arm the innocent soul of the young prince against his dangerous advisers." In a letter directed to Tübingen twenty years later he still held the attitude of the collegiate preachers responsible for the violent procedure of the Catholic party against his coreligionists: "In Styria certainly the whole trouble emanated from the fact that Fischer and Kellin spoke in purposely malicious and injurious words from the pulpit." Assuredly it was more than bad taste when, as Kepler relates in the same letter, the fanatical Balthasar Fischer in his fight against the worship of Mary made fun in the pulpit of the lovely image of the Virgin of Mercy by spreading out his robe and asking if it were seemly for women to crawl under it; still more improper, however, would it be if one painted monks under the cloak of the Virgin Mary. A similar criticism of his own preachers is to be found in a letter Kepler addressed to the margrave, Georg Friedrich von Baden, ten years after these events: "Some of the appointed teachers confuse the positions of teaching and ruling, want to be bishops and have an ill-timed zeal with which they tear everything down, defiantly relying on their prince's protection and power which they often lead to dangerous precipices. This has long ago ruined us in Styria. One could have often sent more modest and more exemplary people to us in Styria, or could have shown the young people in the universities the manner and way to behave in such dangerous places without offending one's conscience and with the necessary cleverness of serpents so that the rulers who are of a different faith may not be disquieted." This is a distinct slur against his former teachers in Tübingen, from where Kepler still heard ringing the titles "ravenous werewolf, Antichrist, Babylonian whore," as the Pope was there designated, just as indeed Maestlin now also sees the devil at work in Ferdinand's action. "We see," writes Maestlin in answer to Kepler's reports "with what raging fury the devil incites the enemies of the church of God, as though he wanted to devour it completely." To this, the way of thinking of his former student compares advantageously. At that time in his purely private notes the latter remarks of himself, "I am just and fair towards the followers of the Pope and recommend this fairness to everyone." He surely erred in one point if he actually did believe, as the cited remarks suggest, that the followers of the Augsburg Confession in Graz would have been left in peace if they had been less provocative. Ferdinand, in his powerful position, would then also have found a way to carry out his plan for the re-establishment of the Catholic religion in Styria. A thirty-years' war, with all its terrible horrors and devastations, had to rage over the German lands before it was understood that one cannot and may not suppress freedom of conscience by outside pressure and force. Even today, admittedly, this point of view is not universally accepted.

The peace-loving attitude, which Kepler displayed in such a sympathetic manner in those disturbed surroundings, had its foundation not only in his character and his noble way of thinking which both respected the conviction of his opponents and conceded to others a freedom which he demanded for himself. Even more, this attitude is connected with his position in regard to those dogmas about which Catholics, Lutherans and Calvinists disputed with each other. Not that he had taken the standpoint that dogma was immaterial and that it made no difference what people believed if only they lived correctly. This attitude has, indeed, been ascribed to him, but without reason. This shallow point of view, which thoroughly misunderstands the relation between faith and life, is a product of a later time, which turned its back to Christianity. Rather, Kepler was convinced that there is only one truth and acknowledged the obligation to penetrate it with all the force of his spirit. As we have shown, in his early spiritual struggles he had arrived at a conception of the doctrines concerning ubiquity and the Eucharist, which deviated from the Augsburg Confession in which he had been reared; regarding ubiquity, he leaned toward the Catholic doctrine, but regarding the sacrament, toward the Calvinist. Whereas he had previously concealed his dissent, he now felt forced to emerge from his reserve. It is reasonable to assume that some of the preachers and teachers who had been expelled did not like having their colleague and fellow religionist turn from them and alone receive permission to return to Graz, while they themselves had to bear the lot of exile. Was it not natural to suppose that he had bought this favor by concessions to the Catholic party? This interpretation is suggested by Kepler's later confession that he had felt compelled at that moment "to lighten his conscience" and had begun, in all modesty, to reveal his scruples to the exiled church members. One lightens one's conscience if something weighs on one. What weighed on Kepler was the realization that he could not agree completely with the people on his side, neither in attitude, nor in dogma. He confessed this to them. Yes, he had made concessions to the Catholics, as to the Calvinists. His conscience required this; he could do nothing different. He had to go his own way, the way which his conscience prescribed for him, whether or not that pleased others. If he thereby achieved an advantage with the ruling party, good. "I did not want to risk my fate in regard to this article [about ubiquity] in which the Papists were wronged." So he spoke to one side. The Catholics should not, however, believe that he belonged to them. No. Therefore, then, in the agitation of his heart he clearly confessed before their prominent follower, Herwart von Hohenburg: "I am a Christian. I have absorbed the Augsburg Confession from what I was taught by my parents, in repeated study of its foundation and in daily trials; I hold fast to it. I have not learned to play the hypocrite. I am in earnest about religion; I do not make a game of it. Therefore I am also in earnest about its practice and about accepting the Sacraments." Thus the man, who with pious zeal sought God, stood not above the parties but between them and it pained him that he had to get along without the consolation of belonging entirely and without reservation to one group. This was a sorrow which accompanied him throughout life.

Of the confessions by means of which he lightened his heart, little has come down to us; perhaps they were mostly oral. There exists one piece of writing in which he exposed in metric form his conception of the sacrament of Communion. More is gained from letters of his repeatedly cited friend, Zehentmair, to whom he had expressed himself in great detail. Unfortunately, every one of the letters which Kepler wrote to him is missing. However, since Zehentmair in his answers repeats the thoughts of his correspondent before he takes a position regarding them, one learns something from these. Once a fragmentary poem by Kepler was mentioned there. It contains much that is worthy of note "about the Papal church which in all Europe proceeds harshly and with hostility." Kepler should send the missing part; everything would be carefully preserved so that he should not be in danger. There is mention of a long letter of Kepler's which was essentially a *dissertatio philosophica*. Therein, from a lofty viewpoint, he had evidently developed his thoughts about the conditions and the political measures of the church. Zehentmair praises his friend, who has the rare combination of a rich and profound intelligence with great piety and who has a great ability to distinguish truth from falsehood. Zehentmair had been marvelously moved and uplifted by his friend's admonition about the humbled state of the church and about the general vexation. Who would judge otherwise concerning God's Providence and mercifulness? Yes, it is true, and every Christian must understand and find it obvious that, from the beginning of time, the lot of the church has been that it grows only under the cross and in persecution, that outside power is harmful rather than useful to it. Thus is it now also. The organization and community of the church are not essential. Thus by contemplating the wonderful guidance of God he arrives at the same decision as his friend: Suppose God withdraws from them the outer means of salvation, the Word and the sacraments, through which the visible congregation of the church grows together into a body, as well as the protection and help of the high lords. This would take place to make us believe in Him alone, perceive the force and power of the Word without human assistance and, as is proper for soldiers of Christ, learn to fight and conquer in the greatest weakness with the help of the Holy Spirit.

Kepler consoled himself with the same thoughts with which he tried to support and edify his friend. His situation after his return to Graz made this necessary. Being deprived of the services of his creed bothered him. He complains thus: "Driven out of our lands are the men through whose intercession I hitherto had intercourse with God; others, through whom I could have intercourse with God will not be admitted." Indeed, preachers still remained here and there in the castles. However, if a subject of the prince asked for the sacrament and one of these preachers complied with the request, then that preacher was banished. In addition, there were the external worries. The school at which Kepler had worked had been discontinued. To be sure, he still retained his none too high salary. However, the outlook for a raise which he had harbored was gone. "How can I presume in my bitter mood, when so many able men live in exile, to demand something more for my idle speculation." Do not the councilors think that of all teachers, the mathematician would have been least missed? Should I therefore depart from Graz, so he asked himself. But his wife clung to her possessions and to the hope for her paternal property. Money disputes with his wife's relatives caused many days of vexation and annoyance. He would also have had to leave behind his little stepdaughter, to whom he was attached. His father-in-law, the child's guardian, would have liked to estrange her from him anyhow. She has a fortune of some 10,000 gulden from her father; Kepler obtains seventy gulden yearly for the child's maintenance, and in addition he gets the yield of a vineyard and a house. All that would disappear. Besides there would be the danger that the child might soon be introduced to Catholicism. He concludes that he should remain for the present and hold out. His Tübingen teachers, to whom he even now feels closely tied and whom he asks for advice, are of the same opinion. No matter how they esteem their former pupil because of his outstanding talents, they still cannot use him in Tübingen, which, of course, they do not tell him.

7. Further scholarly work; studies connected with the notion of "world harmony"

The school inspectors, who were very well disposed toward their mathematician and happy that he remained with them, expressed the wish that he might apply his present philosophical leisure for the furtherance of the mathematical sciences. Kepler needed no inducement. Although he happened to live at a time when, as he says, anyone, no matter how gifted for intellectual work, ran the risk of having his sharpness of intellect blunted, his zeal slackened, and his enterprising will diminished, nevertheless Kepler's extraordinary energy, his intense drive to investigate, overcame all obstacles. He dragged into the domain of his studies a host of scientific questions, now stimulated from outside, now gushing forth from within. Herwart von Hohenburg willingly loaned him the books which he wanted but could not get in Graz. The reading started a flow of his own thoughts. "He who distinguishes himself by intellectual agility has no inclination to concern himself much with reading the works of others; he does not want to lose any time." Nevertheless, he still had to learn. A fine scent guided him in assimilating that which he later needed for his high attainments and in following the right tracks which promised new discoveries. He did not drop a question which seemed important to him; he always took it up from a new angle. His letters which gave details about his works are partly long, learned treatises. Even in the letters which tell about the events described above, the notices are always tucked between scholarly demonstrations.

Naturally, his book, and what was connected with it, still kept him busy. The plan which he had drawn for the continuation of this work, which, after all, was thought of and entitled a "forerunner," discloses the thoughts which spun round in his head. He wanted to write four cosmographical books. One about the universe, about the stationary parts of the universe, about the locus of the sun and the fact that it is stationary, about the arrangement of the fixed stars and the fact that they are stationary, about the unity of the world and so forth; a second about the planets which, in addition to a repeated elaboration of the basic idea of the *Mysterium*, should introduce research about the motion of the earth, about the relationship of the motions according to Pythagoras, about music and so forth; a third about the heavenly bodies individually, especially about the earth, about the origins of mountains, rivers and so forth; a fourth about the relationship between heaven and earth, in so far as these interact, about light, the aspects and the physical principles of meteorology and astrology. The plan never was carried out in this form, because Kepler's scientific activities developed along different lines. Yet one finds research concerning the above-mentioned subjects in other arrangement and in other connections in various of his later works.

It has been mentioned before that in these years he was much concerned with the construction of a planetarium, which was to illustrate his invention. It is regrettable that it was never executed.

The letters which he received in connection with his "world mystery" forced him to take a stand. An aggravating incident which happened to Kepler should be mentioned here. Carried away by his enthusiasm when his invention was successful, he had written about it to, and asked the opinion of, the then imperial mathematician, Reimarus Ursus, who had been commended to him. In youthful exuberance, Kepler bestowed on him the highest praise and placed him above all the mathematicians of his time. Ursus was silent, but without Kepler's knowledge published that letter in 1597 in an astronomical work in which he carried on a controversy in the sharpest terms against Tycho Brahe, with whom he was quarreling regarding the discovery of the so-called Tychonic world system; Brahe had accused him of plagiarism. Thus Kepler now stood between Tycho Brahe, with whom he had just entered a relationship which was important for him, and Tycho's opponent Ursus, whom Kepler had praised to the sky, although Ursus in no way deserved the praise. Assuredly, the latter had risen from swineherd to imperial mathematician, but he did not have the honest manner of Eumaeus[1] and could exhibit no special scholarly achievements. Brahe remonstrated with Kepler, who naturally tried to justify his action since he did not want to trifle away the favor of the great man. This delicate task was accomplished skillfully, without embarrassment. The guileless novice, which he had previously been, could learn a lesson from this experience; he knew now that not all men of science, no matter how high their titles, had the same pure intentions which inspired him and which he had also assumed for others. The matter, however, was not disposed of in one letter, even if the personal motive was cleared up; it recurred in many letters and later when Kepler worked with Tycho, the latter asked him to busy himself still further with the refutation of Tycho's most bitterly hated opponent.

What Tycho Brahe said about his "Mystery of the Universe" was more important than this fight. Besides the reservation by which he had judged the fundamental ideas, Tycho expressed a number of objections respecting individual figures which were used in Kepler's world plan. Indeed, the construction underlying this plan was not exact. The reason why the regular solids did not fit exactly between the planet spheres was sought by Kepler in the inaccuracy of the distances of the planets from the sun, taken from Copernicus. Only more exact observations could help to solve the problem. He did not have instruments at his beck and call. Tycho Brahe alone possessed observations such as he needed. Kepler yearned to get a look at these. No king, he says, could give him anything greater than instruments and access to good observations. How could he manage to become acquainted with the observational results of Brahe, who was so critical and who yet did not know how to make the most of his whole treasure of figures? "I did not wish to be discouraged, but to be taught. My opinion about Tycho is this: he has abundant wealth. Only, like most rich men, he does not know how to make proper use of his riches. Therefore, one must take pains to wring his treasures from him, to get from him by begging the decision to publish all his observations without reservation." Yet Kepler still had to be patient and defer the decision about the discrepancies in his world plan.

In still another domain he wanted to learn more about Tycho Brahe's experiences. The motion of the moon had been presented only inexactly and unsatisfactorily by the previous theories. To make

[1] ED. NOTE. Eumaeus, Odysseus' swineherd, gave his master a kind reception although Odysseus returned disguised as a beggar.

progress in this, Kepler zealously observed the solar and lunar eclipses and compared the observations with the calculations previously carried through on the basis of the Copernican theory. He reached an important positive conclusion. He was the first to trace the hitherto unknown so-called "annual equation" of the moon's motion, which is based on the fact that the period of the moon's revolution is somewhat greater in the winter than in the summer. His attempt to attribute this phenomenon to physical causes, letting the "vis motoria" of the sun compete with the "vis motoria" of the earth, shows that he followed an entirely new path hitherto untrod. The phenomenon of the reddish light of the moon at the time of lunar eclipse induced him to think carefully, especially about optical problems. Likewise Brahe's observations that the apparent diameter of the lunar disc at the time of solar eclipse is about a fifth smaller than that of the full moon at the same distance from the earth gave Kepler much food for thought. By this observation, for whose explanation he referred only generally to an "optica ratio," Brahe had arrived at the false assumption that total solar eclipses are, in plain terms, impossible. It remained for Kepler to discover the optical law on which that phenomenon rests. For, a few years later, he clarified for the first time the effect of small apertures on the appearance of images.

The main disagreement between Tycho Brahe and Kepler concerned their attitude toward Copernicus. The former rejected the new system of the world on theological grounds especially[1] and explained the motions of the planets by a hypothesis which appears as a reconciliation between Ptolemy and Copernicus. He let the planets Mercury, Venus, Mars, Jupiter and Saturn circle the sun. At the same time, the sun, together with its companions, was circling the earth, which remained stationary in the center. He also followed Ptolemy with regard to the rotation of the heaven of the fixed stars. Kepler completely rejected this system, which was presented in similar form at the same time by others, such as Roeslin and Ursus, and which met with approval in wide circles.[2] He wanted to have nothing to do with such a patchwork. He saw in it an inadmissible compromise. "Since we astronomers are priests of the highest God in regard to the book of nature, it befits us to be thoughtful not of the glory of our minds but rather, above all else, of the glory of God. He who is convinced of it,

[1] ED. NOTE. Tycho also raised objections founded on a consideration of the possible parallaxes of the fixed stars and the planet Mars. See J. L. E. Dreyer, *Tycho Brahe*, Edinburgh: Black, 1890, pp. 176 ff.

[2] ED. NOTE. In antiquity a similar geo-heliocentric system was presented by Heraclides of Pontus. Tycho also resented Roeslin's proposal of the system and considered it a plagiarism from his own. See Hellman, *op cit.*, p. 164.

does not lightly publish something other than what he himself believes in, and does not boldly change something in the hypothesis unless the phenomenon can thereby be explained in a more certain manner. Nor does he strive to excel those great lights, Ptolemy and Copernicus, etc., with the glory of new discoveries." Ardently venerating Copernicus, he modestly expressed his relation to him as follows: "Because I am absolutely convinced of the Copernican theory, a solemn awe prevents me from teaching anything else, be it for the glory of my mind or for the pleasure of those people who are annoyed at the strangeness of this theory. I am satisfied to use my discovery to guard the gate of the temple in which Copernicus celebrates at the high altar."

What worried Kepler most in his support of Copernicus was the latter's requirement that the sphere of the fixed stars possess an immeasurable diameter because no reciprocal displacement, no parallax, is shown when the earth travels about the sun. He refused to accept an actual infinite space. If he had to believe, he says, that the distances of the fixed stars can by no means be determined relative to the distance of the sun, then this one argument in the support of Copernicus would be more trouble for him than the concurring opinion of a thousand generations. To get to the root of the matter, he himself then engaged in observations and requested similar ones from Galileo, Tycho Brahe and Maestlin. He wanted to find out whether little differences could not be observed in the altitudes of the polestar at the time of the winter solstice and at the time of the equinoxes. To be sure, his instrument was very crude, hewn from a few beams. To Herwart von Hohenburg, who asked him about it, he jokingly answered that his observatory had arisen from the same workshop as the huts of our ancestors. One can well believe that the result was negative or at least very uncertain. It was a far cry from this crude observation to Friedrich Wilhelm Bessel who, in 1838, using an ingenious method, was the first to succeed in carrying out such a parallax measurement.

Other inquiries in which Kepler engaged related to chronology which, at that time, was the favorite hobby of numerous scholars. One of these questions concerned the chronology of the Old Testament. After painstaking exegetic procedures, in association with Maestlin, he sought to make the dates of the historical books agree, for the purpose of ascertaining the number of years which have passed since the first day of creation, and to determine the relative position of the sun, moon, and planets at the moment of their creation. This constellation, surely, must have been a particularly distinguished and symmetrical one! Herwart von Hohenburg was also a friend of chronological investigations. In his chronological studies he strove to clarify a place in Lucanus in which that Roman poet in his work about the civil war between Caesar and Pompey described in detail a fantastic constellation. In order to fix the date at which such an arrangement of the stars could have occurred, he turned to a number of scholars, among them Kepler. To please the important noble, Kepler took great pains with these calculations only to establish finally that the passage in question must have been an instance of poetic trifling with astrological rules. A further demand of Herwart's was concerned with the report of an ancient writer that Venus had been occulted by Mercury in the year 5 B.C. Barely had Kepler emerged from the tricky calculations which these questions required, when his patron came with a new request, the fulfillment of which caused no less trouble. Herwart desired an accurate determination based on historical sources of the date of the birth of Emperor Augustus and the construction of a corresponding astrological birth diagram for the purpose of explaining certain texts which had been handed down. This request, too, had to be fulfilled. Although, for the most part, all these studies gave Kepler more work than pleasure, and even at one point exhausted his patience, he still drew gain from them. He made himself familiar with the literature of the ancients, he got practice in certain astronomical calculations and penetrated into the confused relationships of the Roman calendar, all of which proved useful to him in his later studies.

Another advantage of corresponding with the Bavarian chancellor should not be overlooked. The latter exchanged letters with many scholars and, as would be expected in his position, possessed far-reaching connections, even at the Imperial Court. Since he everywhere recommended and praised the young district mathematician from Styria, he made him known in wider circles and so helped to prepare for him the way out of the narrowness of his Graz surroundings into a larger world. Moreover, it would be wrong to believe that the correspondence with Herwart had dealt only with these chronological researches. He was very glad to be informed about the other scholarly labors of his protégé and followed them with intense interest. Nor did this experienced man fail to give advice in difficult situations. He was a sober thinker, who refused to believe in astrology. With one critical word he always drew Kepler back to *terra firma* and forced him to an objective testing of his thoughts, whenever the wings of speculation had carried him away.

However, the range of questions which Kepler investigated reached still further. For in his letters there is much discussion about the magnetic declination and about the experimental arrangement which he employed in researches on the phenomena of magnetism. The inclination of the ecliptic and its presumable change in the course of time inspired him, in default of exact data, to "philosophical" reflections. The observations made by the Dutch on their famed northern excursions of 1594–1596, where they saw the sun rise several days earlier than they had expected by calculation, proposed a riddle which he wanted to solve. Finally, he started his weather notes which he carried on from day to day throughout a few decades. They were supposed to help him clarify the influence of the stars on the weather.

All the researches mentioned hitherto were carried on simultaneously, though they tended in different directions. Indeed, they supplied individual building blocks for later works, although at that time they did not yet lead to a completed construction. Yet, in the summer of 1599 just when his little daughter died and the clouds in the Graz sky gathered more threateningly, his efforts were concentrated on a single idea so as to establish the exact ground plan of one of his principal works. It is the idea of harmony. And the work which he traced out in its main parts but which was first to appear in mature form two decades later is his *Harmonice Mundi*, his "World Harmony." In those months, main portions of this book came into being, if not in final form, at least in outline and contents. Although we shall offer a more exact analysis of this famous work in a later chapter, nevertheless something must be said now about these basic thoughts, which were his favorites and which accompanied him during his whole life, consoled him, spurred him on, and charmed him, while drawing nourishment from his other successful astronomical investigations.

In the famous tenth chapter of Copernicus' book I, where the new world picture is sketched in short strokes, Kepler read the sentence "We find in this arrangement a marvelous symmetry of the world and a harmony in the relationship of the motion and size of the orbits, such as one cannot find elsewhere." In what does this symmetry consist, this harmony in the visible world? On what is it founded? How does man come to recognize it? God has not created anything without a plan and in His wisdom and goodness has made the world most beautifully. It carries in itself the features of the omnipotent creator and is His copy. To man, however, God gave a rational soul and thereby stamped him in His own image. In accordance with the whole character of his mind, Kepler felt himself called and driven to establish that sentence of Copernicus' by the trinity of the concepts prototype, copy, image, to unfold it in all its width and depth.

When we say harmony we think first of all about music. The sensation of euphony which various tones awake, whether they advance in succession according to set intervals or sound together simultaneously, belongs to the most direct experience of the human soul. Music which is built on these original experiences is better able than all words to express the basic emotions of human hearts. Forgetting itself and detached, the soul, affected and overcome by the power of the tones, sinks to the depths from which it is born. In blessed rapture it raises itself, carried by its wings, into the purest heights, in which it divines its final home. By sensuous means, music reveals a transcendental world in which everything is as it should be, in which will and law agree and truth in its beauty reveals itself to the perceiving mind. Whence comes this magic power, we know not, we only experience it. Music is given to man as a gift from heaven. To inquire into the physical condition for the realization of the individual tones and chords is a different matter which has nothing to do with the shading of emotion. The first discovery that two strings of equal tautness and quality produced a melodious sound, if their lengths were kept proportional to certain small whole numbers must have acted on men like a revelation. So the octave is sounded to the keynote when this relationship is 1:2, the fifth with the relationship 2:3, the fourth with 3:4, the major third with 4:5, the minor third with 5:6, the minor sixth with 5:8, the major sixth with 3:5. Does not this deal with a wonderful relationship? What does the immediate feeling of a pleasant consonance have to do with relationships of numbers? And why do not other numerical ratios such as 5:7 produce a harmonious sound? Between the realms of tones and of numbers which to a naïve mind lie far apart there is apparently a connection rooted in some remote reaches of the spirit.

That was the world of thought into which Kepler plunged. It had first been unlocked by the Greeks who, following a strong intellectual tendency, founded the science of harmonics, which passed as part of mathematics and occupied a central position in their system of education. Even if not entirely correctly, tradition attaches this accomplishment to the person of Pythagoras. Plato had presented his theory of harmony in the *Timaeus*. He attempted, by fantastic speculation based on the four first numbers, to set up an ideal scale, in which only the octave, fifth and fourth counted as proper consonances. Starting from these, he undertook to explain *a priori* the improper consonances of the whole and half tones. Kepler received particular stimulation from Proclus, the Neo-Platonist, whom he was already studying zealously. He felt that this passage was spoken from the soul: "Mathematics contributes the most for the observation of nature by revealing the well-ordered structure of the thoughts, according to which the all is formed...and presents the simple original elements in their harmonious and uniform composition, out of which the whole heaven was also founded, by taking the corresponding forms in their individual parts." The ancient theory of harmony was first transmitted by Boethius, the famous statesman and philosopher at the court of the Ostrogoth emperor, Theodoric. In the Middle Ages, Boethius' work on music was just as authoritative for the theory of harmony as was Ptolemy's *Almagest* for astronomy. Since Boethius' time harmonics was taught in the quadrivium together with astronomy, geometry and arithmetic.

Whereas, however, the Pythagoreans indulged in a confused, barely understandable number mysticism concerning the foundation of the harmonious relationships, Kepler consciously went his own way right from the start. "I do not wish to prove anything by the mysticism of numbers, nor do I consider it possible to do so." His conception of the nature of mathematical existence, his view of the foundation of geometrical concepts and figures in the divine nature here also formed for him the starting point of his profound reflections as they had done earlier when he introduced the regular solids into his cosmography. Man does not experience the many-sidedness of all possible geometrical forms by perception but, roused by his sensory perceptions, he finds them in his mind. "God wanted to make us recognize them, when he created us after His image, so that we should share in His own thoughts. For what is implanted in the mind of man other than numbers and magnitudes? These alone we comprehend correctly, and if piety permits us to say so, this recognition is of the same kind as the divine, at least in so far as we in this mortal life of ours are capable of grasping part of it." Here we already have the same thought that he expresses several years later in the lapidary form: "Geometry is one and eternal, a reflection out of the mind of God. That mankind shares in it is one of the reasons to call man an image of God."

Now he had to demonstrate that in the geometric forms one could find the above-mentioned musical harmonies in terms of special numerical relationships in order to explore, as he says, "the true nature of musical harmonies." It is quite certain for him that these "world shaping relationships" must be sought in the plane regular figures. Their "knowableness," that is the possibility of constructing them with compass and rule, provides the first distinction between these figures. Thus the three-, four-, five-sided and so forth are "knowable," but the seven-, nine-, eleven-sided are not. The latter polygons, as not knowable, have no existence. They are in no way a part of the divine plan of world structure. By now separating all cases in which the side of a "knowable" polygon cuts a section of the circumscribed circle in such a way that this part forms a relationship to the remaining part corresponding to a knowable figure, he succeeds in setting up a genealogy of basic harmonies which corresponds to the above-named seven musical harmonies. By this he believed he had reduced that which determines the foundation of harmoniousness in music to those special regular solids which have their origin in the divine being.

But not only in music, so Kepler with bold fantasy speculated further, do those "world-forming" relationships arise. "Nature loves these relationships in everything that is capable of thus being related. They are also loved by the intellect of man who is an image of the Creator." So one finds them in the meters of the poet, in dance rhythms and in the beats of music, perhaps also in colors (in the angles of refraction of individual colors of the rainbow), in smells and tastes, in the limbs of the human body, in architecture and above all in heavenly phenomena. Indeed, were they not to be especially expected precisely here where the most sublime order and regularity are seen? He believed he could trace them in the heaven in two ways, in the aspects and in the speeds of the planets.

The so-called aspects were among the innumerable requisites of the astrologers. They inquired what angle two planets in the zodiac form with each other, "how they look at each other," and attributed special significance to the angles of $0°$, $60°$, $90°$, $120°$, $180°$. Attention was also paid to the particular signs in which the planets then stood, if they were watery, fiery, etc., and if the planets were strong or weak in the houses in which they were. A distinction was made accordingly between good and bad aspects. Kepler rejected most of this theory. What he retained was the question of the angles which two planets form with each other. He believed in an effect on the "sublunary" world of nature, that is, on all the beings beneath the moon, "when the light rays of two planets make a convenient angle here on earth." Now such "convenient angles" are for him precisely those which arise when the zodiac is divided according to the above-mentioned harmonic relationships. This effect will, however, not be produced by the planets and their light rays by themselves or according to their position in the houses, but rather in such a manner that the animated sublunary nature, by virtue of an innate geometrical instinct, perceives this harmonic constellation. It is thereby unconsciously agitated with the result that the animated beings achieve with greater zeal and increased activity that for which they are created and ready. In order to explain this effect in its full range, especially on the weather, Kepler also gives the earth a soul. "The natural philosophers may say what they want, there exists in the earth also a soul." What effect can a geometric relationship or a harmony have on the earth? He answers this question by an example: "It is the custom of some physicians to cure their patients by pleasing music. How can music work in the body of a person? Namely in such a way that the soul of the person / just as some animals do also, understands the harmony / is happy about it / is refreshed / and becomes accordingly stronger in its body. Similarly the earth is affected by a harmony and quiet music. / Therefore there is in the earth not only dumb, unintelligent humidity / but also an intelligent soul / which begins to dance when the aspects pipe for it. / If strong aspects last / it carries on its function more violently by pushing the vapors upwards / and thus causes all sorts of thunderstorms: while otherwise / when no aspects are present / it is still and develops no more exhalation / than is necessary for the rivers." To strengthen this theory of aspects, Kepler emphatically refers to experience. "It is experience which furnishes faith in the effectiveness of the aspects, experience which is so clear that only someone who has not tested it himself can deny it." He knows himself safe against superstition. He is fully cognizant of the thousandfold interplay of matter, circumstances and causes which one could not know before. As a consequence, in his general astrological prophesying he would be no more influenced by the heavenly signs than by that which physiognomy, temperament and sickness predict. For the brooding scholar, heavenly influence is only one of the causes which determine the perpetually changing conditions and behavior, the individuality, the ups and downs in mood and action of animated beings. This cause is founded on the nature of the soul, in so far as the nature of the Creator, eternally concerned with geometry, is mirrored in it.

Yet not only with the aspects do the "world-forming" relationships appear in phenomena; Kepler finds them also in the velocities of the planets. It is the old idea of the harmony of the spheres in new form which lends enthusiasm to the new Pythagoras when he is in the swing of his soaring speculation. "Give air to the heaven, and truly and really there will be music," he announces triumphantly. But since there is no air, there appears a "concentus intellectualis," a spiritual harmony, "which pure spirits and in a certain way even God sense with no less enjoyment and pleasure than man experiences when listening to musical chords."

What is his alleged discovery, the "iucundum theorema," about which he reports elatedly in his letters? In his *Mysterium Cosmographicum* he had, as we have seen, already remarked that the rate of increase of the periods of revolution of the planets is greater than that of the paths; if the distance from the sun is doubled, the period of revolution is more than doubled. The relationships of the velocities which Kepler assigns to the individual planets to do justice to this phenomenon are derived from musical intervals and thus from his basic geometric relationships. By repeated trials he succeeds in accommodating all these basic harmonies. Nor does he lack reasons why in an individual case exactly this and no other interval is to be put between the planets. Should the calculation somewhere not come out right, then Pythagoras ought to rise again to instruct him; but he does not come, "unless his soul has migrated into me." And when Herwart objects to his ideas, because the whole theory is based on hunch and conjecture, then he rejoins: "Not every hunch is wrong. For man is an image of God, and it is quite possible that he thinks the same way as God in matters which concern the adornment of the world. For the world partakes of quantity and the mind of man grasps nothing better than quantities for the recognition of which he was obviously created." The distances of the planets from the sun, which Kepler calculated according to his new theory, admittedly agreed with the Copernican as little as those obtained in the *Mysterium Cosmographicum*. His endeavor to explain the eccentricities of the planet orbits, that is the distances of the centers of the orbits from the center of the universe, by *a priori* reasoning on this basis, did not lead him to his goal. He himself felt that he did not get any further with this method of trial and error. He needed more exact data from observation. Again he looked toward Tycho Brahe, who alone could provide him with these. "I wait for Tycho alone; he should communicate to me the properties and arrangements of the paths and the differences of the individual motions. Then, so I hope, I will some day erect a magnificent structure, if God keeps me alive." On December 14, 1599, he communicated to Herwart von Hohenburg[1] his plan for the work. It was to be divided into five parts. He wanted to finish it as soon as possible but fate had other plans for him.

8. *Kepler's oppressing situation and his visit to Tycho Brahe*

Having spoken of Kepler's activity as a scholar, we now turn to his first visit to Tycho Brahe in Bohemia, his final banishment from Graz and his move to Prague. These events are so important and decisive, both in determining the course of his life and in shaping the development of astronomy, that a detailed description of them seems indicated. Though they also brought him painful experiences, they led him to

[1] ED. NOTE. *Johannes Kepler Gesammelte Werke*, XIV, 100 (letter number 148).

the summit of his accomplishment and of his fame and prepared the ground on which, in truth, a new science of the heavens arose. Without Brahe's observations, Kepler could never have found his planet laws, and he was the only one of his time who, having those observations, could have made this wonderful discovery. Anyone who likes to contemplate the profounder relationships in human life and history witnesses a unique drama in the ever memorable meeting of the two men, the genius in observation and the genius in theory, both alike in their enthusiasm for the wonders of the heavens but entirely different in their way of thinking, in their character and their conduct. Kepler, with his pious nature, recognized and respected an act of Providence in this meeting. Fortunately the extant documents furnish such rich sources that it is possible to give a detailed account.

While Kepler became absorbed in peaceful research and bent his ear to the heavenly harmonies, his surroundings were torn by hideous dissonances and were made ready for combat. The regulations of the Counter Reformation proceeded in a more severe form. The shepherds had been driven away; now the herds had to be destroyed. Ever harder was the screw turned. After the expulsion of the preachers from Graz, the Protestant citizens proceeded to visit the divine service of their faith on the neighboring estates of the nobility and to receive the sacrament there. This was now made punishable and the people were forced to have their children baptized as Catholics and to be married according to Catholic ritual. Kepler himself received an order for a fine of more than 10 talers for evading the city clergy at the death of his little daughter. This fine was reduced by half at his request, but he had to pay the other before he could bury the child. Naturally, it was not very long before the expulsion of all Protestant clergymen still in the land was also ordered. All who received them were threatened in body and property. Attendance at other than Jesuit schools was forbidden; the installation of Catholic clergymen in the collegiate church was demanded. Anyone who sang hymns in the city, or read collections of homilies, or Luther's Bible, made himself liable to banishment from the city. Heretical books had to be rooted out and destroyed; barrels and chests, which held books, had to be opened and examined in the presence of the archpresbyter. A watch was set at tolls and gates to keep heretical books out of the city. All these and similar ordinances produced the most violent excitement. In turning down complaints reference was made to the forced conversions previously carried out in Saxony, Württemberg and the Palatinate. There were serious riots in city and country. Threats whirred through the air. Rumors circulated arousing fear that soon there would be no place for a Lutheran in the city and that someone who wanted to emigrate would no longer have the right to take away, to trade or to sell his possessions. That these rumors were not unfounded was soon to become evident.

In such circumstances, Kepler peered uneasily into the future, when he turned from his studies back to reality. It was just common sense to examine the possibilities of a change of residence, so that he would not be taken unawares by events. His earlier intention to hold out became untenable. By this time circumstances were such that he had to say to himself: "No matter what fate might await me if I move elsewhere, I know for certain that it will not be worse than that which threatens us here so long as the present government continues." What so far had held him in Graz had been consideration for his wife and her possessions. But now he says to himself: "I may not regard loss of property more seriously than loss of opportunity to fulfill that for which nature and career have destined me." The school inspectors tried to hold him; they were well disposed toward him. But the councilors, on whom his position ultimately depended, did not all have the same understanding. Some of them scorned his mathematical speculations because they were useful to no one. This is not the time to study but rather to wage war, they thought. Kepler was oppressed by the knowledge that they left him his stipend, not because they expected profitable activity from him, but rather out of mercy and fear that they might make themselves unpopular elsewhere in the empire.

But whither should he turn? To return to Württemberg, to enter upon a clerical post, as he had firmly intended at the time of his departure, was "for most weighty reasons" not possible for him. "I could never torture myself with greater unrest and anxiety than if I now, in my present state of conscience, should be enclosed in that sphere of activity." We know that he no longer agreed with the Tübingen orthodoxy. He contemplates a philosophical professorship in a university and is confident that after the passage of some time he would be one of its ornaments. In his need, he turns to Maestlin for advice and, in touching words, describes his great distress. His old teacher should advise him whether there were good prospects for his career in Tübingen, or if he had better consider another university. If Maestlin cannot promise anything, he should at least tell him how expensive maintenance would be for a family in Tübingen. He made careful inquiries with Herwart wishing that he, Herwart, were master of his destiny or at least interpreter and mediator between him and the master of his destiny, because he had every reason to hope for the best of everything from his patron. But Herwart happened not to be the master of his destiny and needed to be discreet in this delicate situation out of regard for his position. Maestlin, however, answered miserably that he did not know how to advise him. "Had you but consulted wiser and politically more experienced men than me, because in these matters I am as innocent as a child." All he was still able to add were the prices of grain and wine. Indeed, in Tübingen one wished the former pupil all luck, only he should not seek this luck in Tübingen. The professorial college, linked by the bonds of kinship and marriage, would have felt itself threatened in its secure tranquility by this young firebrand. Certainly, he did not become any more acceptable by having written to Tübingen: "Who knows what kind of situation will await you after the events here? With you, too, is the harvest of God's wrath grown to maturity." It is not pleasant to hear something like that.

In these trials the eyes of this forlorn man fell naturally on Tycho Brahe who had, indeed, invited him. Here gleamed a star of hope in the surrounding dark. Here he could find not only what he needed for the solution of the uncertainty in his astronomical inquiries. Here also a better chance to look around for a new dwelling place and sphere of action would offer itself. So such a visit would bring a twofold gain. He had learned from Herwart that in the meanwhile Emperor Rudolph had called Brahe to the court as mathematician and allowed him a glittering salary of 3,000 gulden. The great astronomer had come to Prague in June, 1599. Because he could not find peace for his researches there, the generous emperor had placed the Castle of Benatky at his disposal. This is located twenty-two miles northeast of Prague in a pleasant region on the Iser. In August he had moved there. For Kepler there now remained only the finding of an opportunity to travel to Prague without cost. Soon such an opportunity appeared. At the beginning of January, 1600, Johann Friedrich Hoffmann, Baron of Grünbüchel and Strechau, a member of the Styrian diet and councilor to Emperor Rudolph, was about to return to the Prague court. He was very kindly disposed to the district mathematician and also had already come into contact with Tycho Brahe. He was ready to take his protégé with him and personally introduce him to the imperial mathematician.

In the meanwhile, in December, in a very friendly communication, Brahe had repeated his invitation to Kepler and expressed the hope he would come to Prague not because he was obliged to do so by the disfavor of fate, but rather of his own free will and because he desired joint studies; in any case he would find Tycho ready to advise and help him and his family. But Kepler had started for the Bohemian capital on January 1, before this letter arrived. Upon his arrival he stayed in the city for a few days as the guest of Baron Hoffmann. Brahe was overjoyed when he learned at Benatky of Kepler's impending visit and promptly on January 26 wrote him a welcoming letter[1]. "You will come not so much as guest but as very welcome friend and highly desirable participant and companion in our observations of the heavens." On February 4, Kepler rode out to Benatky with one of Brahe's sons and an assistant of Brahe's. Baron Hoffmann also gave him a specially written recommendation in which he said that the newcomer was beyond all praise and for many reasons worthy of favor and protection.

That day there came face to face for the first time the two men who had searched for each other from afar and each of whom expected much from the other. One was Tycho Brahe, the recognized prince of astronomers, who at fifty-three felt old age creeping up, but who with the fabulous vitality of an extrovert was still able to assert himself. Concerned with his age, he wanted to bring his plans to a safe haven before his career was closed. He distinctly perceived that for the conquest and working up of his prodigious material he needed an assistant with young and different abilities. The other was Johannes Kepler, the newcomer in the world of scholars, who at twenty-eight, in happy self-assurance, stood at the start of his career and only a short while before had rendered the first proofs of his extraordinary ability. He proudly carried in his head novel and completely original ideas and with the impatience of youth sought precise information from Tycho. Searchingly and hesitantly, they approached each other, in order to bridge the gap between them. Brought up as a noble, the Dane possessed the freedom and superiority furnished by noble birth and property. In his youth, accompanied by a tutor, he had traveled and become acquainted with the world. He could associate with kings and princes and his name was of consequence wherever he went. In the face of resistance, which increases the higher one's demands and claims, he maintained and upheld, he even stubbornly exaggerated, his rights. His haughty nature extorted subordination and adaptation from everyone who was dependent on him. He did not much care that his overbearing nature made enemies for him. Nor did it matter to him that, at his departure from Denmark, the reproach was made by church people that he had received no sacrament for eighteen years. What a contrast did the young Swabian present. His background was narrow

[1] ED. NOTE. *Johannes Kepler Gesammelte Werke*, XIV, 107–8 (letter number 154), especially: "Aduenies non tam hospes quam Amicus gratissimus nostrarumque in Coelestibus contemplationum, per ea quae nunc ad manus habeo, Instrumenta spectator et socius acceptissimus."

and middle class. He had studied on scholarships and become acquainted with the world more from books than by free intercourse. All he understood and knew was dependence on others, and of necessity he was compelled to defend his inner freedom against the forces which threatened it. He considered himself unfortunate when he found himself in disagreement with those who belonged to him, and was very desirous of regularly receiving the sacrament. As a little man who continued to receive his 200 gulden annually because his superiors took pity on him, he faced a great man, whom the imperial favor (in spite of the opposition of some councilors) had granted an income, such as none at the court, neither counts nor barons, drew even after a long period of service. These personal contrasts were comparable to those of their immediate surroundings. Tycho Brahe was accustomed to having many people about him. Not only did he have a numerous family, two sons and four daughters, but he was also surrounded by a staff of assistants and students who helped him in his work. All met at the common meals which dragged on interminably. A loud and lively traffic filled the rooms of the castle. In addition, this restless man had various structural alterations made, so that everything should be as he wished and was accustomed to. As a consequence, there was a coming and going of craftsmen and messengers, and there were discussions with officials which, to his annoyance, naturally did not always go smoothly. Amidst this bustle, Kepler may often have thought of the quiet of his study in the Graz Stempfergasse.

Among Brahe's co-workers there were two in particular with whom Kepler now and later again had dealings. One was a Westphalian nobleman, Franz Gansneb Tengnagel von Camp. He had lived with the astronomer for two years at Uraniborg and departed with him. Although he never earnestly busied himself with astronomy, he was arrogant toward the other assistants; he occupied a firm position in the family because he was about to marry one of Tycho's daughters. It was he who later provided some troublesome hours for Kepler. The other was Christen Sörensen Longberg, or Longomontanus, as he is called. He had previously worked eight years on the Island of Hven with the master, and thereby earned his confidence. Now he had come to him again just a few weeks before Kepler. Because he was very familiar with the situation in Brahe's house and with all the work, Longomontanus had some advantage over the novice. Jealously guarding his position, he let Kepler observe this. Some years later, he reproached Kepler offensively for occupying himself with the question of the moon's motion with which he himself had been entrusted by Tycho. Kepler replied that conditions were not the same with astronomers as with smiths, where the one made swords, the other wagons.

Naturally, in Benatky the observations could not be continued on the same scope as formerly at Hven, even though Brahe in his restless zeal urged and pushed for something always to be done. Most of his instruments were not yet set up, and the important big ones still had to be sent from Denmark. Assuredly, Kepler took little active part in the observations because he was not suited to this on account of his weak vision. But so much the more eagerly did he participate in the theoretical works, which were taken up according to the established plan. When Kepler arrived, Longomontanus was occupied with the theory of the planet Mars, which offered special difficulties. However, because Longomontanus met with some success in representing the motion in latitude yet did not succeed with the longitudes and was stuck, the new arrival, at his own wish, undertook this difficult task, while Longomontanus was assigned the theory of the moon. We shall soon hear what significance this distribution of work had for Kepler. He had been led to Bohemia by the hope that he would immediately learn finished values for the eccentricities of the planet orbits and for the distances of the planets from the sun. Had he obtained these values, he would have been able to test his discovery in the *Mysterium Cosmographicum* and his harmonic structure. But this hope proved ill founded. For the most part, these values still had to be calculated from the actual observations. In addition, Tycho Brahe turned out to be very miserly and reserved in the communication of his results and the disclosure of his observational material. He gave the inquisitive questioner, as the latter reports, no opportunity to participate in his experience, except that while eating, in the discussion about other matters, he one day mentioned the apogee of one planet, the next day the nodes of another. Only when he saw how his new assistant treated the theory of Mars with bold courage, did he disclose the observations of this one planet to a greater extent. Soon Kepler intelligently realized the condition of Brahe's work; he knew what was there and perceived what was still lacking. His verdict about this is to the point: "Tycho possesses the best observations and consequently, as it were, the material for the erection of a new structure; he has also workers and everything else which one might desire. He lacks only the architect who uses all this according to a plan. For, even though he also possesses a rather happy talent and true architectural ability, still he was hindered by the diversity of the phenomena as well as by the fact that the truth lies hidden exceedingly deep within them. Now old age steals upon him, weakening his intellect and other faculties or, after a few years, will so weaken them that it will be difficult for him to accomplish everything alone." With this insight, Kepler came to understand the task which lay before him. He considered himself summoned as the architect who should erect the new structure; he felt the power in him to accomplish the task which no one else in Germany could carry out.

Because of this condition of affairs Kepler saw himself constrained to stay longer than he had originally contemplated. If he did not want to lose the main object of his journey, he must, as he well saw, remain with Brahe one or two years. He thought that in that time he would be finished with what he needed to carry out his plans. His purpose would, indeed, also have been served had he been allowed to copy Brahe's observations. However, as he well understood, Brahe would not permit that, because they represented a treasure on which he had spent his life and his fortune. On the other hand, Kepler considered it an advantage during a longer visit to wait from afar for further events in Styria and hoped that by then a satisfactory solution for the choice of his future residence might also be found. In order to clarify his own thoughts and to create a basis for discussion with friends whom he wished to ask for advice, he wrote down his pertinent considerations in great detail. How can I obtain the consent of the councilors to so long an absence from my position of employment, he asked himself. Should one ask the emperor's intercession? Should he himself seek it? Should Tycho Brahe write? Each proposal had its pros and cons. If the emperor were to attend to it, then his father-in-law's mouth would be shut. The latter considered it disdainful to emigrate from Styria with wife and daughter, as though driven away by poverty, the most disgraceful and worst crime of the time. However, the emperor would turn to the archduke, not to the diet. The diet might take that badly. However, if he were to turn to this body, then he would do it in the tone of a command. What would then be the attitude of these representatives? And who recommends him to the emperor? Tycho? He would prefer someone else because it appears that Tycho is not in the good graces of all those who matter. He could not very well write himself. The diet might believe that he sought personal gain which was not the just conclusion. He merely wanted to invest his salary in the best possible way, as long as the school no longer existed. Should Tycho petition, full justice would not be done to the authority of the diet. He might be turned down or completely discharged, which would weigh on him because of wife and father-in-law with their retinue. Now, if he had consent, economic difficulties would present themselves. He had to bring his wife along to Prague, too, because he could not leave her for such a long time without damaging his reputation. Who is answerable for the cost of the trip? Who takes care of their property? Should he neglect these for the sake of astronomy? He would do that had he not been appointed administrator of this property. Furthermore, should he call for his wife himself or have her brought by one of her relatives? The first would be best. A further question: Where should he live? Prague would be suitable for his studies. Here the nations came together, here he could best care for his affairs. Here also a friendly tone prevailed among the German speaking people, so that his wife could find comfort as recompense for the absence of her relatives and friends. To be sure, everything is expensive in Prague. In Benatky there are only a few Germans; one would be lonesome there. Tycho's dwelling is crowded, the family commotion big. Kepler did not want to involve his own family in this. They were accustomed to quiet and simplicity. On the other hand, collaboration with Tycho required proximity to him. Finally, the main question: Would it not be better for Kepler to be in the employ of the emperor and in the bargain dedicate himself to Tycho, than to be solely dependent on the latter? However, the emperor paid wages irregularly. Kepler's family would often get into financial difficulties which his wife would consider extreme misery. If he obligated himself entirely to Tycho, would the latter not take too complete possession of him? Such a situation would be detrimental to his vocation and his studies and might be taken in bad part by the college councilors. Yet perhaps this would be prevented by the form of the consent. At any rate, he would have to set certain conditions, if the matter depended solely on him. If Kepler's wife wanted to live at Benatky, Tycho would have to place an entirely separate dwelling at his disposal. Tycho would have to arrange for wood and declare himself ready to provide a definite quantity of provisions, meat, fish, beer, wine, bread. In any case, Kepler could not be satisfied by general promises. Furthermore, Tycho might impose on him only such astronomical tasks as were necessary for the publication of the planned works. Besides, Tycho must not prescribe time and matter for his studies, but must place confidence in him. Since Kepler would need no spur, rather a rein, in order to avoid consumption, Tycho must leave him free time during the day if he had worked into the night for him, as often occurred. Tycho would have to pay him 50 gulden four times a year for his work and would have to procure a salary for him from the emperor. Out of this, Kepler, from time to time, would repay Tycho for what he had received from him.

In addition to this memorandum, Kepler composed still another, setting the requirements differently for a collaboration with Brahe.

On the advice of Longomontanus, he entrusted this to Johannes Jessenius, the Wittenberg professor of medicine, then likewise staying in Prague. Jessenius was on friendly terms with Brahe, and Kepler also came in close contact with him. The draft was not to be surrendered to Brahe; rather, it was to form the basis for discussions, which Jessenius was prepared to hold with him in Kepler's name. But it passed into Brahe's hands anyhow, although he must have been hurt by the way certain parts of it were expressed. Since certain difficulties soon arose when the two men lived together, it was all the more necessary that the conditions be settled exactly. As early as March, Brahe wrote to Baron Hoffmann about these difficulties, expressing the hope they would be cleared up easily.[1] Further particulars were not given in the letter. One can, however, easily imagine what oppressed Kepler. In Graz, however much he was cramped and bound in other respects, he had enjoyed complete freedom in his studies. Now he had to fit an organization which to him was strange and sometimes uncomfortable. Without knowing or intending it, Brahe, in the lordly manner which he showed toward his assistants, had made Kepler, too, feel his dependency more severely than he could endure. Yet, Kepler considered himself above being an amanuensis. He became more resentful the more he perceived on the one hand Brahe's superiority in the conduct of life and on the other hand his own superiority over both the other assistants and the master himself in questions of theoretical astronomy. Besides, the noisy surroundings in Brahe's house, and the style of living with which Kepler was entirely unfamiliar, were thoroughly repugnant to him. In the document mentioned above he showed this distinctly, saying that it would no longer be possible for them to live together satisfactorily, because the endless confusion in the household exasperated him and tempted him to be abusive beyond measure in talk and argument. So, near the outset there arose a latent tension which soon exploded, nearly completely severing the bond between the two.

On Wednesday, April 5, in Easter week, a discussion about the conditions of their collaboration took place between Brahe and Kepler in the presence of Jessenius. There appears to have been a sharp conflict, although Brahe exhibited considerable willingness to meet the conditions set by Kepler.[2] At any rate, Brahe believed it advisable to write down his remarks at the discussion immediately and to have them

[1] ED. NOTE. Printed in *Tychonis Brahe Opera Omnia*, edidit I. L. E. Dreyer, VIII (1925), 254–6.

[2] ED. NOTE. See Kepler's written explanation of April 5, 1600, printed in *Tychonis Brahe Opera Omnia*, VIII, 296.

certified by Jessenius. When Kepler explained that he wanted to return to Prague the next day, Brahe attempted to hold him for a couple of days more, because he expected an answer to arrive shortly from the imperial court, where he had taken steps which showed good prospects in regard to Kepler's employment. Kepler, nevertheless, insisted on his plan and departed on April 6 with Jessenius. Shortly before, there was another altercation. At his departure Kepler showed signs of contrition, and apologized for not being able to control his feelings. Brahe whispered in Jessenius' ear that he would like to have a written apology from Kepler and Jessenius promised that, on the way, he would persuade the latter and reproach him for irascibility and lack of self-control, behavior which did not become a respectable and learned man. However, Jessenius' exhortation missed its effect. On the very same day, or the next, Kepler sent a sharp letter out to Benatky, getting off his chest everything which had accumulated. The letter is no longer preserved, but there is a letter[1] from Brahe to Jessenius dated April 8 in which Brahe told about Kepler's letter and expressed his opinion most sharply about Kepler's unseemly conduct. Kepler, who liked to compare his own irritability to that of a mad dog, had, indeed, behaved like one. Brahe, also carried away by his anger, declared he did not want to battle with someone who talked so impudently and who had so little balance in his attitude. He wanted to have nothing further to do with Kepler and wished he never had had anything to do with him. He said nothing about the cause of the argument. When he says in the letter that Kepler could excuse his unbridled sallies and arrogant sneers neither by Brahe's wine nor by Brahe's neglect, then it may be surmised that Kepler was irritated by a disdainful treatment at table.

Thus the break seemed to be complete. Only, as is usual with such attacks, the mood of highest excitement fortunately did not last. When the violent outburst was over, Kepler suddenly saw the whole quarrel from the other side. His choleric wrath, high as it previously had flared, changed suddenly to deepest remorse. Just as previously the injustice to him seemed greater than it was in reality, so now the guilt for his uncontrolled behavior loomed larger in his eyes. He was deeply repentant and thoroughly ashamed of himself. He reviewed all the benefits which Brahe had bestowed on him, even enlarged them and reproached himself with base ingratitude. He was deeply despondent that God and the Holy Spirit had abandoned him so very much to his attacks of impetuosity and to his sick soul. He implored God in His

[1] ED. NOTE. Printed in *Johannes Kepler Gesammelte Werke*, XIV, 112–14 (letter number 161) and in *Tychonis Brahe Opera Omnia*, VIII, 298–9.

infinite goodness for forgiveness and strove hard, as he says, for a Christian reconciliation. Some time later[1] when Brahe came to Prague, Kepler sent him a letter most movingly expressing all these thoughts and feelings and indulging in self-accusations and assertions of his repentance and good will. He accepts all the blame for the disagreement, withdraws his indictments and asks in the name of God's mercy for pardon for his offenses. He promises to unite himself to Brahe by every manner of service, to gratify him in all honorable matters and to prove by deed that his feeling is and always was different from what one could have concluded from the unbridled state of his heart and body in bygone days. After this change in Kepler's attitude, Brahe was easily reconciled; although he did not admit it openly, he may have recognized the justice of some of the complaints. The quarrel which could have become so disastrous was settled. Baron Hoffmann and Jessenius served as true friends in the reconciliation. It was a painful emotional experience through which Kepler lived in those weeks when he was practically helpless at the mercy of his violent outbursts of temper and one after another overstepped the bounds of anger and remorse. This is explicable only by assuming an abnormal frame of mind, just as he himself used the miserable condition of his health as an excuse. The man who, as he once said, "had brought along a delicate disposition to his profession" suffered greatly because of his separation from his family and his concern about his uncertain future.

After the storm had passed and the air had cleared, Kepler betook himself to Benatky and again took part in the work. He had stayed in Prague three weeks, but now he longed for home. Again he cast about for an opportunity to go home at the lowest possible cost. A relative of Brahe's, Friedrich Rosenkrantz, who intended to travel to Hungary, was prepared to take him along as far as Vienna, but was not yet ready for the journey. So he still had to wait through the month of May. He eagerly used the time for studies of his own. Arrangements for a longer collaboration were once more discussed. However, there was no further written agreement. The main content of the settlement consisted in Brahe's promise to make every effort to obtain a decree from the emperor by which Kepler should be summoned to Bohemia for two years to aid in the speedy publication of Brahe's planned work. The main condition was the willingness of the Styrian representatives to consent and to continue to pay Kepler his salary for this period. To the salary was now to be added an increase of 100 gulden, for the granting of which Brahe likewise promised to intercede with the

---

[1] ED. NOTE. See *Johannes Kepler Gesammelte Werke*, XIV, 114–16 (letter number 162), or *Tychonis Brahe Opera Omnia*, VIII, 305–7.

emperor and which he himself would guarantee. Kepler promised so to perform his work that he should pay regard above all to the glory of God, then to that of Tycho Brahe and only lastly to his own. To carry out these stipulations, at the next opportunity Brahe negotiated with the imperial vice-councilor Corraducius. He found him most co-operative. Corraducius promised to ask the emperor to intercede with Archduke Ferdinand and the Styrian representatives for their approval. Brahe was willing to assume payment of the moving expenses. At first the question of where Kepler should live remained open. Whereas Kepler wanted to stay in Prague, Brahe put great value on his remaining near him. If Kepler did not want to live in Benatky, then he would provide him with a suitable lodging in the neighborhood. Yes, Brahe thought he would prefer that Kepler stay in Styria rather than come to Prague; indeed, then he could exchange opinions with him by letter. In the background lay a certain mistrust on the part of Brahe toward Kepler. He did not want Kepler to make too many private contacts; particularly he feared Kepler might join with his hated enemy, Ursus, against whom Brahe was then bringing a law suit. Such a union, in any case, he wanted to avoid. The same suspicion was also apparent when Brahe inquired of Herwart what Kepler had written him about Brahe's system of the universe, the subject of the fight with Ursus. The question entirely solved itself the same summer or fall by Brahe's giving up Benatky and settling in Prague, at the wish of the emperor, who wanted his mathematician close by. Besides, the worry about Ursus, whom he persecuted most harshly to the last, ceased when the latter died in August of the same year.

9. Plans and work after Kepler's return

Finally, on June 1, Kepler, in the company of Friedrich Rosenkrantz, set out on the homeward journey, equipped with a warm and flattering letter of recommendation by the imperial mathematician. The joy of reunion was soon dampened. The letter of recommendation in no way made the desired impression on the councilors, who in the meanwhile were kept busy by the measures of the Counter Reformation. To be sure, they praised the zeal of their mathematician, but gave him, as they worded it, the advice, in reality the order, under threat of dismissal, to push astronomy aside for the time being and dedicate himself to medicine and turn his mind to the common good, away from speculations, which are beautiful rather than useful in this time of need. Because he had proved by his five months' absence in Bohemia that he could get along all right without his home, he was to travel to Italy in the fall, there to prepare himself for the profession of physician. Kepler had believed and hoped that after his return he could peacefully await a favorable decision from Bohemia. Now he saw himself in new difficulties because of this changed attitude of the councilors, since the above-mentioned main condition of his agreement with Brahe and consequently the agreement itself threatened to fall through. What was he to do? Under these circumstances, his plan to stay in Bohemia receded into the background. To follow it would have meant the sure renunciation of his salary. Foregoing the visit was easier for Kepler because he had learned at Brahe's how hard it was really to come into the possession of the money granted by the emperor; even Brahe himself had difficulties in this respect. Thus, even if everything else had been in order, Kepler would have had to fear exposing himself and his family to want on that score. He really considered learning medicine. At that time, medicine and astronomy, or rather astrology, were not so very far apart; even Copernicus and Tycho Brahe practiced the art of healing and busied themselves with the preparation of medicaments. However, it was impossible for Kepler to give up his astronomical researches now. His mind resisted this to the utmost. His new thoughts needed to mature and reach the light. What he had begun with success must not come to a standstill, especially not now, when the use of the Tychonic observations promised him the richest advancement in his undertaking. He must not leave the crop to perish at the moment when it was at its finest growth. So he considered dedicating himself for a few more years to his astronomical researches; but at the same time, in order to save his salary and to satisfy the representatives, he would study medicine.

But still another path—to devote himself completely to the science of the heavens—seemed to open before him. Could not Archduke Ferdinand use him as a mathematician, just as Tycho Brahe was appointed by the emperor? The difference in creed would not necessarily be a hindrance here, any more than the difference in religion of the emperor and his mathematician. Kepler had fancied something like that for some time. "I would pay readers in order to take care of my eyes which already grow feeble and to save time. I would send messengers here and there to obtain books and gather the advice of learned men. I would build instruments. I would appoint others for observing because I am less suited thereto. People would also be appointed for calculating." Again he turned to the Bavarian chancellor for advice and help without, however, openly expressing his scheme with regard to the

Archduke Ferdinand. But since he appealed to the mutual connections between the courts in Munich, Graz and Prague, Herwart von Hohenburg must have understood what he was aiming at. The latter was, to be sure, a better politician than Kepler and therefore advised him to go to Bohemia and to rely on Brahe, who "would not neglect to take such measures as would help his plan come true."

In order to introduce and recommend himself to the archduke, or at least to produce a favorable attitude, in case the emperor should turn to Ferdinand in connection with the leave of absence, Kepler, immediately after his return, composed an astronomical essay in the form of a letter to the ruler of his country. In it he asked him for his favor, and also pointed out the great gains which the emperor, Ferdinand's cousin, had obtained for the science of the heavens by the summoning of Tycho Brahe. (He did not neglect to state the latter's high salary.) He explained that his own astronomical studies would come to an end if he lacked the archduke's good will. He made the solar eclipse of the impending July 10 the occasion for this composition. In addition to the calculation of this eclipse, the essay contains longer arguments about the theory of the moon's motion in which he discloses significant new ideas. He had not yet been able to see Brahe's material about the moon but had heard orally his methods and bases for computation of eclipses. It had become clear to him that he had to follow a different path regarding lunar theory. Brahe stubbornly insisted that the motions of the planets and the moon ought to be represented only by uniform circular motions resting one on another. Therefore he had introduced ever new epicycles in his theory of the moon, the motion of which exhibited especially striking and complicated inequalities. Previously, in Bohemia, against Brahe's violent contradiction, Kepler had opposed this theory because it seemed to him that "simplicity is more in agreement with nature." Consequently, he accepted a non-uniform motion of the moon in its orbit. Here he was guided by an entirely new and highly significant physical foundation for the phenomena of motion. "There is a force in the earth which causes the moon to move." (*In Terra inest virtus, quae Lunam ciet.*) However, this force becomes weaker, the further the moon withdraws from the earth; it will therefore move slower at a greater distance. No such theory had been heard before. We shall soon see how Kepler was led to his great discoveries by such conceptions. Besides this new theory, his essay also refers to the new method by which he wanted to arrange the observations of the eclipse. It rests on the principle of the image through small apertures. Although this procedure for observing was not new, up to then it had been applied only in very rough form. Now, however, Kepler had conceived and constructed an apparatus, with the help of which he could obtain exact numerical values of the progress and size of an eclipse. He hoped, thus furnished with the observation of the eclipse at hand, "to remedy any imperfections in Brahe's lunar theory and to be able to test his allegedly certain conclusions by the all-revealing experience."

Assuredly, the success which Kepler promised himself from the dedication of this essay failed to materialize. Ferdinand appears to have accepted it graciously and to have rewarded it with a gift. But there the matter rested. Lunar theory and eclipses were not at that time of overwhelming importance to the archduke. But Kepler could forget his precarious position in his happiness at the heavenly spectacles. He erected his new instrument in the market place in Graz and with it, on July 10, observed the eclipse which he had previously calculated. In the days that followed, he energetically worked at evaluating his observations. Suddenly, on July 22, he clearly saw the optical reasons explaining the apparent diminution of the lunar disc at the time of solar eclipses. He found the laws which are valid with the pictures of small openings and thus overcame one difficulty which had caused Brahe much trouble. He intended to assemble his results as soon as possible for printing.

10. Sharpened Counter Reformation measures and exile from Graz

A few days later there followed a blow which tore him from his studies, put an end to his deliberation regarding a future dwelling place and brought catastrophe. On July 27, 1600, the archduke published a decree which ordered all burghers and inhabitants of the entire precincts of the city of Graz—likewise all doctors, procurators and all noble persons, with the exception of the old members of the gentry and nobility—at the penalty of 100 ducats to be summoned by the city magistrate to the church July 31, at six o'clock in the morning for examination regarding their faith. On this day the archduke appeared in the church with a big retinue. After a sermon by Bishop Martin Brenner of Seckau, the Reformation commissioners seated themselves at a table in the middle of the church. Then those who were present, more than a thousand citizens and officials, were called up man by man and asked about their creed. Anyone who was not Catholic, or did not pledge himself to become Catholic before long and to go to confession and communion, was banished and obliged to quit the country in a short period of time after payment of 10 per cent of his assets. The activity of the commissioners lasted three days. On August 2, Kepler also appeared before them and because he refused to become Catholic he was banished. The list of the sixty-one men who shared this fate is still extant. The fifteenth entry is Hans Kepler, with the appended note "should quit the land inside of six weeks three days." On August 12, he was discharged from the service of the district, in case "he, contrary to expectations, can no longer be held in the land." The dismissal followed a petition which he himself previously had delivered with a request for a written reference and a gracious termination of his employment. Such a termination allowing him a half-year's salary was granted him by the diet and he was paid on August 30. On September 4, the councilors gave the requested reference, praising him highly for his work as professor; they regret that he had not been able to remain undisturbed in his position and recommend him most warmly.

Among the extensive acts for the Counter Reformation there are three, all dated August 3, 1600, and all reporting to the archduke that Kepler, along with others, had declared himself ready to become Catholic. These writings originated with Ferdinand's secretary, the Capuchin Father Peter Casal, the Chancellor Wolfgang Jöchlinger, and the official procurator, Angelus Costede, a member of the Reformation commission. On examination, all these reports are seen to rest on the assertion of a Capuchin Father Ludwig, who evidently had had a discussion with Kepler after his refusal. That the rumor spread so rapidly and was similarly reported to the archduke from three sides proves the importance attributed to the position taken by the district mathematician. Several biographers concluded from these reports that Kepler had wavered in his belief or had pretended to waver. This interpretation must be emphatically opposed. Throughout the whole period he showed such determination and resolution that one cannot possibly believe that he had become undecided from one day to the next. In the days immediately after August 3, he himself had sought dismissal from the representatives. In the reference from the councilors it was likewise confirmed that he was expelled "because of steadfast public avowal of the pure Augsburg Confession." To maintain, however, that he had played the part of waverer, perhaps in order to gain time, would be to insult the memory of that man who for his whole life was most bitterly and most sacredly in earnest about his religious belief and had shown often enough that he was also ready to make a sacrifice for his faith. That report is much more readily explained in another manner. A whole series of precise statements, which will be cited later, testify that Kepler always interpreted the "Catholic" church as the fellowship of all men who had become children of God through baptism. When he spoke of the Roman Catholic Church he usually used the expression papist-religion. Thus he later oftentimes explained he belonged to the Catholic Church. Certainly, it was in this sense that he had declared himself as a Catholic before Father Ludwig, so that that report rests on a misunderstanding. There is no doubt about his sincerity which despite all cautious intelligence in decisive moments distinguished itself by courageous confession and faithful conviction.

The same people who reproach Kepler for a wavering attitude also raise and attempt to answer the question why Kepler was not excepted from the expulsion this time as he had been in 1598. However, we are now dealing with a universal regulation, especially designed against district officials. Therefore, such a question need not be investigated; on the contrary, it would have been more to the point to look for a reason had Kepler been excepted. In connection with their question, those authors now maintain that Kepler's expulsion was justified, because he secretly composed "consoling letters and little tracts" and circulated them clandestinely among his coreligionists; this distribution of "pamphlets" could have been interpreted by the Catholic party as "preachers' agitation." The nature of these so-called pamphlets and little tracts is apparent from our earlier arguments, reporting everything which the extant documents contain in this regard. Anyone who examines this evidence without prejudice will have to admit that only malevolence could have found "preachers' agitation" therein. In that disturbed period, Kepler found it necessary to unburden his heart to his friends in his troubled conscience and to discuss with them the difficult situation in which they were placed by the persecution. He suffered under the knowledge that he did not agree with the people who stood closest to him. His observations, of course, were partly merely a justification before his own coreligionists. He wanted to seek and give solace, not to agitate. If this is culpable, then the blame is not on him who does this but on him who makes such action punishable. All his life the role of a religious agitator was alien to him.

The bad situation in which Kepler saw himself placed by the expulsion left him no time for lengthy consideration. He had to act. Financial regulations created difficulties. Because the archduke had decreed that no one might lease to a Catholic property which he had not sold within the space of time fixed for him, real estate prices sank steeply. Also all household utensils had become very cheap. Since his wife's whole fortune was in real estate, great loss threatened him. It seems, nevertheless, that an exception was made and he received permission to lease.

Likewise, he received a partial reduction of his exit tax which consisted of 10 per cent of the wealth taken along; he needed to pay only half. This, too, was supposed to be restored after a later decree by the archduke. The order was, however, never carried out. Kepler had tried to obtain remission of the tithe referring to the restitution which had been granted him after the expulsion two years before. Had he not returned at that time, he would have escaped the tax. He was allowed to take his stepdaughter, Regina, with him; Jobst Müller, the grandfather, as guardian, raised no further objection although he and the part of his family remaining in their native land certainly had accepted the Catholic faith (his son Michael's eldest daughter later became a Dominican nun).

The greatest worry of the expelled man was the question "where to go." Since his salary ceased, his arrangements with Brahe were no longer applicable. He immediately informed the imperial mathematician about this turn of events. The latter answered without delay.[1] The emperor had only just, in an audience in which Brahe had the opportunity to report on his arrangement with Kepler, nodded in agreement. Even if everything would now have to be adjusted anew, he should come immediately with or without wife and household goods. Brahe would do everything in his power. "Do not linger; hurry here with confidence as rapidly as possible." Indeed, Brahe now needed Kepler all the more because his principal collaborator, Longomontanus, had recently left him for good and returned to Denmark. Nevertheless, this generous offer could not completely dispell Kepler's old scruples. Should he surrender himself entirely to the good will of Brahe and begin the journey into the unknown? It was not feasible to write to Herwart. With the close connection between the courts of Munich and Graz, the Bavarian chancellor could not intervene in this ticklish situation. He had previously made known his advice, to go to Bohemia, and shown readiness to give assistance there. In September Kepler wrote[2] once more to Maestlin. He described his misfortune and communicated his intention of going to Linz with his family, leaving them behind in that city, and traveling alone to Prague, in order to see there what could be done. If the inconveniences there should be great, he would return to Linz and proceed on his way on the Danube to his native Swabia "provided God extends my life that long." He would enter a medical career in the hope of receiving a "professiuncula" in Tübingen. Because it was too late for an answer to arrive before his departure, he asked that a letter be sent to Linz. He was ready. Now

[1] ED. NOTE. *Johannes Kepler Gesammelte Werke*, XIV, 145–9 (letter number 173).

[2] ED. NOTE. *Ibid.*, pp. 150–2 (letter number 175).

that the step had to be taken and there was no more turning back, he felt within himself that consolation which greatness of soul was able to give. "I would not have thought that it is so sweet, in companionship with some brothers, to suffer injury and indignity for the sake of religion, to abandon house, fields, friends and homeland. If it is this way with real martyrdom and with the surrender of life and the exultation is so much the greater, the greater the loss, then it is an easy matter also to die for faith."

On September 30, 1600, Kepler left the inhospitable city with his wife and stepdaughter; two wagons carried his household goods. He may have remembered the past years with sadness as he drove along that autumn—his friends and associates, his first book which announced his discovery, the founding of his household, the happy and sad hours spent in the Stempfergasse. Graz lay behind him.

III

IMPERIAL MATHEMATICIAN
IN PRAGUE

1600–1612

---

1. The first months

WITH Kepler's expulsion from Graz begins the most important period of his life. What he undoubtedly regarded as misfortune became his good fortune. What he undoubtedly considered a hindrance to his researches brought him the richest success. What he undoubtedly experienced as oppression led him to freedom. It was good for him that he got out of Graz. That city was too restricted for a man of his stature. He needed other air in order to develop completely. He needed other people for the riches of his intellect, another pulpit for spreading his ideas and discoveries, other patrons for promoting his works. Even in Tübingen he would have encountered opposition. Prague was the right place for him, the residence of the emperor, where he met with fitting esteem. The title which he obtained sounded well and gave him prestige. Little as he sought his fame in titles and external honors, all the more did the world care about them. Of all the cities in which he dwelt for some time, Prague was the only one where he remained unmolested because of his faith. The twelve years which he passed there were blessed. But he gave more than he received.

First, to be sure, he still had to cross purgatory. Hesitating, he approached the abode which fate had prescribed and prepared for him. His rejection by Tübingen must, so to speak, have shoved him in. It was as though he refused to grasp his good fortune. When he arrived in Linz, no letter from Maestlin awaited him. Until then he had hoped to be summoned to his homeland with some view to a tolerable future. Now this had come to nothing. All that remained for him was to continue on to Bohemia. He gave up his original plan to leave his wife and stepdaughter behind in Linz, because there the fear arose in him that one in the absence of the other might become sick among strangers. He left only his household goods in this city when he proceeded on his journey. En route he came down with a fever. Before he arrived in Prague, he announced himself by letter to Tycho Brahe.[1] Brahe had moved in from Benatky and taken temporary quarters in the inn at the Sign of the Golden Griffen[2] on the Hradcany[3]. To him the distressed Kepler expounded his schemes. As a former recipient of a ducal scholarship, he must first go home to Württemberg. He wanted to see if, by the good offices of its ambassador at the imperial court, he could reach there without great expenditure. He also wanted to find out if, with the duke's permission and recommendation, he could proceed to another university, perhaps Wittenberg, Jena or Leipzig. However, if Brahe could get him a favorable position, he would give it first consideration. For financial reasons, a decision must be reached in four weeks at the latest. He certainly talked too big when writing in this letter that his teachers at the University of Tübingen had previously made him promises that would justify his harboring great hopes from the duke's connections with the university. Was he ashamed to admit that his people left him in the lurch, or did he only want to prevail on Brahe to push his case with so much the greater energy? If he assumed greater benevolence from the duke than from his theologians and advisers, then he was not wrong, as was shown by a later occasion.

Kepler, in wretched physical condition and depressed mood, arrived in Prague with his family on October 19. There he was again hospitably received in Baron Hoffmann's house. A bad winter was in store for him. The intermittent fever (*febris quartana*) wouldn't yield; it tortured him for three-quarters of a year. A bad cough was added, so that he feared consumption. His cash quickly melted away. The change of residence had cost him 120 gulden, a considerable sum in comparison with his current annual salary of 200 gulden. In Prague everything cost four times as much as in Graz. His wife, who was used to a good standard of living, suffered under the retrenchment which she had to impose on herself and under the separation from her family, in a foreign city. Finally she, too, became sick. At last, in

[1] ED. NOTE. *Johannes Kepler Gesammelte Werke*, XIV, 152–5 (letter number 177), from Kepler to Brahe in Prague, dated "[Böhmen], 17. Oktober [1600]." The places given in the *Werke* for Brahe and Kepler have been changed because Dr. W. Norlind of Lund called the attention of Miss Martha List to the fact that Brahe had gone from Benatky to Prague in August and that Kepler was detained by illness somewhere in Bohemia.

[2] ED. NOTE. "Gasthaus zum Goldenen Greif."

[3] ED. NOTE. The Hradcany is a district on the left bank of the Vltava or Moldau river. It is on the hill of the same name which is the site of the castle of the same name.

December, the long-desired letter from Maestlin came. Its content robbed Kepler of his last hope. The Tübingen teacher let it be understood that he could give no advice. He couldn't hold out a prospect of any sort for a professorship, not even for a very small one. "Only one thing I do industriously; pray for you and yours." Kepler was shaken by this letter. "I cannot describe what paroxysm of melancholy your letter has occasioned me, because it destroys all hope of going to your university. So I must stay put until I either get well or die," he wrote back. Once again, a few weeks later, he besieged the old man's heart. "I need solace," he begged. "The love for my homeland pushes me toward you, whatever its future destiny may be. Once already I have gone under, when the world toppled about me; I have no fear." Yet Maestlin remained silent and for more than four years did not answer Kepler's letter, no matter how urgently the latter repeatedly asked him to do so. Hafenreffer, whom Kepler likewise had approached, sincerely meant it when he assured Kepler that his prosperity and welfare were nearest to his heart day and night. However he, too, could give only the inadequate advice, that Kepler should repair to Tübingen as steward with a few nobles, and there await a further opportunity. Also the attempt to reach Saxony, whither Kepler had evidently turned through the good offices of the Saxon ambassador, failed. The court preacher Lyserus gave him empty solace and advised[1] him to turn toward Wittenberg where, as contrasted with an earlier time, mathematics was in a sad state.

Brahe was happy to have Kepler with him again and eagerly strove to secure Kepler's position. Kepler himself admits that if Brahe could do as he wanted, no one would be more satisfied than he. Yet it was not done solely by a gracious nod of the emperor's. Indeed, such a gesture set the state machine in motion. But this worked so slowly that anyone dependent on waiting until it gave out hard cash fared badly. So, within the four weeks which Kepler had set, no decision in any direction had been reached. Nor is anything said anywhere in the extant documents about fixed agreements with Brahe or with the emperor. In spite of all pains, the situation, which Kepler from the very beginning had wanted to avoid at any price, had arrived—he was completely dependent on Tycho Brahe's mercy and had to rely on him for everything.

He took as much part in the astronomical works as was possible in his bad state of health, even if not everything by which he sought to satisfy his thirst for knowledge met with Brahe's approval. Meanwhile

---

[1] ED. NOTE. *Johannes Kepler Gesammelte Werke*, XIV, 159–160 (letter number 181), dated Jan. 19, 1601.

the latter's instruments had been set up in and near the imperial country seat, Belvedere, which Ferdinand I had built. Observational activity was limited in comparison with the industry which once had prevailed at Uraniborg. At the end of February, 1601, Tycho Brahe moved to the house previously occupied by the late vice-chancellor Curtius from whose widow the emperor had acquired it for 10,000 taler in order to put it at the astronomer's disposal. Evidently Kepler also moved into that building at the same time. The house stood in the very spot on Loreto Place where later the Czernin palace rose. In addition to his planet studies Kepler was assigned a work by Brahe which did not make him particularly happy. He was to compose a defense of Brahe against Ursus, whom Brahe did not want to leave in peace even in the grave. Kepler undertook this task, elegantly eliminating the personal motive. He attempted to refute the reproaches of Ursus, while sticking to the subject. The composition was never completed and was first published in 1858 by Christian Frisch in his complete edition of Kepler's work. It is of great interest, especially in respect to methodology, since in it Kepler explained the notion of the astronomical hypothesis and rejected the interpretation that as such it is only a matter of obtaining correct conclusions by calculation.

Because his health did not improve and his pecuniary resources became ever more scanty, Kepler, in the spring of 1601, made plans to travel to Graz. Since his father-in-law, the old Jobst Müller, died just at that time, Kepler wanted to look after his wife's interest. At any rate, it was his object to turn his wife's fortune into cash. He remained away from the end of April until the beginning of September. No objections to his visit were raised by the government, although he had been expelled from the city. The journey agreed with him, although he seems not to have accomplished much else, since he reports about an "iter inutile." The attacks of fever ceased after he had stayed in Graz a few months. He met old friends, was a guest everywhere, as he wrote his wife. His spirits rose again. We learn about an ascent of Mt. Schöckel from which, by measurements toward neighboring lower mountains, he tried to ascertain the curvature of the earth; on the ascent he encountered an unusual storm, which he described in detail. From a letter by his wife to her "beloved master of the house" (the only one of hers to survive, because Kepler used the unwritten portions of the paper for astronomical drawings and calculations), he learned the news from Brahe's house and the household concerns that gave her trouble. In it, naturally, there was talk of the unpleasant money worries. Brahe had promised to help out Kepler's wife with money during his absence. Now as she believed she had to complain in a letter to her husband that she was getting too little, he immediately wrote Brahe an angry remonstrance. Brahe, much vexed over Kepler's "rude and caustic" words, did not write back himself, but assigned this business to his pupil Johannes Eriksen who executed this commission in a truly thorough manner. Kepler should calculate what his wife had already received from Brahe, and put up with the total; in the future he should behave more sensibly and with greater moderation toward his benefactor and have more confidence in him. However, that was exactly what roused and was bound to rouse Kepler's touchiness; he did not want to receive benefactions, but a suitable recompense to which he could lay claim for what he did.

Thereupon, agreement was soon re-established. Nevertheless, in spite of good will on both sides, the two men had not settled their differences permanently. We know what pressed heavily upon Kepler. Brahe, in Prague, always felt like a stranger in a strange land and suffered severely when recalling his homeland and the events which, not without blame on his part, to be sure, had removed him from there. Only with difficulty did he get along with the persons around the emperor. "He was not the man," Kepler reports, "who could live with anyone without very severe conflicts; let alone with men in a high position, proud advisers of kings and princes." His health showed itself unequal to these agitations and the great cares which dragged him hither and thither. "He always resembles a lost man, but always somehow extricates himself. His success at this is to be wondered at, if one considers the means employed which must rather have led to death." Although just fifty-four years old, he was, as Kepler says, beginning to become childish (*puerascere*). In spite of all capriciousness he was, however, basically good-natured. What was particularly difficult for Kepler to endure and what cramped him in his work was the lack of confidence with which Brahe still guarded his observations, so that none at all should reach other hands. He showed him his choicest observations, but only inside his four walls and just said to him: "Get to work." If Kepler troubled with observations other than those which were precisely left to him, this was taken as unseemly prying. As early as his first visit to Benatky Castle, on that fatal April 5, 1600, he had signed a declaration by which he promised to keep strictly secret all information which Brahe gave him from his treasure of observations, as is fitting for a philosopher. Kepler condemned the "evil shameful custom of the times, of somehow imposing upon the useful and glorious work of other people for one's own benefit, of depriving the true author and of ascribing the proper glory to others." He himself made light of that lack of confidence in himself. "I am moved by an exceedingly powerful desire for knowledge of the heavens and I cannot stop myself from communicating my thoughts to the masters of science so that by their hints I immediately make progress in our divine art." It was not youthful exuberance which spoke thus; Kepler followed this view as long as he lived. In order to further science, he likewise exerted himself to induce other astronomers such as Maestlin, and Magini in Bologna, a rival of Tycho Brahe's, to offer their observations to Brahe in exchange for his. In this, however, he met with no success. All these circumstances made the collaboration of the two men more difficult and burdensome. But, on the other hand, one must emphatically point to the generosity which Brahe always had shown toward the young scholar. He was the only one who did not leave him in the lurch after his expulsion and who eagerly befriended him. No one perceived and acknowledged this more than Kepler himself.

2. Tycho Brahe's death and Kepler's appointment as imperial mathematician

Shortly after Kepler had returned to Prague at the beginning of September well and strong, Brahe introduced him to the emperor. The ruler congratulated him on his recovery and gave him the glorious commission of collaborating with Brahe in compiling the new tables of the planets which the latter planned. Brahe took this opportunity to ask the emperor to allow him to call them the *Rudolphine Tables*. Kepler was pleased at the favorable course which his affairs took. Hardly, however, had a few weeks passed, when an event occurred which again at one blow brought about a change in his life. On October 24, 1601, after a short illness, Brahe died of a bladder ailment. Kepler was with the family at the deathbed. Shortly before his passing, the dying man turned once more to his esteemed assistant to whom he wanted to leave his scholarly estate and begged him to carry through the presentation of the motions of the planets by Brahe's plan, following his hypothesis, not only according to the Copernican one. On November 4, the corpse was laid to rest with great pomp in the Tyn Church where an Utraquist divine service was still permitted. Jessenius gave the funeral oration. Kepler composed a longer elegy about his departed master, which appeared in print as an appendix to the funeral oration. Although he could not offer up his conviction of the correctness of the Copernican theory as a sacrifice to the last request of the dead man, nevertheless he did not neglect to show repeatedly how far the phenomena of the motion could also be presented according to the

Tychonic theory, and gave animated expression at every opportunity to his reverent and thankful acknowledgment of everything for which he was indebted to Brahe and his observations.

Two days after Brahe's death, Barwitz, an imperial adviser, came to Kepler and brought him the glad tidings that the emperor had decided to transfer to him the care of the instruments and incompleted works of Tycho Brahe. He was informed of a salary and he was ordered to make application for the corresponding sum. Therewith he was appointed as successor to the man on whom he had hitherto been dependent. He was imperial mathematician himself. He had a position such as he had long hoped for and dreamt about. Now he could work freely. The rich store of observations was at his free disposal. For he was dependent thereon, since he had to complete Brahe's works and especially had to carry out the great tables, for which barely any preparations had been made. That here lay a source of vexations and difficulties he was, to be sure, soon to discover. But at any rate, Kepler, with good cause, took charge of the observations in which were put down Brahe's lifework. The great observer had fulfilled his task; he was called away. Now the great theoretician took over—he, who from these observations had to solve the secrets which they concealed. So the sadness over Brahe's passing mixed involuntarily in Kepler's breast with the joy over his own promotion. Hafenreffer and Rollenhagen,[1] the well known poet of the "Froschmäuseler" (The Frogs and Mice), sent congratulations. Herwart particularly took joyful part in the favorable turn in Kepler's fortune. He was convinced that there was no mathematician other than Kepler in all Germany, indeed in all Europe, who was able to succeed Brahe. He voiced this conviction directly to Barwitz, the imperial adviser, to whom he wrote: "I, as one informed about these matters and having also some experience, know very well that at this time as far as one can judge from the works that have been published (*ex operibus editis*) no one can be found who can be compared both in intellectual power and in mathematics (*et ingenio, et fundamentis artis Matheseos*) with this Master Kepler, let alone be preferred to him, so that I have no doubt whatever that when it is brought to the attention of His Majesty most graciously and most humbly he will not let him go for any amount of money." While he said he knew a position for Kepler in the university in Lauingen, yet it would be best for all parties if he were to remain in Prague. Only one would then also have to "direct things in such a manner that he" would be "reimbursed for his past as well as his present expenses." Herwart was quite at home in questions of salary and particularly knew

[1] ED. NOTE. Georg Rollenhagen, 1542–1609, German satirist and clergyman.

how they were handled at the imperial court. Therefore, right from the beginning, he advised Kepler: "My friend should not be satisfied with a small and pinched salary but should ask for a large and comfortable one. And the estimate of it should not be based on bodily needs (*pro quantitate corporis*) nor on the "living standard" (*et tenuitate uictus*) but on the greatness of your mind (*sed pro magnitudine animi tui*) and of the subject matter (*et rei subjectae*) and you should make an effort to receive an adequate down-payment at once." How justified Herwart's warning was, soon became apparent. Advised by men near him, Kepler left the fixing of his salary to the emperor, who granted him an annual sum of 500 gulden, beginning October 1, 1601. However, he had to dance attendance for months before at last, on March 9, 1602, he received the first payment. The sum appeared very trifling compared to that which Brahe had received. However, it must be borne in mind that, for the observations which constituted the wealth of the departed one, the latter's heirs made great demands which had to be complied with. Soon Kepler moved into a new dwelling in the New Town opposite the Emaus cloister, nearly an hour's distance from the castle, at which he frequently had to present himself.

The deeper one penetrates into the events from Kepler's first expulsion up to his appointment as imperial mathematician, into the motives which guided the people concerned, into the temporal relationships which they bore to each other, into the significance which they possess for the history of astronomy, that much more clearly does one recognize the hand of a higher guidance. In order that the paths in life of the two great astronomers who uniquely supplemented one another could unite, it was necessary that both be displaced from their widely separated residences, in order to meet at the court of an emperor whom history reproaches for having neglected the affairs of government for the sake of his astrological and alchemical bents. Kepler himself expresses his conviction of the rule of a divine decree in these events when he writes: "If God is concerned with astronomy, which piety desires to believe, then I hope that I shall achieve something in this domain, for I see how God let me be bound with Tycho through an unalterable fate and did not let me be separated from him by the most oppressive hardships."

3. Astronomia Nova; *and the second and first planet laws*

Even in the midst of the hardships and afflictions in which his life abounded, Kepler was very seldom forsaken by his remarkable ability to plunge into and bury himself in studies and speculations. Similarly, during the suspension and fear of the last year, his ever active mind made keenest use of the pauses, which care and illness allowed him, to pursue his scholarly researches. If we examine these researches, we immediately arrive at the beginnings of his most renowned accomplishment, the discovery of the planet laws. We see him in the months of his first visit to Tycho Brahe and then later when concerned with other works, occupied above all with the foundation of the glorious structure of his *New Astronomy*. It is not, however, as though he had laid this foundation according to a precise preconceived plan. The work that was being formed grew out of him in accordance with the unfathomable laws of gifted creation. Assuredly, he appears as the active one, the agent, the calculator, the meditative one, the designer, the constructor, but he was the sufferer, the stimulated one, the hunted one, since his genius directed his mind, led his hand, showed him the trail which he must follow, called him back when he made a mistake, spurred him on and left him neither rest nor quiet until everything was completed and he, who had carried it out, finally regarded with amazement the work in which he had succeeded. For anyone whose mode of thinking is not so unassuming that he operates only with complete results but who is able to derive pleasure and profit from considering the multiplicity of the divine manifestations of life, it is uncommonly fascinating to follow, in this unique example, the separate phases in the development of revolutionary opinions in the models of nature, and to follow step by step the work of the genius who introduced this revolution. In the example at hand, this is all the more possible since the extant sources give us all the information desired. Obviously, for various reasons, this undertaking cannot be completely carried out within the bounds of a biography. The difficulties of the material are not insignificant in the case before us because of the astronomical and mathematical knowledge which must be assumed. However, a biographer would fulfill his task badly, if he were to let these difficulties restrain him from giving the reader something more than only the last formulations of the final results. In the introduction to the great work, in which Kepler informs the world of his brilliant conclusion, he compares his voyage of discovery with those of a Columbus and a Magellan in whose narratives we find great entertainment. Whereas in reading we take no part at all in the hardships of the travels of the Argonauts, the reader of Kepler's works would get an opportunity to trace the obstacles and thorns on his paths of thought. However, so he asserted, this is the common lot of all mathematical books. As some people find pleasure in that, others in something else, so there will also be some who, having overcome the intellectual difficulties, will be filled with joy to have before their eyes at one time the whole series of his discoveries.

It was already noted that previously, during Kepler's first visit, the elaboration of the theory of Mars was assigned to him by Tycho Brahe. Because he did not find there, as he had hoped, ready values for the sizes which he needed for corroboration and correction of his harmonic speculations, he had to set about calculating these sizes himself. Now then, what happened at closer range with these calculations? It is common knowledge that Mars, like the two other "upper" planets, Jupiter and Saturn, advances in the ecliptic from day to day from west to east until, in about 780 days, it has completed one synodic revolution. When it gets near to being in opposition to the sun, thus some time before it culminates at midnight, it is stationary, just as though it wanted to ponder, even moves back a bit, in order to continue its journey in the old direction sometime after the opposition. It is known that Copernicus had demonstrated in a startlingly simple manner that this remarkable loop was a reflection of the motion of the earth, from which we observe without perceiving this motion. Even if no account is taken of this loop, nevertheless another irregularity is still perceptible. The times between two oppositions, that is the synodic periods of a planet, are not exactly equal, as the observations of several centuries have established in detail. Because people, in their curiosity, want to know where the planets would stand at a particular time in the future, the calculation of this irregularity had to be taken into consideration. How was this possible? In the presentation of these motions of the heavens the ancients began with the principle that a natural retrograde motion must of necessity be a uniform circular motion. Supported in particular by the authority of Aristotle, an axiomatic character was given to this proposition, whose content, in fact, is very easily grasped by one with a naïve point of view; men deemed it necessary and ceased to consider another possibility. Without reflecting, Copernicus and Tycho Brahe still embraced this conception, and naturally the other astronomers of their time did likewise. In order to master that irregularity mathematically, the center of the universe (the earth according to Ptolemy, the sun according to Copernicus) was assumed to be somewhat away from the center of the orbit. The distance between the two points was called the eccentricity, the circular path the eccentric, the axis connecting the centers of the universe and of the orbit the line of apsides, and the intersections of this straight line with the orbit the apsides (according to Copernicus perihelion or nearest to the sun, aphelion or farthest from the sun). Even if the motion in the orbit proceeds uniformly, it still appears to an observer in the center of the universe as irregular, namely quicker at perihelion, slower at aphelion.

Yet since this simple aid did not manage to save the appearances, as it was expressed, additional assumptions were made. Only with the orbit of the sun (according to Ptolemy and Tycho Brahe) or of the earth (according to Copernicus) was it considered possible to get along without such supplements; here a simple uniform circular motion on an eccentric was retained. What those supplementary assumptions rested on is only hinted at: Ptolemy assumed a point on the line of apsides (equalizing or compensating point or *punctum aequans*[1]) from which the motion on the eccentric should appear uniform, so that in reality it is non-uniform. Copernicus, whom Brahe followed in this, sought to reach the same end by the superposition of two uniform circular motions. For what follows it is not necessary to go into this in greater detail. Only note that some astronomers in Kepler's time saw the chief merit of Copernicus precisely in the fact that his theory does greater justice to that axiom of uniform circular motion than does that of Ptolemy.

To develop the theory of Mars meant, consequently, to calculate the position of the line of apsides and the value for the eccentricity. Since a circle is defined by three points, to solve this problem it was necessary that three points on the planet's orbit be known. These are obtained from the observations of the opposition, because (to use the words of Copernicus) at an opposition it is immaterial whether one observes from the moving earth or the stationary sun, since in this configuration planet, earth and sun lie in a straight line. Now Tycho Brahe had a series of ten such Mars oppositions from the years 1580–1600 (later in 1602 and 1604 Kepler added two more). They formed the material which Kepler had before him when he set to work on the task set for him. Naturally the result had to be the same each time, no matter which group of three oppositions was taken as a basis if, yes, if, the assumptions as to the form of the path and the form of the motion were correct. Let it also be noted that the calculations were entirely carried out solely with the mathematical aids supplied by the geometry of Euclid and trigonometry.

This scanty sketch of the fundamental ideas of the earlier planet theories marks the runway from which Kepler started his flight into new regions. From the very beginning he set to work at his task with optimistic impetuosity. He believed that in eight days he would master the difficulties which had stumped Longomontanus. He even made a bet that he would accomplish this. When it did not go that fast, he

[1] ED. NOTE. Equant.

kept hoping from day to day that he would reach a happy conclusion. He was madly bent on his calculations. The purpose for which he wanted the results receded into the background. Immediately after returning from his first visit to Benatky, he wrote to Herwart: "I would already have concluded my researches about world harmony, had not Tycho's astronomy so shackled me that I nearly went out of my mind." It was clear to the honorable seeker after the truth: "Those speculations may not *a priori* run counter to obvious empirical knowledge, rather they must be brought into agreement with it." Now it was a question of testing the empirical knowledge, of cornering nature and forcing her to answer prudent questions. At the new task, new abilities developed in him. The Kepler who speculated made way for the Kepler who computed and weighed critically. That he went straight to Mars, was a most propitious piece of good luck. For since this one of the three outer planets has by far the largest eccentricity, it alone did not fit the earlier theories and so made possible the new discovery. "I consider it a divine decree," writes Kepler, "that I came at exactly the time when Longomontanus was busy with Mars. Because assuredly either through it we arrive at the knowledge of the secrets of astronomy or else they remain forever concealed from us."

Now how did Kepler take hold of this work? Certainly he must have started by the traditional method. However, the vain attempts of Brahe and Longomontanus had already demonstrated that the conventional procedure was found wanting if all the assumptions made there were retained. Therefore, he had to drop or change one or another. In the first place his criticism was directed toward the supplementary assumptions discussed above. From the very beginning he rejected the superposition of two uniform circular motions, as introduced by Copernicus. Introducing an equant (or *punctum aequans*) was more satisfactory because then the motion of the planets in reality is irregular and appears uniform only from this mathematical point. This will be seen to have fitted into his fundamental concept from the very start. Previously, without any evidence, a very precise assumption about the position of this point had been taken as the start for the calculations. Kepler abandoned this procedure and left the position of the point on the line of apsides open. Since he thus introduced one more degree of freedom, the task naturally became more complicated. Instead of three points of the path, as hitherto, he now had to make use of four so as to be able also to calculate the position of the equant. Accordingly, he selected a favorable quadruple of four observations of oppositions, by the use of which he carried through the very cumbersome calculation. The solution of the task was possible only by a procedure of approximations. Not less than seventy times, as he informs us asking for sympathy, had he had to carry out the entire series of difficult separate calculations which the solution required before everything agreed sufficiently for him to be content. And the result? He checked the path so calculated against the other available observations and saw that for all, the calculation fitted well with the observation within the limits of accuracy of two minutes which conformed to the Tychonic observations. Since those were distributed over the whole ecliptic, he had good cause to conclude that he possessed a means of calculating the position of Mars for any desired moment within those limits of accuracy. How he could triumph with such a result! Was not his problem thus solved?

Indeed, anyone else would have been satisfied, but not Kepler. He wanted to be absolutely certain of his results and accordingly sought further confirmation. Like a possessed collector, he sat before the accumulated treasure of observations and with the eye of a connoisseur selected a few rare pieces which enabled him to calculate the eccentricity of the orbit directly, in a highly original manner. However, he obtained, not a confirmation, but a contradiction which was so large that at the maximum there was a difference of eight minutes for the planet's position. The triumph was too early. Such a difference was not to be neglected. Here observation confronted observation, both indubitable. Logic decided: there must be an error in the suppositions regarding the form of the orbit and the form of the motion. One or the other or both assumptions were wrong. So much Kepler clearly stated. "These eight minutes showed the way to a renovation of the whole of astronomy." Kepler was undaunted. The sincerity and purity of the purpose which guided him in his inquiries is expressed when he seals the negative result with the incomparable lovely words: "After the divine goodness had given us in Tycho Brahe so careful an observer, that from his observations the error of calculation amounting to eight minutes betrayed itself, it is seemly that we recognize and utilize in thankful manner this good deed of God's, that is we should take the pains to search out at last the true form of the heavenly motions."

Now the scene changes. On the stage appear two thoughts, which had long been standing behind the wings and were barely able to wait for their turn to play their parts. After all, they had advanced once before. Both figures sharply criticized antiquity, both for the same reason. Copernicus in his picture of the universe had, it is true, placed the sun in the center. However, since he relied entirely on Ptolemy whom he esteemed highly, he had, in presenting the planet theories, always assumed as center of the universe, not the sun itself but rather the center of the earth's orbit, which was somewhat to the side of the sun, and referred all his calculations to this. Tycho Brahe, in his system, had made an assumption corresponding exactly to this. Because both of them, on the strength of these assumptions, erected their planet theories on oppositions to the so-called mean sun instead of to the true one, inaccurate figures naturally entered the calculations from the very beginning. Very early Kepler rightly took exception to that. He required that all values be referred to the true sun.

The second thought which he introduced concerned the earth's orbit. Copernicus had assumed that the earth moves uniformly in her circular orbit; he had not found any of the supplementary assumptions necessary here as contrasted with the paths of the upper planets. In this regard, too, he followed Ptolemy as did in turn Tycho Brahe. Why, Kepler now asked, should a different theory be valid for the earth than for the other planets? It was not solely a deduction by analogy which put this critical question into his mind. Behind this question stood an important positive thought. It was the same thought from which the previous consideration had also grown: The sun is the seat of a force which moves the planets in their course, and, what is more, the motion is so much the quicker the nearer the planet is to the source of the force. If that is the case, then the sun itself, the body of the sun, must be the middle point of the whole planet system, not an empty point like the center point of the earth's orbit. And if the effect of the force decreases and increases according to the distance, then the earth, in its eccentric orbit, must also move faster when nearer the sun, slower when farther away.

In the sun there is the seat of a moving force. This was the great new guiding thought which from now on shone in front of him in his inquiries and led him to the discovery of his laws, the great theme which he henceforth varied to the utmost and tried to found out of observed facts with all the consummate skill with which he was able to utilize the observational material. He now wanted to abandon the old beaten track and adopt new ways. He was no longer willing to be satisfied with a kinematic and pure geometric presentation of the motions; he wanted to explain these by their causes. As he rightly said, if the earlier masters, and Copernicus and Tycho as well, always had proceeded *more Ptolemaico mutatis*, so he now intended to clean house, getting rid of the entire furnishing of epicycles, and demonstrate the planetary system as governed by inner laws, regulated by physical forces. Even in his student days, he had such thoughts and we have seen similar physical considerations emerge several times. Now the time had come to introduce them systematically into the science of the heavens, to shape astronomy into celestial mechanics. With this Kepler had set himself a difficult task. The mathematical resources of his time no longer sufficed to accomplish it. Also, we see how in the solution of his problems, which later was relatively simple to accomplish with the tools of mathematical analysis, he slaved unceasingly, without quite reaching the goal which he had set for himself.

However, before introducing his physical notions into the theory, Kepler still wanted to prove empirically that his supposition in regard to the motion of the earth was correct. More exact information about it was therefore necessary, because of course all observations are made from the moving earth, so that an error made in regard to this motion necessarily also creeps into the working up of the observations. For this reason Kepler perceived in the correct theory of the earth's orbit the "key to a more deeply penetrating astronomy." As daring and rich in fantasy as he was in his speculation about the universe, just as thoroughly and carefully did he now proceed, taking no step without gathering authorization and confirmation from the observations. Indeed, while following his Mars researches, one almost gets the impression that sometimes he deals with individual tasks and proofs out of pure delight and pleasure in the observations.

But now, how could he get more exact knowledge of the earth's orbit? In any event, not by conscientiously employing the old methods; even though Brahe, also assuming a uniform circular motion, succeeded, with the aid of his accurate observations, in doing justice to the phenomena within the limits of accuracy set by him. Now here again Kepler's inventive genius was active and suggested an ingenious trick. Hitherto the point of view had been from the earth to Mars; now Kepler wanted to follow the earth in its course from a point on the orbit of Mars "as from a watchtower." He, so to speak, transposed his eyes to a particular position of Mars' orbit and from there found out directly the relative values of the distances from sun to earth. Since the sidereal period of Mars was accurately known, such points of time, when it was in the same positions in its orbit, could easily be specified. Kepler chose three. Since, naturally, at these points of time the earth occasionally was to be found at various sufficiently accurately known positions in its orbit, he succeeded in calculating, by elementary geometrical means, the relative distances of the earth from the sun for these three points of time. But in this he mastered the assumed circular orbit of the earth and could calculate the distance of the sun from the central point of the orbit, that is the eccentricity of the earth's orbit. Out of this was derived a value for this eccentricity which beyond doubt corroborated his surmise that the theory of the earth's motion is the same as that of the upper planets, namely that the earth, too, moves non-uniformly in its orbit. This extremely ingenious procedure was still further profitable for his later researches; to wit, it also provided him with the relation between the radius of the earth's orbit and the distance of Mars from the sun at that place in its orbit which was in question, that is a relative distance of Mars from the sun.

Now, however, the moment had come to introduce his physical conceptions. The procedure just now sketched, as a more exact inquiry proved, had demonstrated not only the irregularity of the motion of the earth in its orbit, in general; from it could also be derived a measurement for the points at which the earth has its greatest and smallest distance from the sun. It was shown that at these places, that is at aphelion and perihelion, the speed of the earth is inversely proportional to its distance from the sun. This measurement he immediately extended to the whole orbit and thus advanced the general proposition which he had already had in his mind for a long time: The rate at which the earth moves in its orbit, as a consequence of the force issuing from the sun, is inversely proportional to its distance from the sun. And he introduced still another generalization. What holds for the earth, holds also for the other planets. Naturally, as Kepler well knew, experience still had to prove whether these inductive conclusions were admissible.

But how can one calculate with this proposition, that is solve the problem set by astronomy, to ascertain the place of the planet in its orbit at a given moment? This was a difficult matter for Kepler. (The motion of a point which travels on a circle in such a way that its rate is inversely proportional to its distance from a given eccentric point leads, according to modern analysis, to an elliptical integral.) Yet Kepler was not frightened away. He divided half of the circular orbit beginning at one apside into 180°, calculated the distance to the sun of each one of these little graduated arcs (letting the semi-diameter of the orbit equal one) and added these 180 numbers. The sum gave him the measure of the time it takes the earth to travel around half its orbit. If he wanted to calculate the time when the earth had moved 50° from the apside, he added the first fifty values of the distances. The ratio, then, of this sum to the previous one is the same as that of the time sought to half the period of revolution. Thus was solved the problem of computing the time which it takes the earth to reach a given point on its orbit. The inverse problem, to be sure, of calculating the position of the earth at a given moment of time, could only be solved by interpolation with the help of a table, constructed in accordance with the previous procedure.

Now, however, calculating with the sums of the distances was exceedingly bothersome. And Kepler immediately looked around for a suitable short cut. He himself tells about his next step: "Since I was cognizant of the fact that there are infinitely many points on the orbit and correspondingly infinitely many distances, the thought came to me that all these distances are contained in the plane of the orbit. For I remember that once Archimedes also divided the circle in the same manner into infinitely many triangles, because he tried to find the ratio of the circumference to the diameter." So now Kepler, tempted by this consideration, not mathematically indisputable, replaced the sums of the distances by the corresponding areas and succeeded in finding the time it takes a planet to pass over a particular section of its orbit. He did this by measuring the area bounded by the rays from the sun to the end points of these sections. In that way he obtained what is today called the second planet law: *The radius vector describes equal areas in equal times.* Since the areas, which he thus introduced, were easy to calculate, he thenceforth used this proposition as the working hypothesis for further research. He was nevertheless completely cognizant that the two propositions, the distance proposition and the area proposition, are not identical. Immediately after he had accomplished the change from the one to the other, he himself pointed out the difference with mathematical precision; the applications he made in this connection constitute a specimen of his clever mathematical way of thinking, as well as his accuracy and thoroughness. The difference in the results of the two propositions in their application to the motion of the earth was irrelevant considering the limits of accuracy of that time. Only experience with other planet orbits of greater eccentricity could establish which proposition was the correct one.

Although calculating those areas was simple, Kepler still could only solve the problem of calculating the position of a planet for a given point of time indirectly as previously, because the statement led to a so-called transcendental equation. Every mathematician well knows the great significance of this famous "Keplerian problem" in the further development of the theory of functions.[1]

---

[1] ED. NOTE. This problem was well stated by Robert Small in *An Account of the Astronomical Discoveries of Kepler . . .*, London: Mawman, 1804, p. 296, as follows: "Having the area of part of a semi-circle given, and a point in its diameter, to determine an arch of the semi-circle, and an angle at the given point, such that the given area may be comprehended by the lines including the angle, and by the required arch: or, to draw from a given point in the diameter of a semi-circle, a straight line dividing the area of the semi-circle in a given ratio." Geometers have been unable to achieve a rigorously accurate solution. For a modern solution with calculating machine, see Jens P. Møller, "On the Solution of Kepler's Equation," *Festschrift für Elis Strömgren*, Copenhagen: Munksgaard, 1940, pp. 163–74.

These researches clarified the form of the motion for Kepler. Now it was a question of testing the other assumptions of the earlier theories, those regarding the shape of the orbit. He began with the information that the orbit cannot possibly be circular. This truth he demonstrated by referring to the distances of the sun to Mars, which were known to him from his earlier detailed researches. If the orbit were circular, then he would always have to come up with the same orbit whatever triplet of distances he were to use. Now, however, the negative was demonstrated, since a different result appeared depending on the choice of the three distances. But how to proceed now? His physical concepts which hitherto had brought him rich returns now enticed him on an extremely burdensome and very long detour. As soon as it had become clear to him that the orbit departs laterally inward from a circle, he believed it possible to furnish a physical cause for this phenomenon. He prepared a certain mechanism showing the motion due to the force issuing from the sun. The mechanism led to an egg-shaped orbit with the blunt end at aphelion and the point at perihelion. Conquering this picture of the motion mathematically gave Kepler infinite trouble. He calculated the breadth of the "moonlets" which lie between a circular orbit and his oval one. He sought to ascertain the area of the ovoid. Then again its perimeter. He attempted to solve the problem now with the sums of the distances, now with the areas. For all 180° he recalculated the distances of the sun to Mars provided by his mechanism. For in no other way could he complete his integration problem. When the result did not agree with the observation he changed his preliminary statement; at least forty times, so he remarks, he carried out such a calculation for all 180°. If only he had succeeded in ascertaining the area of an ovoid by geometry, without its being necessary to reckon repeatedly "in smallest divisions." Yes, if the orbit were a perfect ellipse, he wrote at that time to a friend, so the problem would already have been solved by Archimedes and Apollonius. Only the picture of the motion, with which he had fallen in love, did not admit of such a thing. Then what was the fault in these disagreements? Kepler sought it in his area proposition, he sought it in an erroneous use of the proposition of distances. The only place he did not seek it was where it lay, in his picture of the motion. Only when all possibilities were exhausted did he decide, with heavy heart, to desist from this. He had again triumphed too soon. Later he joked about the over-great haste, with which he had taken hold of his problem. "Hasty dogs bear blind young."

After this lack of success, Kepler once more took up where he had let himself be pushed aside from the correct path. He began to calculate distances and, indeed, very thoroughly. So, at last, he had marked off a great many points on the orbit of Mars. Indeed, he now had the parts at hand but lacked the picture which comprehends and puts these parts together. Excited, he was on the lookout for a solution. Now was it chance or a good fairy which set him on the correct track? He had calculated the width of the "little moon," which his oval produced. In his trials he had found that this width may be only half as big. The number stuck in his head. Then he accidentally hit upon the idea that precisely in the ellipse, whose eccentricity is equal to that of the orbit of Mars, the difference between the semi-major and the semi-minor axis was half as big as the width of that "little moon." It was for him, as he says, as though he were awakened out of sleep and saw a new light. It was clear to him that the rule, by which the distances change from point to point of the orbit, proved correct precisely for this ellipse. The question of the shape of the orbit was solved. What held for Mars, must also hold for the other planets. The law was announced: *Planets move in ellipses with the sun at one focus.*

It was a steep and long path which Kepler had to retrace in order to scale the summit which he had seen from the distance. Some reader to whom the subject is foreign may already be nearly out of breath from the attempt to become acquainted with this path as it has been described in the previous statements of particulars. And yet here everything is put in the simplest form, all mathematical detail being left out, and only the main lines shown. Anyone who penetrates deeper into Kepler's exposition finds himself transplanted into a confusion of calculations and deliberations. What sounds exceedingly simple in our arguments divides into difficult single problems, for whose solution Kepler himself had to contrive a method because up to then no one had set or carried out such problems. It is necessary to force a way with him through the thicket of his numbers, to share in his detours, to overcome the difficulties of his abstruse style. Yet it is worth the trouble. The power of the logic which impels him forward is captivating, the ability with which he masters every difficulty is admirable, the rich flashes of ideas which streamed in on him are pleasing, every new outlook which he opened is enjoyable. His prodigious industry, his inventive genius, his mathematical sense, his unhesitating sense of fact are to be marveled at. The same man who came to Prague to complete his *a priori* structure of the universe we now see calculating, for months, for years, because the observations required it. It is always the observations which chain him, which he forces to answer his questions. The problem which he had mastered should be made clear. The numbers giving the position of Mars at the times of the observations are on many pages in Brahe's journals. A confused muddle! Kepler brought order out of this chaos. He had hunted out the laws uniting these numbers, so that they no longer stand together unrelated but rather each can be calculated from the other. In this connection one circumstance still deserves special mention: that is the limits of accuracy of the Tychonic observations. That these limits were narrow enough so that Kepler could not afford to neglect those very important eight minutes, we have already seen. But had they been considerably narrower, he would certainly have been caught in a fine meshed net, because in many of his calculations he would no longer have been permitted to overlook certain inaccuracies, as was necessary for the progress of his research. Thus, theory and practice harmonized remarkably with one another.

It is a new land which is glimpsed from the position next to Kepler on his summit. He left far behind him not only Ptolemy but also Copernicus and Tycho Brahe. Perhaps it seems that it makes little difference whether the planet orbit is a circle, or an ellipse deviating little from the circular shape. Yet Kepler's prodigious step forward consists precisely in the fact that with his ellipse proposition he had overthrown for all time the two-thousand-year-old axiom, according to which every motion retrograde in itself must of necessity be a uniform circular motion. By that step he had made the orbit free for a new development of astronomy. And nothing is more difficult in science than to set aside such deep-rooted opinions. People who have not read Kepler often tell the story as though Kepler had found his laws in a purely geometrical way, so to speak by trials. It is naïve to believe that a fortress as strong as that axiom indicates can be taken by such means. No, everywhere in the solution of the problem confronting him, physical concepts were in the background and drove him forward. They became more and more intimately intertwined with his astronomical thinking. In 1605 he wrote in a letter: "I admit that for at least five years past I have used for physical considerations at least half the time left me by the affairs at court." When he was reproached for having a passion for innovation because he wanted to mix together such heterogeneous sciences as astronomy and physics, he explained: "I believe that both sciences are so closely bound with one another that neither can achieve perfection without the other."

Nowadays we are so accustomed to seeing mechanical forces operating in the planetary motions that it is difficult for us to think that it was once different. And yet Kepler ran up against rejection and lack of understanding on all sides. Maestlin, Fabricius, Longomontanus and others shook their heads. Even many years later Maestlin advised his former pupil to leave physical causes and hypotheses entirely out of the question and to explain astronomical matters only according to astronomical method; geometry and arithmetic alone are the vibrations of the knowledge of the heavens. It is Kepler's greatest service that he substituted a dynamic system for the formal schemes of the earlier astronomers, the law of nature for mathematical rule, and causal explanation for the mathematical description of motion. Thereby he truly became the founder of celestial mechanics. The goal that he pursued he summarized clearly: "My goal is to show that the heavenly machine is not a kind of divine living being but similar to a clockwork in so far as almost all the manifold motions are taken care of by one single absolutely simple magnetic bodily force, as in a clockwork all motion is taken care of by a simple weight. And indeed I also show how this physical representation can be presented by calculation and geometrically." If magnetic force is here replaced by the designation attractive force and the limitation "almost" is omitted, then with these words the great problem of classical celestial mechanics is formulated. In historical accounts it is repeatedly stated that it was Galileo who founded the Copernican theory physically. While fully appreciating Galileo's accomplishments in the domain of mechanics, it must still be emphatically pointed out that he completely failed to comprehend the idea of a celestial mechanics. In none of his works did he take notice of Kepler's planet laws although he certainly knew them. Not once in his famous *Dialogue* about the systems of the world, which appeared a quarter of a century later, did he speak of them, although they surely should have played a central part. Yes, as though Kepler had spoken into the wind, Galileo praised Copernicus in this work, because he understood how to present the planet motions by uniform circular motions; he sticks here throughout to the old Aristotelian distinction between "natural" and "violent" motion. So it was Kepler first of all, not Galileo, who freed astronomy from the bonds of Aristotelian physics.[1]

[1] ED. NOTE. Although it is not advisable to go into detail at this place concerning Galileo's failure to mention Kepler's laws of planetary motion, it is important that the reader's attention be drawn to certain recent contributions to the literature concerning this subject even though they can be treated only superficially here. In this regard see:

Erwin Panofsky, *Galileo as a Critic of the Arts*, The Hague: Nijhoff, 1954, especially pp. 20 ff.

Alexandre Koyré, "Attitude esthétique et Pensée scientifique," *Critique*, Sept.–Oct., 1955, pp. 835–47, which is a critical review of the Panofsky pamphlet listed above. (Prof. Koyré's earlier *Etudes Galiléennes* (1939) is cited by Prof. Panofsky in the 1954 pamphlet.)

———, "A Documentary History of the Problem of Fall from Kepler to Newton; *De Motu Gravium Naturaliter Cadentium in Hypothesi Terrae Motae*," *Transactions of the American Philosophical Society*, new ser., vol. 45, part 4, 1955.

Erwin Panofsky, "Galileo as a Critic of the Arts: Aesthetic Attitude and Scientific

Thought," *Isis*, XLVII (1956), 3–15, which is an "abridged and somewhat revised version" of the author's pamphlet listed above and which uses suggestions made by Prof. Koyré in his review.

Edward Rosen, Review of Panofsky, *Galileo as a Critic of the Arts*, 1954, in *Isis*, XLVII (1956), 78–80.

Erwin Panofsky, "More on Galileo and the Arts." *Isis*, XLVII (1956), 182–5, discussing Prof. Rosen's review.

As is known by all three of the authorities cited above, there is available (Galileo, *Opere*, Ed. Naz., XI (Florence, 1901), 365–7) a letter dated July 21, 1612, from Frederico Cesi to Galileo which refers to the ellipses of Kepler as a matter of common knowledge. Since Galileo was not ignorant of Kepler's laws, why did he ignore them? As Prof. Panofsky says (1954, pp. 24–5) "At the very beginning of the *Dialogue*, Galileo unequivocally endorses the belief...in the perfection...of the circle not only from a mathematical or aesthetic but also from a mechanical point of view. According to him the qualities of uniformity and perpetuity, reserved to rectilinear motion in post-Galilean dynamics, exclusively belong to the circular movement which Huygens and his successors have taught us to consider as vectorially accelerated." Prof. Panofsky goes on to say (p. 25) that Galileo says that before the world was created rectilinear motion may have had some use but that thereafter only circular motion is naturally appropriate to the bodies constituting the universe. Prof. Panofsky believes that the "haunting spell of circularity," borrowing a phrase from Prof. Koyré, "made it impossible for him [Galileo] to visualize the solar system as a combination of ellipses." "Kepler, on the other hand," says Prof. Panofsky (1954, p. 26), "did break the 'spell of circularity'...he considered the rectilinear...movement as privileged as far as the physical world is concerned:..." As Prof. Rosen points out (*Isis*, XLVII, 79), Kepler abandoned the traditional circular motion because no circle would fit the observations at his disposal. These were the Tychonic observations on whose accuracy Kepler justifiably relied. Prof. Panofsky points out that both the famous astronomers attempted to support their celestial mechanics by a comparison of the motions of the stars with those of the human body, and in this also arrived at opposing views, Kepler believing that all muscles move in accordance with the principle of rectilinear movement and Galileo that all human movements can be reduced to a system of circles.

Prof. Koyré says (1955, pp. 842–3) that Galileo rejected the Keplerian ellipses for the simple reason that they were ellipses, and that for Galileo the Keplerian astronomy was an astronomy of Mannerism. Galileo very probably had the same aversion for Kepler's symbolism and use of cosmotheological reasoning that he had for the allegorism of Torquato Tasso (1955, p. 846).

Prof. Rosen (*Isis*, XLVII, 79) finds a reason for Galileo's neglect of Kepler's laws in Kepler's obscurity, prolixity and mysticism which were so repugnant to Galileo that he had no desire to seek out "the nuggets of real gold hidden away in Kepler's heap of dross."

The question, which was more modern, Galileo or Kepler, brings out the fact that both men had prejudices, although different ones. Kepler was able to substitute celestial dynamics for celestial cinematics because he clung to Aristotle's interpretation of motion as a "process," believing that the planets would cease to move if the force emanating from the sun ceased to act on them. Galileo, on the contrary, considered motion as a "state." Prof. Panofsky concludes his 1956 article with the statement that "as Galileo ignored—and, in a sense, was bound to ignore—Kepler's ellipses, so did Kepler ignore—and, in a sense, was bound to ignore—the principle of inertia quite clearly (though restrictedly) stated in Galileo's *Second Letter on the Sunspots* of 1612." It may be trite to add what Prof. Caspar, because of his desire to emphasize the importance of Kepler's overthrow of the age old axiom of circular motion, did not include here, that the Newtonian synthesis rested upon the works of both Galileo and Kepler, that both the Keplerian celestial dynamics and the Galilean dynamic physics were needed.

An unusually interesting and comprehensive treatment in English of the physics and metaphysics of Kepler's universe should be mentioned here because it distinguishes the different aspects of Kepler's approach to the structure of the universe: Gerald Holton, "Johannes Kepler's Universe: Its Physics and Metaphysics," *American Journal of Physics*, XXIV (May, 1956), 340–51.

Of course Kepler did not reach the high goal of celestial mechanics which he was the first to set up and perceive. It was reserved for Newton's genius, by stating the law of gravity, to crown the structure which Kepler had begun, and to prove clearly that the planet laws follow as necessary consequences of this general law of nature. Yet various of his remarks show how close Kepler actually came to this law, for example, when he says the magnetic, that is the attracting, force of the sun spreads itself out like light and at another place proves that the intensity of the light diminishes with the square of the distance, or when he categorically asserts: "If one would place a stone behind the earth and would assume that both are free from any other motion, then not only would the stone hurry to the earth, but also the earth would hurry to the stone; they would divide the space lying between in inverse proportion to their weights."[1] That is an unprecedented speech in a time when the Aristotelian theory of weight was still universally recognized. Here the clever researcher positively has at hand the idea of universal gravitation. However, in the physical presentation of the planet motions he did not follow this concept but pictured the motion differently. He divided it into two components, a circular revolution around the sun and a deviation along the radius vector. The first motion is taken care of by the sun. The force issuing from the sun spreads out in the plane of the ecliptic and grows weaker with the distance away. Now the sun, since it rotates (an assumption which Kepler made *ad hoc* a few years before the actual discovery of axial rotation), by means of its likewise rotating rays of force, pulls the planets around in a circle. The phenomenon that the rate of the orbital velocity is less than the rate of rotation of the sun is to be explained by the inertia of the planet body which by nature inclines to rest. Stimulated by the very important work on magnetism published in 1600 by William Gilbert, the Englishman, Kepler explains the deviation in the radius vector by imagining the planet bodies polarized, that is consisting of parallel magnetic filaments one end of which is pulled by the sun while the other is being repelled. A vital force is supposed to hold these filaments continually in a parallel position, and perpendicular to the line of apsides. Now if the rays of force of the sun pull the planet around away from aphelion, where the effect of the sun is the same on both poles of the planet, then the end of the filaments, which undergoes an attraction, lies nearer to the sun than the other. Thus the planet steers toward the sun and, indeed, it does so until it reaches perihelion.

---

[1] Ed. Note. *Johannes Kepler Gesammelte Werke*, XV (letter number 358), from Kepler to David Fabricius, from Prague, Oct. 11, 1605, pp. 240–80, especially p. 241. Similarly, see *ibid.*, III (*Astronomia Nova*), 25.

From that point on, the reverse takes place. The deeper reason why Kepler, in explaining the mechanism of planetary motion, could not penetrate to the knowledge which we owe to Newton is obviously that he lacked the conception that a mass remains in uniform straight line motion if no external forces act on it.[1]

In the summer of 1605 Kepler had collected those researches about the orbit of Mars, which we have followed above. The first part had already been finished during Tycho Brahe's lifetime. The area proposition was introduced soon thereafter, in the year 1602. Other works occupied the year 1603. Carrying through the oval hypothesis and rejecting it took up nearly the whole following year. In December, 1604, once more weighed down by thoughts of death, he considered depositing his manuscript, as far as it went, at the University of Tübingen. But he soon recovered from his depression and abandoned this plan. Final success was near; at Easter time, in 1605, he discovered his ellipse proposition. Justifiably he could give the work he had composed the proud title: *Astronomia Nova αἰτιολογητος Seu Physica Coelestis, tradita commentariis de Motibus stellae Martis.* It is the first modern astronomy book.

There were grave obstacles in the way of publication. The first difficulty Kepler encountered emanated from Brahe's heirs. After Brahe's death a strained relationship had developed between them and Kepler. This immediately came to the surface when, executing the imperial commission to publish the works left by Brahe, Kepler tackled the completion of the printing, already well along, of Brahe's great *Progymnasmata.* To this book which dealt with Brahe's solar and lunar theories as well as with the fixed stars and the new star of 1572, he composed an appendix[2] and here and there he made improvements. Now, to his annoyance, the "Tychonians," without his knowledge, had various notes printed, which he had written down for private use. They also kept him from correcting proof, so that many errors remained. The main subject of the fight between the two sides was Kepler's use of the Tychonic observations. After Tycho's death, the right to possess them was transferred to his heirs. The emperor wanted to acquire them along with the instruments and offered the heirs the

[1] ED. NOTE. See the footnote before last, concerning recent literature. Kepler wrote to Fabricius (...*Gesammelte Werke*, vol. XV...241), in a letter dated Oct. 11, 1605, that by nature every body inclines to rest, "Quodcunque materiatum corpus, se ipso aptum natum est quiescens, quocunque loco reponitur."

[2] ED. NOTE. In a letter to Magini, Kepler explicitly stated that he was the author of this appendix, "Appendicis ad Progymnasmata ipse author sum:..." *Johannes Kepler Gesammelte Werke,* XVI, 279–80 (letter number 551), dated from Prague, Feb. 1, 1610, especially p. 279. The appendix, which was published as part of the book, can be read in *Tychonis Brahe Dani Opera Omnia,* III (1916), 320–3.

sum of 20,000 talers for them. Naturally, however, the imperial treasury had no money to satisfy them. Indeed, in the course of the year, they received a few thousand talers. This, however, was not sufficient to cover the accumulated interest. On the other hand, Kepler could not carry out the imperial order without free and un-hampered use of the observations, of which, after all, he had taken charge immediately after Brahe's death. Besides, as a follower of Coper-nicus and an opponent of the Tychonic system of the universe, Kepler's researches led him further and further from the theoretical point of view of the man without whose observations he could not have carried out his investigations successfully. Out of this arose, spontaneously, the reproach on the part of Brahe's heirs, that Kepler was not using the observations in the sense intended by the man who had acquired them by many years of toil, and that he was looking out for his own fame and advantage. His chief opponent was Brahe's son-in-law, Tengnagel, who, in order to keep the upper hand, repeatedly promised a publica-tion of his own on the basis of the observations, although he was not at all qualified for such a task. Indeed, he preferred not to be called a mathematician. Contrasting character traits made an amicable agree-ment more difficult. Kepler appropriately characterized the way his opponent wanted to guard Brahe's treasure by comparing him, follow-ing Aesop's well-known fable, with a dog in the manger who cer-tainly eats no hay himself but also lets no one else near it. The chief plan to be realized was the working out of the *Rudolphine Tables*, whose completion was very important to Tengnagel and his family, partly because of the paternal glory, partly because of the sound of money clinking which they hoped it would bring. Whereas they were not at all aware of the difficulties of this problem, Kepler, on the other hand, fully understood that only someone full of self-confidence and willing to risk his scholarly calling could hope to solve this problem rapidly. To him it was clear that, before the work on the tables could be approached with a view to success, he first had to solve the prob-lem in the midst of which he was already embroiled, and free the old planet theories from the faults clinging to them, that is found a new astronomy.

In the course of these disputes, the emperor's father confessor, the prelate Johannes Pistorius, was selected as the man to whom Kepler should now and then report on the use of his time and about his studies. Pistorius was favorable to Kepler and in the circumstances it might not have been too unpleasant for Kepler to consent. It was, however, less pleasant when in the year 1604 in return for the relinquishment of the Tychonic observations he had to agree in writing not to publish anything based on them without Tengnagel's approval until the *Rudolphine Tables* were completed. This made the publication of the commentary on Mars dependent on Tengnagel's consent. Now the latter wanted to undertake the elaboration of the tables himself and orally promised the emperor that they would be completed within four years. However, he was in no hurry with this work, and hardly did or could set about them in real earnest. Consequently Kepler found himself in a disagreeable position. This did not improve when Tengnagel soon after was named imperial appeal counsel and joined the Catholic Church, further increasing his influence at court. Tengnagel, after taking office, could think less than ever about carrying out the work he had promised. Thus he had Kepler in his power. Accordingly, the latter foresaw further disputes because Tengnagel's sole object was to guard the fame of his father-in-law, whereas Kepler, on the contrary, had the freedom of research in view. In fact, Tengnagel threatened to prevent the printing of the commentary on Mars when Kepler fancied himself free of his agreement after his opponent failed to fulfill his promised term of four years. Nevertheless, an agreement was finally reached because Kepler declared himself ready to insert a preface by Tengnagel at the beginning of his work.

The second obstacle to publication was the delay of printing for financial reasons. At the end of 1606, Emperor Rudolph granted 400 gulden, "because for the extension of the fondness of patronizing astronomy, which is our custom and that of our predecessors in the Austrian House, we did not gladly leave untouched the previously mentioned book, in which so many glorious secrets of nature are included." However, since Tengnagel's approval had not yet been given, the printing could not be started immediately. In the meanwhile Kepler spent the money "in a great part otherwise and for household needs" because the payment of his salary stopped. With the remainder he got the printing under way in 1608 at Ernst Vögelin's in Heidelberg. Since the sum still at his disposal did not suffice and Kepler wanted to travel to Heidelberg, he had to ask the imperial Maecenas for further financial aid. As a consequence the latter bestowed an additional 500 gulden. Since everything moved slowly, the printing was not completed until the summer of 1609, while the author was in Heidelberg. The emperor had denied Kepler the public sale of the book and ordered that "he give no one a copy of it without our previous knowledge and consent." He reserved the ownership of the entire edition evidently because it was composed by Kepler in the pursuance of his office and printed with imperial money. In this order there is also implied a recognition of the great importance attributed by the emperor to the book which he wanted to distribute himself. However, since the imperial treasury remained continually in arrears with salary payments and the emperor's situation had in the meanwhile become so precarious that he could trouble himself but little more with such things, Kepler tried to recover his losses and in the end sold the whole edition to the printer. In make-up, the book corresponds completely to the importance due to the *New Astronomy*. In big folio format and lovely print, it is the most magnificent of all the works which Kepler published. Because only a small edition was printed, it is today by far the most expensive of the great astronomer's first editions.

4. Astronomia Pars Optica

It is reasonable to suppose that the planet orbits with their secrets and whims had so filled Kepler's thoughts that no room remained for other scientific research. And yet we see him at the same time busy with another comprehensive complicated question. This certainly touched upon the former but still, for the most part, had its own form and meaning, namely the subject of optics.[1] When he went from Graz to Prague, he already carried a sizable portion of the questions within him; others were aroused by the association with Tycho Brahe as well as by the exigencies of his work on the planet orbits. One can recall the eclipse observation which Kepler had made in July, 1600, with his own instrument, devised and constructed for this purpose, as well as the successful detailed considerations he made in this connection regarding the laws for pictures by a pinhole camera. Because of unfavorable circumstances he did not carry out his original intention of publishing immediately what he found here. But the problem which was here set had established itself in his thought and developed further. However well instructed a person was in general about the occurrence of solar and lunar eclipses of his time, there still arose a host of separate questions which needed explanation and solution as soon as he concerned himself more closely with these phenomena, often observed since antiquity. Yet such a dependable astronomer as Tycho Brahe had denied the possibility of a total solar eclipse. The many observations

[1] ED. NOTE. The optical work of Kepler, in the setting of its time, is well handled by Dr. Vasco Ronchi, director of the Istituto Nazionale di Ottica in Florence, in "L'ottica del Keplero e quella di Newton" in *Atti Della Fondazione G. Ronchi*, anno XI N.3 (1956), pp. 189–202. See also Vasco Ronchi, *Optics: The Science of Vision*, translated from the Italian and revised by Edward Rosen. New York: New York University Press, 1957, pp. 40–51, 263–5.

zealously gathered by Kepler both from ancient literature and from contemporary reports frequently did not correspond exactly with each other or would not agree with the calculations as to the passage of time and the size of the obscuration to the same degree which a good theory demanded. There could be various reasons for these disagreements. They could stem from the fact that the numerical values, on which were founded the calculations for the sizes and distances of the two heavenly lights, were inexact, or from the fact that the phenomena of the motions of the sun and moon had not yet been completely mastered. The cause could, however, also lie in too crude an observational procedure that did not take into consideration certain external circumstances and relied too much on estimates instead of on exact methods of measurement. In addition, still other questions appear, with which Kepler had been partly occupied previously. Whence comes the reddish light of the moon during a total lunar eclipse? What is the explanation of the reported luminous appearances around the sun at total solar eclipses? The question concerning the diminution of the lunar diameter during solar eclipses was already recalled. So many questions, so many problems. They all made Kepler restless and forced him to formulate a great plan. He wanted to write a book exploring and presenting the sizes and mutual distances of the sun, moon and earth, primarily supported by the phenomena of eclipses. Since he knew from a report by Theon of Smyrna that the great astronomer of antiquity, Hipparchus, had composed a work (since lost) dealing with the same subject, he wanted to give his book the title *Hipparchus*. However, the first step toward carrying out this plan was to examine and clear up those optical questions without whose solution the necessary exact and trustworthy foundation to the work could not be securely laid.

A further series of questions on optics, connected, indeed, with the above, but playing a part in all astronomical observations, are grouped about the subject of atmospheric refraction. The importance of this is perceived when it is remembered that it is precisely by the refraction of its rays in the atmosphere that the sun, when on the horizon, is raised by an amount approximately equal to the diameter of the solar disc, so that the sun seems to touch the horizon with its lower rim when in reality it is immersed just beneath the horizon. Certainly, the amount of refraction decreases with increasing altitude; nevertheless, it is still so great that it must always be taken into consideration if the precise star positions, determined with refined instruments, are to have meaning. Brahe's improvement of the art of observation consequently aspired to more exact knowledge of refraction. He himself clearly realized this and took the trouble to determine it empirically by comparing observed star altitudes with those which had been calculated with certainty. Nevertheless, his results were not satisfactory. Curious to relate, he believed that the amount of refraction was dependent on the distance and brightness of the source of light, so that he set up three different tables of refraction for the sun, moon and stars. The first two of these differed only slightly. On the other hand, the last disagreed very considerably with the other two. In addition, he believed it impossible to determine any refraction whatsoever for the sun at an altitude of more than 45° or for the stars at as little as 20° and over. Since Kepler, for his researches on Mars, was very anxious to have at his disposal the most accurate possible star places, freed from refraction, it is evident how important for him, in this respect, too, must have been the question under consideration.

In order to make himself completely familiar with what science knew about these optical matters, he plunged into what was, in his time, the most important work on this subject. It was the one composed by the Silesian-born scholar, Witelo or Vitellio, in about 1270 and printed three times in the sixteenth century. Although this work contains but few independent investigations by the author, it nevertheless presents an apt and inclusive digest of what had previously been accomplished in this domain by men of science, in particular by Ptolemy and the Arab, Alhazen. By using this work as a starting point, Kepler put together everything he knew concerning optics. As was his way, he explored the subject in all directions. One question led to another. With true Keplerian thoroughness, the entire subject was torn up and overturned as with a ploughshare. So, in the very first years of his Prague stay, a book of no less than 450 quarto pages came into existence, a book by which the author, to be frank, earned the universally recognized epithet of founder of modern optics. It bears the title: *Ad Vitellionem Paralipomena, quibus Astronomiae Pars Optica traditur.* Kepler completed the main portion of the work on it in the year 1603, during which he interrupted his Mars researches. At New Year's in 1604 he was able to present the finished manuscript to the emperor.

The book is highly significant because in it Kepler, with an abundance of new thoughts and insights, with his clear grasp of the problems and proofs, prepared the ground for a new treatment of optics and solved, if not all, at least a good portion of, the tasks which he set for this science, in such an exemplary fashion that today we still build upon the foundations laid by him. Corresponding to the provinces of the problems from which he started, the work falls into two parts.

Whereas he deals with the questions of pure optics in the first part, the second part to some extent appears as a text book of astronomical optics. He begins with a chapter on the nature of light, in which he attempts by metaphysical speculation to trace the characteristics of light as a means created by God for forming all things and bringing them to life. Even if we find these considerations, which attracted Goethe's special attention, somewhat odd, still they offer the principle of photo measurement, that the intensity of light decreases with the square of the distance. On the foundation of experiment and experience rest his further successful investigations into the laws of pictures by a pinhole camera and into the foundations of catoptrics and the place of the picture produced by plane and curved mirrors. To be sure, in spite of his penetrating research and efforts, he did not attain complete success in the question at the core of the search for the universal law of refraction. He did, it is true, discover a very good approximation-formula for it with which he was able to calculate an improved table of refraction. But the discovery of the familiar law of refraction was reserved for Willebrord Snel[1] and Descartes, who met with better luck in carrying Kepler's considerations to their conclusion. The most brilliant morsel in the first part is the appended section about the process of vision. Today everyone knows that this occurs in such a way that the light rays which reach the eye from the subject are refracted by the substances filling the eye and thus an inverted image forms on the retina. Only the minority remember that it was Kepler who first correctly recognized and explained this process. Similarly, he was the first who could reduce the operation of eyeglasses in nearsightedness and farsightedness to the laws of optics. He could also successfully establish the importance of seeing with both eyes for the perception of distance.

In the second, or astronomical, part, he first gives lengthy details about the various light phenomena for the individual heavenly bodies, about the shadows of the earth and moon, about the then only very inaccurately known parallaxes, that is displacements to which the sun and moon are continually subject from the point of view of an observer on the earth. Then he heads for his main theme, the skillful observation of the apparent diameters of the heavenly luminaries and of eclipses. God, according to Kepler's conviction, had so regulated the paths and sizes of the sun and moon that by the occurrence of eclipses not only should man be invited to contemplate the divine

---

[1] ED. NOTE. For a discussion of the spelling of Snel's name, see George Sarton, *The Appreciation of Ancient and Medieval Science during the Renaissance (1450–1600)*, Philadelphia: University of Pennsylvania Press, 1955, xiii.

work which the kindness of the Creator lets be perceived but also should be prepared to investigate them more profoundly. "These eclipses," he calls out, "are the eyes of astronomers, these diminutions of light signify for science growth in abundance, these faults illuminate the soul of mortals by most highly valuable and artistic pictures. O what a great argument in praise of the shadow to be recommended to all people!" Happy in the subject matter, he spares no pains in showing how one can accurately establish the details of the phenomena by observation and evaluate the observations by calculation. Here he is not dealing with the above-mentioned main subject of the planned *Hipparchus*; indeed it is not the subject matter of optics. It was supposed to be carried into effect in a special work which, however, was never published in the planned form.

Just as the *Astronomia Nova* is of great importance for the early history of infinitesimal calculus because of the numerous new problems to work out "according to smallest divisions" (to which not enough attention was being paid), so the mathematician is much the richer for the *Optics*. Since here Kepler, while dealing with curved mirrors, speaks about conic sections, he shows how their various forms continuously follow from each other. There is found, he claims with the aid of an illustration, that there is a transition from the straight line through infinitely many hyperbolas to the parabola and from these through infinitely many ellipses to the circle. Thus on the one side the parabola has two open figures, the hyperbola and the straight line, on the other side two closed ones, returning on themselves, the ellipse and the circle. Here is expressed an entirely novel view of these geometrical images.

Kepler dedicated both works, the *Astronomia Nova* and the *Optics*, to his imperial master. If the latter valued his mathematician, he had every reason to do so. To found afresh two sciences in a few years was in reality an incomparable feat.

5: *The religious situation in Prague*

After looking at the researcher at work and becoming acquainted with his results, it might be advisable to glance at the circumstances of Kepler's life in Prague. It was the old Prague, so rich in historical memories and magnificent buildings. At that time, as the emperor's residence, it formed for a decade the focus of European politics. At this advanced post the German nationality came in contact with the Slavic east, which was continually kept in agitation and commotion by political and national antagonisms and also especially by religious ones. For example, a few years later the tragic religious war, which for thirty years afflicted and devastated Germany, supposedly began there.

Kepler was expelled from Graz for reasons of faith. Just as in that city his denomination had decided his fate, so, all through his life, the religious state of affairs in the place where he dwelt was very important to him. His conscience did not permit him to attach himself completely and without reservation to any of the confessional parties which fought each other, but he was nevertheless inspired by a strongly pronounced religious spirit. Now, in Prague, the confessional situation differed from that in Styria and also from that in his native Swabia. Whereas in Württemberg it showed a closed unitedness, and in Graz was characterized by two distinct sharp fronts, the affairs in Bohemia were more complicated. After the very severe religious and economic convulsion, which the land had suffered through Hus and the Hussite wars, the more moderate followers of Hus had finally made peace with the old church on the basis of the Basel-Prague compacts[1] which granted them communion *sub utraque specie*, that is under both forms. On the other hand, the extreme elements considered themselves severed from the old church in doctrine and later joined the so-called Unity of Brethren. Thus, in Bohemia, even before the appearance of Luther, there were three camps: the Subunites, as the Catholics were called, the Utraquists and the Bohemian Brethren. Then when the waves of the reformation beat against the land, the multiplicity of creeds became still more variegated. Especially with the nobility and in the cities, the Augsburg Confession won numerous followers. In 1575 these united with the Bohemian Brethren in the Bohemian Confession, which in essential matters agreed with the Augsburg Confession, but nevertheless showed a strong Calvinistic touch. The antagonisms grew sharper when in the year 1556, shortly after the founding of the Order, the Jesuits entered Prague and settled in the Clemen monastery which was assigned to them. They were led by Peter Canisius, distinguished by his zeal and his model conduct, a member of the most prominent order in Germany. With the greatest energy and by a clever use of all the means at their command, in order to win friends in all ranks for themselves and for their cause they worked successfully for the reversion of the populace to Catholicism and for the strengthening of the position of the old church. On the other hand, not least important was their far-reaching educational activity by which they

[1] ED. NOTE. November 30, 1433, published July 5, 1436. (See Carl Joseph Hefele, *Conciliengeschichte*, VII (Freiburg im Breisgau, 1874), 568–74, 618–626.) These compacts were formed under the Council of Basel which lasted in one form or another from 1431 to 1449.

could obtain great influence. Since, in their Clementinum, they founded an academy which rapidly rose to prestige and power, there began a competition with the Carolinum, the heart of the university founded in 1348 by Charles IV, which had developed into a bulwark of the new doctrine. This competition was so much the more dangerous to the latter because its scholarly importance in spite of a few strong forces had at that time sorely declined and the Clementinum enjoyed the favor of the emperor. Then, too, Maximilian II had taken an uncertain and divided stand in the Confessional quarrel whilst it was not known to which of the opposing parties he belonged at heart. Then his son, Rudolph, who in 1575 had taken over the Bohemian and in the following year the German imperial throne, true to the education he received in Spain, had immediately joined the Catholic side and proceeded to take corresponding measures. The influence and the prominence of the archbishop rose. Besides the Jesuits, Capuchins also came to the city. They likewise made a point of zealously attending to the goals of the Counter Reformation. The antagonism between the throne and the majority of the ranks who belonged to the new faith grew sharper. In a religious edict of 1602, thus barely two years after Kepler's arrival, it was decreed that only Catholics and Utraquists were to be tolerated in the land.

Under such conditions, did not Kepler have grave fear that the fate which he had suffered in Graz was repeating itself here in Prague? Could he hope that he would be able to make his way safely through all this confusion? Was it not rather to be presumed that he would again be confronted with the need to make a decision? In fact the fuel which had collected in such quantity would necessarily explode as soon as a vexatious occasion struck a spark, especially in view of the passion characteristic of the Bohemian populace. And yet we see that Kepler was able to attend to his investigations undisturbed during the whole time he spent in Prague. Neither from his letters nor from other documents do we learn anything about anyone's having put difficulties in his way because of his faith. Indeed, he seems to have been urged to take the same step which Tengnagel had taken, and to become a Catholic. In 1605, Bartolomeo Cristini, the mathematician to the duke of Savoy, could report by letter to the astronomer Magini in Bologna that he had learned from the apostolic nuncio in Prague that there was hope that Kepler would soon join the Catholic Church. Still there is undoubtedly the same misunderstanding here that was shown above regarding Kepler's declaration before the Reformation commission in Graz.

6. *Emperor Rudolph II and Kepler*

The basis for Kepler's not being persecuted because of his faith during his time in Prague, indeed not once being bothered, is to be sought, first of all, in the person of Emperor Rudolph II, who was carved out of different wood from his relative Ferdinand, in Graz. With the former, sympathy for all kinds of arts and sciences outweighed the concern for the importance of the confession, so that he did not ask the numerous scholars and artists whom he drew to his court about their doctrinal belief. Distinguished by a delicate and big-hearted manner and bestowing his diverse exertions and generous support and promotion on many activities of human spirit and industry, he had, during his thirty-six years' rule, been able to create a flowering season for Prague, unlike that which it once enjoyed under Charles IV, it is true, but none the less pronounced and fruitful. The fact that he summoned Tycho Brahe and Kepler to his court, so that Prague became the cradle of the "New Astronomy," meant for all time a brilliant title of glory. His political conduct can also be thanked for the fact that in the confessional confusion, fraught with disaster, peace was maintained above all and the catastrophe was postponed for several years.

Certainly, the imperial throne did not radiate bright resplendence, but rather a discreet glow in which it began to glimmer. Not determined energy, which firmly holds the helm of the government fast and guides the ship safely through the storm, produced the result, but rather deliberate reserve, which shrinks from daring grasping and can put off decision, avoided much failure. Not on a large scale did urgent enthusiasm bring art and science to a flowering; rather it was done by a sick soul's capricious zeal for compiling, which throws itself on a thousand shining or iridescent things which are before it, and by the inquisitive bent which meditates on the hidden relations in nature and tries to distill miraculous materials out of retorts. This is the figure of the strange, slightly built, unmarried eccentric on the throne, who looked into the world with big eyes. It is known that his shyness, increasing with advancing age, forced him for days at a time to shut himself off entirely from the outer world. Here—irresolute, idle, meditative, suffering—he gave way to his moods or, unconcerned about everything which happened outside, devoted himself to his favorite tastes, contemplated his collections, cut, wove, constructed mechanical apparatuses and clocks. His whole love was directed toward his art and treasure room, for whose enrichment he made the largest expenditure and set in motion his agents in all important places in Europe. Everything imaginable was amassed there, pictures in great number (mediocre, and highly renowned by Dürer, Titian, Brueghel), coins, carvings, sculpture, textiles, gems in enormous quantities, wax figures true to nature, monsters, clocks, mechanical apparatuses, pigskin bindings. The search for curiosities and rarities was no less decisive in the choice than the feeling for beauty and value. Exotic animals and plants of the most diverse type were kept in the marvelous gardens. To satisfy his zeal for collecting, the emperor drew artists and craftsmen from all over to Prague, painters, sculptors, copper-engravers, goldsmiths, artists in glass, stone polishers, brass-founders, mechanics, clockmakers. It is significant that only the large scale art of architecture was almost entirely neglected. Zealous attention was also given to music. Paired with the emperor's passion for collecting, was his predilection for occult science. Numerous alchemists practiced their dark profession in secluded laboratories; they enjoyed the special favor of the ruler, who took an active share in their experiments and offered them great resources. They immediately formed a separate fraternity at the court and were able to use their ruler's whim for their own benefit. Besides alchemy, the emperor also favored astrology. He attentively followed the phenomena in the heavens, believing that the fate of men is determined and ruled from there. However, it was not as if the sovereign did not also appreciate true science and its new knowledge. Right and wrong, real and unreal were grotesquely confused in the goals of his diverse efforts.

It is obvious that under such conditions, state affairs must have suffered, all the more so since the emperor jealously saw to it that none of his rights be taken from his hands. The councilors were in a difficult position, due to their ruler's unpredictable manner. Envoys from foreign courts often had to wait for weeks at a time for an audience. The only ones through whom one could, in the end, still penetrate to the emperor and accomplish something with him were the grooms of the chamber, who naturally often utilized their power shamefully.[1] The historian will be interested to learn that, in spite of the shortcomings of the administration, Kepler, in a letter written in 1598, while still in Graz, that is from afar, attributed positive values to the emperor's political actions. His opinion dates from the time when

---

[1] AU. NOTE. An example of this is offered by the attempt of Brahe's heirs to gain possession of the money granted them by the emperor for the instruments and observations. For that purpose they turned to Ruzky, the groom of the chamber, and promised him no less than 5,000 talers if he made good his promise to see to it that the debt be paid out of definite receipts due. In fact, this scamp, as the documents show, once diverted the sum of 1,000 talers from the imperial treasury and later the sum of 2,000 talers.

Turkish help was considered and haggled over at a Reichstag, which the emperor, by the way, most rarely convened. Kepler wrote: "It looks as though the emperor possesses a certain Archimedian manner of motion. It is so gentle that it barely strikes the eye, but in the course of time puts the entire mass into motion. There he sits in Prague, understands nothing about the military profession, but yet, without authority (as one previously believed) accomplishes wonders, keeps the princes submissive, makes them accommodating, obliging, liberal, checks a monarch who for so many centuries had been formidable, makes him pliable by the long drag of war, without himself suffering such a great injury as to outweigh the ravages to the enemy. In this manner he lays the foundation for a position of full power, so that only the subjection of the Turks still seems to be missing."

Life in the city of Prague naturally received color and form from the court. The great court city with many dignitaries and officials, the numerous envoys and agents, who resided continually in the city or arrived because of extraordinary motives, lent brilliance and bustle to the public life and increased the wealth. Thousands of hands bestirred themselves and created lovely or useful things. Craft and trade flourished. Good times came to the Jews in the Prague ghetto. Adventurers streamed into the city from afar to make use of the opportunities it offered. Questions of science, art, religion, politics were eagerly discussed. National and religious differences also had this in their favor, that they spurred all parties to positive accomplishment. The cultivation of occult sciences diffused an iridescent veil of smoke into the intellectual atmosphere. It ruled life and activity everywhere. In the last analysis all this was due to the solitary man on the Hradcany, who seldom left his castle and there was hardly capable of managing himself. Such are the strange rules that govern the interaction of human forces.

That was the court at which Kepler served as imperial mathematician and those were the surroundings in which he spent his Prague years. When he went through the streets or appeared in the imperial antechambers in fashionable court costume with standing lace collar, he was outwardly one among many. But in importance he was unique. No name among the many could be compared with his. He towered far above all. The thoughts which occupied him, the problems with which he wrestled have already been shown above. And there it also appears very clearly that he did not stand in the twilight of the semi-scientific or perverted efforts, which prevailed in the neighborhood of the emperor. He speaks only very rarely of alchemy. It may be concluded from that little that he rejected the art of making gold. He had set himself a high goal and tried with the clear light of reason to illuminate the secrets of nature. He expressed this effort aptly in one of his calendars:[1] "I may say with truth that whenever I consider in my thoughts the beautiful order, how one thing issues out of and is derived from another, then it is as though I had read a divine text, written into the world itself, not with letters but rather with essential objects, saying: Man, stretch thy reason hither, so that thou mayest comprehend these things." This challenge, which he set for himself, he also sought to justify in astrology. He took pains to develop this subject in an irreproachable scientific manner as regards goal and method. Kepler had previously turned against Tycho Brahe when that scientist had maintained that in theology and astrology one is not allowed to ask for reasons; that there one must simply believe what authority says about the former and what experience says about the latter. On the contrary, Kepler conceived the researcher's task as being exactly this, to trace back to the causes what is definitely established empirically. Casting horoscopes is for him, it is true, "an unpleasant and at this time very begrimed work," which one permits oneself to use only with great discretion. Still "if it should not be used, neither should it be smothered, but rather does it cry with raised voice and show the divine wisdom appearing in the creation of the world. It still remains that people are distinguished from one another more by heavenly bodies than even by institutions and habit: That when bad aspects occur, the prevailing plagues and illnesses reign most fiercely: That the souls of men become wild at strong aspects and for all sorts of purposes brave and restless: And it is permissible to let a ruler of the vulgar crowd once in a while take advantage of such influences."

Kepler also followed this basic statement in fulfilling the astrological wishes of his sovereign. It was, of course, not exactly easy to satisfy all these wishes and at the same time to meet his own rigorous requirements. The emperor followed the scientific work of his mathematician and noted well which were his unusual abilities. To what extent he made use of his services and how often he ordered him to report, cannot be said. Kepler never gives any information about his personal interviews with the ruler. When in the year 1605 he writes in a letter that business at court deprives him of half of his time, it may be a question of the many errands which he had to run in order to get hold of his salary. Occasionally he tells that the emperor frequently de-

---

[1] He composed four calendars in the years 1603–1606 in Prague. Unfortunately, only the two for the years 1604 and 1605 are extant. (Transferred from text.)

manded written reports from him about some question or other. Several such essays have been preserved. There information is given about the nativity of the Emperor Augustus, about the nativity of Mohammed, and the fate which, according to the stars, is to be expected for the Turkish kingdom, supplementing and criticizing astrological arguments which were presented to the emperor from other sides, about the judgment of the then pending Hungarian question in accordance with the arrangement of the stars, about an astrological calendar which the emperor had read, about a prognosis in the fight in which the Republic of Venice was involved with Pope Paul V, about a pump without valves, which Kepler had devised. Rudolph was very much interested in Galileo's new discoveries with the telescope, which will soon be discussed, and asked his mathematician for a detailed report thereon. Two further propositions concern the so-called "fiery trigon." This deals with an astronomical phenomenon, to which astrologers had always imputed special significance and which just at that time gave rise to the rashest and most absurd prophesying. The astrologers divided the twelve signs of the zodiac into four groups of three each, by putting together in one group the first, fifth, ninth, then the second, sixth, tenth and so forth. With these groups, in order, were associated the qualities fieriness, earthiness, airiness and wateriness. The first group, the "fiery trigon," was considered especially distinguished because it contains the beginning of the zodiac. The division is connected with the shifting of the so-called great conjunction, that of Jupiter and Saturn, which since time immemorial played an important part for the believers in the influence of the stars. The two consecutive conjunctions of these two planets sometimes occur about 117° from each other, so that ten consecutive conjunctions always fall in one of these three trigons. Since there are approximately twenty years between two such conjunctions, the designation fiery, earthy, etc., trigon is also carried over to the corresponding period of two hundred years and so an eight hundred year period, in which the series of the conjunctions repeats, is arrived at.[1] Now since such a period began at the end of the year 1603, a great many works about this subject were printed. The powerful figure of Charlemagne had come on the scene eight hundred years before, and eight hundred years still earlier Christ was born. And so, what sort of epochal event would come now? That was the question which occupied the minds of those who believed in the influence of the stars and thus also most actively that of the emperor.

[1] ED. NOTE. For a fuller discussion of the period between great conjunctions, see Hellman, *op. cit.*, pp. 135–6, n. 37.

7. *The New Star of 1604*

The excitement reached its highest pitch in the fall of 1604 when an extremely bright new star appeared in the immediate proximity of the two planets which were forming a great conjunction, and which Mars, in the meanwhile, had also approached. At dawn on October 11, an imperial official, earnestly concerned with weather observations, came to Kepler in acute agitation with the announcement that on the previous day he had seen a brilliant new star in a gap in the clouds in the evening sky. Kepler hesitated to believe the report. In the days following, the sky was overcast and the mathematician had almost forgotten the tidings when, on October 17, the weather was clear and he saw the wonderful spectacle. In Ophiuchus near the three outer planets, Saturn, Jupiter and Mars, which were all close together, a fourth star had appeared; it competed with Jupiter in brilliance and sparkled in all the colors of the rainbow, like a well-cut diamond turned in the sunlight. How curious that exactly at that time when those planets had a rendezvous and exactly at the place of this rendezvous a new star should appear next to the old trusted wandering ones. No wonder that immediately innumerable pens were set in motion to investigate the phenomenon and interpret this meeting. Qualified men and even more who were unqualified poured forth their thoughts about it and the lovely spectacle was also celebrated in verse. Naturally not only Rudolph but also all spectators waited for the distinguished imperial mathematician to express himself. Kepler immediately published a short report in German. Belief in cosmic continuity, critical intellect, pleasure in storytelling, roguish humor and serious preaching zeal gave him the words.

Yet that was not sufficient for him. He collected all the material that he could get and arranged for continuous observation. In a magnificent book which appeared two years later, and is dedicated to the emperor, he set forth everything, going far back, which he could say about the event. Right at the beginning he turned toward the illness of astrology which had befallen not only some few but the largest portion of mankind. To him the astronomical side of the phenomenon is the most important, the gradually decreasing luminosity, the color, twinkling, distance, and material of the nova. He pointed out that it is part of the world of fixed stars, which at that time could not be taken as a matter of course. In contrast to the opinion that the planets had ignited the new star, he supported the stand that he was here dealing with an agglomeration of heavenly material, which also manifests itself in other phenomena. The causes of such an agglomeration he seeks in an architectonic natural ability inherent in that material. He refers to the analagous creative ability of the earth, which according to the law of spontaneous generation is able to bring forth all kinds of lower animate creatures. He rejects the possibility of the star appearing by accident at the same place and time as the great conjunction; in it he sees God's way which adapts itself to men and makes use of the rules of astrology which are in themselves objectionable in order to exhort men, who are dependent on him, and to inform them of his opinions. But what was the wonderful phenomenon supposed to signify? There were many interpretations. There was talk of a universal conflagration, of the Day of Judgment, of the overthrow of the Turkish kingdom, of a general revolution in Europe, of the appearance of a great new monarch: *Nova stella, novus rex.* Kepler himself let his thoughts rest on the conversion of America, on a universal migration out of Europe into the new world, following the old passage of peoples from east to west, or on the downfall of Islam or even on the return of Christ. Still he himself again pushes all those thoughts aside. To write the long chapter about the new star's significance regarding destiny was for him, as he says, a truly unpleasant, complicated task, charged with painful and annoying toil. "The mind accustomed to mathematical demonstration, on contemplating the faultiness of the foundations resists for a long, long time, like a stubborn beast of burden until, compelled by blows and invectives, it puts its foot in this puddle." He works his way past the numerous reefs with many ifs and buts, yet does not anchor, speaks in the third person, advances the pros and cons, makes careful suggestions about the admissibility of an opinion in accordance with the standpoint from which a subject is approached. Thus Herwart von Hohenburg was correct in expressing his opinion about this in the words "there is that much to be concluded from his Honor's hither and thither disputing in itself, that he himself does not know on what he should rely in this case." Then, too, Kepler breaks off his arguments with a jerk, by referring the readers who want to know the future to the books of prophecy in general circulation and explains that he was not appointed as a public prophet by the emperor but received the job of continuing and completing the renovation of astronomy begun by their master Brahe. As a good and freedom-loving German, he had omitted everything controversial from his arguments, although there was much on the tip of his tongue which would not have been welcomed by any of the contending parties in the state or church. It would be best if, in the presence of the celestial sign, people were to commune with themselves, examine their mistakes and vices, and repent. In a letter of that time, he rejected the absurd and worthless prophecies which were then circulating through the world, using the pithy words: "I do not want to conceal from Your Highness that it seems to me more and more that we are looking for too much art in these things. We should grasp the ox by the horns, the buck by the beard, and so on. Moreover, to speak about these signs, one should value them for what they are worth and in accordance with their effects. If they signify nothing we are acting foolishly when we investigate them. If they signify something, well then, the significance must be so constituted that even the common man can understand it." Apart from many essential arguments characteristic of his picture of the world, in Kepler's book about the new star there still flash so many thoughts of an intelligent and ingenious head, there are so many manifestations of a warm heart, that anyone whose mind is not entirely mired in the one-sided thinking of the later natural science is happy to follow him and does so with enjoyment.

Kepler also added an appendix to his book about the new star. While passing several weeks of the summer of 1605 in Styria, he came upon a newly published tract by Laurentius Suslyga, a Pole, in which the author produces proof that our time reckoning starts four years too late. Kepler was highly pleased and immediately did some thinking on his own about it. About five or six years before the starting date of the Dionysian calendar a great conjunction had also taken place, similar to the one in his own time; that conjunction also introduced the eight hundred year period of a "fiery trigon." Now if Christ had been born four years before the start of that era, does not the star, which after the birth of Christ led the Wise Men out of the Orient to Him, form a highly significant counterpart to the new star which just appeared, which also shone forth at the time of a great conjunction and also at the start of the eight hundred year period? At any rate Kepler closely followed Suslyga's thought processes and, in the main, accepted his conclusions. Because of this thorough review, he later felt the urge to write further on this subject. He was so much the more ready to hold to his thesis on learning that Maestlin had made the same assertion in his lectures even before the appearance of Suslyga's tract.

This example shows how Kepler managed to execute his astronomical commission. The same caution and reserve is also noticeable in his other prognosticas. He always protested when he was expected to prophesy events and fates from the stars. The emperor, despite his greater astrological propensities, was well satisfied with his mathematician. When in the year 1610 he favored him with a grant of 2,000 talers, he bestowed the highest praise on him by declaring in the deed "that we have graciously viewed, observed and reflected upon the faithful, diligent, proper and untiring most humble service which our mathematician, the faithful esteemed Johan Keppler, now in his tenth year here, has, with special pains, most obediently exhibited and shown and daily performed to our agreeable and most gracious pleasure and satisfaction. He henceforth is bid most humbly to do no less."

8. Financial worries

Thus everything would have been satisfactory had Kepler only really received what belonged to him—his salary and the extra settlements. But here there had been great trouble. The emperor's passion for collecting devoured huge sums and plunged him into debt. In addition, the war with Turkey continually made heavy demands on the state treasury. Rudolph approved on a large scale but from lack of funds the paying office of the court treasury always proved slow to pay off the agreed monies. So Kepler had to fight continually to come into his due. This he did very thoroughly as is seen from a writ of the imperial exchequer in which the treasury is admonished to settle Kepler's account "to put an end to this complaining and bothering." But nevertheless this complaining and bothering did not have much result. Besides, in the dedications with which he presented his books to the emperor, he was always able to express his money agonies in newly contrived ways, not without slipping in the remark that he could look after his interests better with other works. To Herwart von Hohenburg, who, to the best of his ability, used his connections to help his old protégé, Kepler wrote in 1607; "My hungry stomach looks up like a little dog to the master who once fed it." According to Kepler, the arrears in his court salary amounted to some 3,000 gulden in March, 1611. For this he did receive an order for over 2,000 talers or $2,333\frac{1}{3}$ gulden on Silesia, but he could recover nothing from there either. Anyhow, the aforementioned charitable gift of 2,000 talers remained on paper. Other court employees must have fared similarly, as one can also read about the same kind of difficulties in other parts of the kingdom. Only that was no solace and the circumstance was particularly bad for Kepler. It was annoying and agonizing for him to be dependent on the revenues from his wife's property. The posts which Kepler actually obtained and from which his great debt is assembled are not easily calculated in detail because the records are not all extant and Kepler repeatedly received extra compensation on special occasions and for special services. But it is well understood that the inconvenience of the mathematician caused not only vexation but also concern for the future of his family. Matters went so far that as early as 1603 he thought about seeking leave and a recommendation to the duke of Württemberg and asking Herwart von Hohenburg to advise him if there was a position open anywhere with an income of 500 gulden. "You ask me about my lot," he wrote in the same year to a friend in Italy. "I live, to be sure, and hope; that is I am stuck fast in the misery of court life in which hope is the anchor, the prosperity, the promised harbor." His feeling of insecurity in his position which, whenever the occasion arose, made him keep a lookout for other provisions, is shown in the words he wrote to Duke Maximilian of Bavaria when, at the instigation of his chancellor, he presented him with a copy of his *Optics*: "Even though at present I live peacefully under the protection of his Imperial Majesty and have an honorable salary which is paid so far as the burdens of war permit, nevertheless caution stands at my side as a continual reminder and from time to time pulls me very harshly by the ear and whispers this speech to me: The harvest will not last forever; therefore build yourself nests." Just as the weak little stem of the grape vine lifts its tendrils to the high growing elm, so similarly he wanted to put his humble request to his Highness to take his studies under his powerful protection. The answer was, certainly, not exactly encouraging, since the duke receipted the presentation of the book with so slight a gift that his chancellor, Herwart, thought it suitable to improve it out of his own pocket.

The lack of resources had a special evil consequence for Kepler. He could not, as he had wished and needed, keep a permanent assistant for his voluminous labors of calculation. The few helpers about whom we know, such as Matthias Seiffart, Johannes Schuler, and Caspar Odontius, occasionally were engaged in his service, but only for a rather short time.

In this manner the miserable insufficiency of money forms the black side of Kepler's court position, the deep shadow which lay over his Prague years, and, as will become apparent, not over these alone. Once he characterized his position and his attitude in fulfilling it with precision. After a silence of more than four years, Maestlin had written to him and had excused himself by saying that he had not had anything of equivalent value to write to his old pupil and had fancied that the latter had climbed to such a high position that he could look down on his former teacher. This peevish man would have found greater pleasure in the role of a protector; he lacked the warmth of heart for sharing in the joy of his onetime pupil's rise. It was then that Kepler gave the lovely answer: "High honors and offices I do not have. I live here on the stage of the world as a simple private man. If I can squeeze out a portion of my salary at the court I am happy not to have to live entirely on my own means. As for the rest, I take the attitude that I serve not the emperor but rather the whole human race and posterity. In this confident hope, I scorn with secret pride all honors and offices and, in addition, if it is necessary, also those things which they bestow. I count as the only honor the fact that by divine decree I have been put near the Tychonic observations."

9. Kepler's patrons and friends

His position as imperial mathematician brought Kepler into contact with the men of intellectual ability in the city as well as with many high court officials and men of nobility, of whom not a few distinguished themselves by their love of science. Here, too, the lovableness of his being, his striking wit, the many-sidedness of his knowledge and intellectual interests gained him friends and patrons. Even though one should not rely too heavily on words of praise and esteem in letters written to him, since they may be but expressions of courtesy, it none the less is apparent from remarks in letters between third parties that his personality enormously impressed everyone who came into close contact with him. The majority of his Prague friends followed the Augsburg Confession or belonged to the Bohemian Brethren. Still he was on friendly terms with the Catholics.

Foreign princes, who came to the court, gladly drew him on to longer conversations. Once the elector of the Palatinate, when he sojourned in Prague, engaged Kepler's services for days. On such an occasion Elector Ernst of Cologne occupied him for a period of eight days with a mathematical question. For this elector he composed a detailed memoir about the systematization of weights and measures, eloquently coming out for the decimal division. Besides, the elector wanted to know how Kepler intended to measure the earth's circumference without carrying out observations in the skies. That was a thought which kept revolving in the mathematician's head. He believed he could accomplish his end by measuring both the angle formed by horizontal sights from two places far apart, but visible from each other, and the distance between them. The elector, who was descended from the Bavarian ducal house, suggested to him that Munich and Freising were suitable places, because one could see out of the windows of the castles from one place to the other. Kepler believed that cosmography would also be helped by this and a determination of measure established for all time, "if only science lasts so long."

Baron Johann Friedrich Hoffmann von Grünbüchel und Strechau, an imperial adviser, has already been shown as one of Kepler's friends by his large-scale hospitality in receiving the mathematician on the latter's move to Prague. He was the spokesman of the Protestant party in Styria. After long travels which even led him to the Orient, he entered the imperial service and bought an estate in Bohemia. He had had himself instructed in mathematics by Ursus and Valentin Otho. His extensive correspondence with scholars of all nations and of all branches of learning characterized the breadth of his education. He especially exhibited his big-hearted generosity as well as his admiration for Kepler's talent by equipping him with two astronomical instruments of Brahe's pattern, a $2\frac{1}{2}$ foot brass azimuthal quadrant and a $3\frac{1}{2}$ foot iron sextant.[1] Even if these instruments lacked the refinement of Brahe's, they were nevertheless of great value to Kepler since, in spite of the imperial order, Brahe's were not placed at his disposal. Because Brahe's heirs did not receive the promised sum, the valuable instruments were kept hidden under lock and key.[2] It was therefore wise foresight that Hoffmann, while Brahe was still alive, had instruments of his own made for his protégé. Kepler used these to observe the planets, the new star, the moon and comets. The oppositions of Mars in 1602 and 1604, which he followed with them, he used in his search for the planet laws in the *Astronomia Nova*. In the eleventh chapter of this work he also cited a series of observations by which, in vain, to be sure, he hoped to obtain the horizontal parallax of Mars.

[1] AU. NOTE. Recently Father Richard Rankl, director of the Kremsmünster observatory, was able to show by good arguments that an instrument in the collection of this observatory is very probably the identical sextant which Baron Hoffmann had given Kepler.

[2] AU. NOTE. The instruments which had performed such useful services for science were actually ruined in the next two decades. At any rate, this was the fate of the larger ones. In 1604 Emperor Rudolph wanted to have them conveyed under Kepler's supervision to a safe storage place in Vienna. However, the plan was not carried out. Christopher Scheiner, a Jesuit father, famed as a mathematician, according to his letter to his fellow Jesuit, Paul Guldin, on a visit to Prague in 1617, was obliged to ascertain that most of the instruments had been completely ruined and no one was bothering about them; only three or four had been found uninjured in an imperial summer house.

ED. NOTE. Recently, Dr. Derek J. Price, formerly of Christ's College, Cambridge, and after 1958 at the Institute for Advanced Study, Princeton, to whom I am indebted for the following statement, discovered several astronomical instruments and books which seem to have come from Brahe. They are in the collection of Baron von Essen at Skokloster Castle, Sweden. During the Thirty Years' War the seventeenth century ancestor of the Baron had been a warrior in Prague. With the exception of a simple brick sundial in the "Herregaard" on the island of Hven and perhaps the Skokloster instruments (if their identity can be firmly established), there are no instrumental remains extant. Tychonic instruments preserved in several museums in Europe are for the most part later copies, reconstructions or fakes and, at the best, of "Tychonic type" only.

Peter Wok von Rosenberg was the leader of the Utraquists. As the last of one of the most powerful and richest families of Bohemia, he led a lavish life in splendor and luxury and squandered colossal sums on his alchemical passion. After Brahe's death, Rosenberg transferred his friendship for him to his successor. Kepler gave evidence of his gratitude by dedicating to him the first work which he published after his change of residence to Prague. It was an essay about the more reliable foundations of astrology.

An entirely different figure, who likewise bestowed favor and friendship on the imperial mathematician, was Wenzeslaus Budowetz von Budow, an aristocratic phenomenon of Czechish nobility. He had remained in the Orient for a long time and had learned Turkish and Arabic and called up Christian Europe to fight against the Turks. A pious man, he inclined to Calvinism, which was to be his destiny. He had especially taken Kepler to his heart and, when not staying in Prague, exchanged letters with him.

In the rank of court advisers, who changed frequently, Johannes Barwitz had the firmest position. He was able to restrain himself longest, and enjoyed soonest the trust of the ruler. Kepler extolls his services to science, especially to the Tychonic astronomy, and to those who carry it on. At the court he so exercises his protection over the representatives of science as to be unsurpassed in this effort. Kepler had a staunch supporter in him, particularly in maintaining his salary claims.

A truly friendly relationship tied the mathematician to Johannes Matthäus Wackher von Wackenfels, a court adviser twenty years his senior. Wackher von Wackenfels was born in Constance and was related to Kepler. The son of parents without means, he had his own talent and his own ability to thank for his brilliant career which ended with admission to the imperial council and to the nobility. Originally brought up in the Reformation creed, he had transferred to the Catholic Church even before coming to Prague. With his many-sided scientific interests and his great conversance in all possible spheres, he was the correct partner for Kepler's scholarly conversations which often were seasoned on both sides with remarks full of humor. At New Year's, 1611, Kepler presented his friend with an essay in which he undertook to trace the reason for the six-angled shape of snow crystals. Composed as a letter to Wackher, the droll writing appears as an exceedingly charming study which not only exhibits the sagacity, the rich fantasy, and intelligent descriptive gift of its author but also pushes into the light his position between the old Aristotelian view of nature and the new physical description and explanation of phenomena.

In the house of the imperial councilor Johannes Polz, Kepler was always a welcome guest. In the letter of recommendation, which he gave a son of this family who was going to Tübingen, he expresses himself in greater detail about his relationship with the family. The father is very fond of him, the mother, like the whole family, is distinguished in Prague by Austrian elegance and genteel, noble breeding, so that it would be due to this family if he ever improves in this, in which he certainly still has far to go. Moreover, here he would always find the best advice in his difficult transactions. The neediness of his own household and his lowly position were no hindrance whatsoever to his coming and going there as it pleased him.

Still another is counted in Kepler's circle of friends, namely Johann Georg Gödelmann, the ambassador of the electorate of Saxony, who occupied himself with history and employed his juridical knowledge on witchcraft. He took over the position of godfather to Kepler's son Ludwig. When Kepler had missed his mark with his weather prognostics for August, 1604, and many "persons of high intelligence and prestige" had advised him to give up calendar making, it was Stephan Georg von Sternberg who encouraged him to continue. The latter had also otherwise "let" him "perceive extensively" his "praiseworthy affection for mathematical studies."

In a relationship of a special type Kepler approached the prelate Johannes Pistorius, who was twenty-five years older, and who, as father confessor and adviser to Rudolph, occupied an important position at court. As we have already heard, Kepler had to give him an account of his work from time to time. However, that was no hindrance to the two men carrying on friendly relations with each other which ended only with the death of the prelate in the year 1608. Their scholarly conversation extended to the most varied subjects because Pistorius as Polyhistor, as Kepler designated him, was conversant in all possible domains of knowledge. Thus he had, in contrast to his friend, who was not willing to approve of it, predicted that some time with the help of lenses one would arrive at a much more refined method of observation. The question of the year of the birth of Christ was earnestly discussed between the two since Pistorius was unwilling to accept Kepler's thesis. But in addition, also, the contrast in religious creed frequently gave rise to lively analysis in which both parties supported their points of view with enthusiastic energy. Pistorius had an unusual career behind him. Born the son of a superintendent in Hesse, he had dedicated himself to the study of theology, jurisprudence and medicine and earned a doctor's degree in the latter. His first occupation after the close of his studies had led him to the court of the margrave of Baden-Durlach as physician-in-ordinary. But soon his theological interests had won the upper hand. He transferred to the reformed creed and later to the Catholic Church and was active as one of the sharpest polemicists of his time, orally and in writing. It is easy to imagine how heated were the discussions when two men like him and Kepler argued over questions of belief. A letter which Kepler wrote in 1607 to Pistorius in Freiburg, whither that prelate had gone from Prague, gives an insight into their relationship. Pistorius had written that he was so seriously ill that he saw the end approaching; still he was of good cheer and was preparing himself for the welcome path which would free him from the vanities of the world and would lead him to the redeemer Jesus Christ and to participation in the heavenly heritage.[1] Kepler, otherwise so peace-loving in religious matters, took this remark as the occasion for such a sharp attack on the Catholic Church, as scarcely seems possible from him and of a kind which he makes nowhere else. After words of courteous and friendly sympathy he suddenly lights upon the word inanities. These inanities, in his opinion, largely occupy the fervent zeal of that party, which drivels of a right of salvation and represents Rome as the sole gate to heaven. From there originate the total submission of the rulers to those who are the indirect rulers and the complete scorn of those who hold fast to liberty. It is also the source of a perverted interpretation of words and deeds, as though a blow thus struck against the enemies of the Roman priest-king would hit the enemies of God, and that which would benefit the strongest party might also be of advantage to God and the church. In addition Kepler reproaches his sick friend with the sins which he had committed, while believing he was rendering God a service by harassing certain people.[2] He summons him as witness before God's judgment seat that it is not out of hate for Pope and Bishops, but out of fervor for God and concern for the prophecies of Christ and

[1] ED. NOTE. *Johannes Kepler Gesammelte Werke*, XV, 412 (letter number 413): "...nec doleo: sed totum me ad iucundam illam viam comparo, quae me liberum ab inanitatibus mundi factum, ad Christum Saluatorem meum et ad partam ab illo caelestem haereditatem ducet."

[2] ED. NOTE. *Johannes Kepler Gesammelte Werke*, XV, 488–92 (letter number 431), Kepler to Johannes Pistorius in Freiburg, dated from Prague, June 15, 1607, p. 489: "...Neque sanè dubito te hac fiduciâ Christi Servatoris, partaeque per illum coelestis haereditatis, hoc contemptu, quod scribis, adeoque, quod interpretor, odio et poenitentia, inanitatum hujus mundi (· quarum ego magnam partem esse puto, studium partium fervidum, privilegia beatitudinis somniantium, Romae coeli portas unicas statuentium, indeque natum in Reges indirectos Regum, merum obsequium, merum contemptum eorum, qui libertatem amplexi sunt, sinistras dictorum factorumque interpretationes, quasi quod his medijs Romani Sacrorum Regis hostibus noceatur, Dei hostibus noceatur, quod vicissim factioni validissimae lucro fit, Deo et Ecclesiae non aliter lucro fierj possit ·) hac inquam inanitatum hujus mundj poenitentia paratum et accinctum, nunquam non aptum esse, beatae illi transitioni faciendae...."

the Apostles that he perseveres in that freedom in which God had permitted him to be born. It is only the distorted teaching of mediocre minds which connects those prophecies with the Roman monarchy or churchly despotism. Kepler maintains that he had not bowed under the Roman yoke of those who both saddle the Christians with indifferent ceremonies and explain the words and commands of Christ in a dangerous fashion, by claiming for themselves the sole right of interpretation. These people make an effort to chain the human intellect which God addresses through his servants. Finally, the letter writer asks the prelate to receive his writing in the same way he had always received his words when they had argued orally about this subject.[1] Pistorius answered this heated outburst simply by expressing the wish that Kepler would leave theology out of the question because he understands nothing about it. He ascribes things to the Catholics which they had never thought. Finally Pistorius assures his friend of his sincere love, much as it pains him that Kepler is blind in matters of belief. It was unnecessary for Kepler to say that he wrote his letter in the innermost agitation of his heart. The reader perceives it all too distinctly. However, it is difficult to say by what means he was so excited without being provoked, why he let himself be carried off to such a sharp attack by the catchword "vanities," why, toward his very sick friend, he neglected to have the tact which he otherwise showed. It can only be assumed that memories of specific earlier conflicts play a part in this. Also it must not be forgotten that Pistorius had been accustomed to use exceedingly sharp and rude words in his polemics. Moreover, Kepler would learn only too soon that those who had brought him up in the ideas which he here expressed, in practice interpreted entirely differently from him the freedom which he so emphatically pleaded. He was to learn that the Württemberg consistory was more popish than the Pope and did not possess that tolerance with which Kepler was treated by the Catholics in Prague and also later in the Hapsburg lands. Accordingly, in the course of time, his views must have grown milder in various respects.

In Jost Bürgi, the unusually gifted imperial mechanic and watchmaker, Kepler had a friend such as he needed. Bürgi was born in Lichtensteig in Toggenburg.[2] After learning the watchmaker's craft, he had come to Landgrave William of Hesse and rendered him indispensable service in his observatory in Cassel not only by producing astronomical instruments but also as observer and calculator. For it

---

[1] ED. NOTE. The letter closes with several scholarly paragraphs on the birth year of Christ.

[2] ED. NOTE. In St. Gall canton, Switzerland.

had become apparent that this man, who had been unable to attend any university, possessed both a rare technical skill and an unusual mathematical gift, so that the landgrave, in a letter to Tycho, described Bürgi as a second Archimedes. In the year 1603 Emperor Rudolph, who had long been casting an eye on him, had succeeded in drawing him to Prague completely. He had been there once ten years before, for a short period of time, to deliver to the emperor in behalf of the landgrave a planetarium equipped with a gilded celestial globe which he had made. Kepler and Bürgi were often side by side at common work and discussion, the mechanic assisting his friend by the dexterity of his hand, and the mathematician assisting the other by his scholarly education, in which he had an advantage. While still in Graz, Kepler had devised an "amusing artistic little fountain" in which, without use of valves, the water is driven up high by the pressure of a feather. After other mechanics had tried without success to execute the model, Bürgi succeeded in constructing, in accordance with Kepler's idea, a "little fountain" which drove the water up to three times the height of a man. Naturally it pleased the emperor, to whom it was dedicated. After that, in his "Progress Tables," Bürgi was the first to make use of the principle of logarithms for the simplification of practical calculation. However, being too cautious and reserved, he forsook the child of his intellect instead of rearing it for the public, as Kepler observes, and first published his tables in 1620. As a consequence, in 1614 John Napier, the Scottish baron, destroyed Bürgi's glory of being the first to publish logarithmic tables. Kepler, for whom each shortening and simplification must have been very urgent, considering the many long calculation tasks which he had to carry out, showed great interest in Bürgi's invention. In the written material left by Kepler at his death there is a manuscript about another mathematical subject, which shows how he helped his friend to put his thoughts on paper in suitable form. In one of his works Kepler gives his opinion about Bürgi: although he possesses no knowledge of languages, he still surpasses many professors in the domain of mathematics; in practical mechanics, indeed, he will be celebrated by posterity as a master-mind, just like Dürer in painting.

It might be noticed that nothing has yet been said about Kepler's connection with the University of Prague. In fact, there is scarcely anything to be found about this in his written remarks. The statement, which can occasionally be read, that he taught there for a while in 1605 calls for supplementary proof. The reason for the surprising fact is to be sought in the state of decline, mentioned before, in which the university found itself at that time. Because the funds over which the university had command were only very scanty, the masters in Charles

College all had to lead a common life in celibacy. The situation which had developed out of this could have nothing attractive for important scholars. Two men with whom Kepler was connected nevertheless stood out. One, Johannes Jessenius, the distinguished anatomist, is already known to us. He had been a sponsor at Kepler's summons to Tycho Brahe and since then remained kindly disposed to him. He moved from Wittenberg to Prague at the same time as Tycho came to Prague but soon abandoned the teaching profession because of its restriction to celibacy. From him Kepler gained anatomical knowledge about the structure of the eye which he needed to explain the sight process for his *Optics*. When Jessenius became rector of the university in 1617 he contemplated getting Kepler for his academy. But the latter had been gone from Prague for a long time and the city in the meanwhile, after Rudolph's death, had taken on a different aspect. It was Kepler's good fortune that nothing came of the summons. Otherwise he would easily have been pulled into that circle several members of which ended their lives on the scaffold a few years later, in the course of the further calamitous development, both political and confessional. Among these was Jessenius, who leaned toward Calvinism.

The second of the men from the university circle to be named here is Martin Bachazek, a predecessor of Jessenius as rector. He represented the sciences but was better known and more deserving as the pedagogue who took great pains to improve instruction in the Protestant and Utraquist circles of Prague, although he did not achieve lasting success under the difficult circumstances. Kepler got along so well with him that in the year 1604 he moved into a dwelling in the same house where the two could see and speak with each other daily. At that time they observed the new star together and Bachazek had a little wooden tower built for this express purpose. He also took part in another observation in the year 1607, which led Kepler to compose his *Phaenomenon singulare*. This contains an exceedingly clear report of the event. In April and May of 1607 Kepler had earnestly observed Mercury in the evening sky. According to calculations, it was supposed to enter into lower conjunction with the sun on May 29. Since a heavy storm arose in the evening and night of May 27, the belief that the aspect could be the cause of this disturbance suggested to him that perhaps this conjunction should be fixed earlier. Yet from time immemorial, the theory of Mercury caused astronomers very special difficulties. Since, besides, he supposed Mercury to be near a node of its orbit, he set to work to observe the sun on the afternoon of May 28. He was just discussing the matter with a Jesuit when the sun came out of the clouds. He immediately cut the conversation short and repaired with Bachazek to the loft of their house, where single sun rays shone through thin cracks between the shingles. He held a piece of paper against such a ray and there, on the little picture of the sun so formed, espied "a little daub, quite black, approximately like a parched flea." He was certain he was observing a transit of Mercury. He was overcome by the greatest excitement. To insure against error, he repeated the observation under other conditions. "And to make sure it was not a mark on the paper, we kept moving the paper back and forth so that the light on the paper moved, and everywhere the little black spot appeared together with the light." Kepler drew up a report and had Bachazek verify it. In his excitement he ran to the Hradcany. On the way he related his experience to the Jesuit. But just then the latter had to pay attention to prayer time and bade the sun to wait. On the Hradcany Kepler had the important message announced to the emperor by a valet. Then he repaired to the workshop of Bürgi, the court mechanic, who was out. With one of Bürgi's assistants Kepler covered a window and let the light shine through a small hole in a tin plate. Again the "little daub" appeared. The watchmaker-journeyman, too, had to place his signature beneath the report: "Heinrich Stolle, little watchmaker-journeyman, my hand." "When the clouds had departed, one had seen how they came in on the little daub, but that the little daub remained stationary in the light while the clouds moved over it." However, Kepler had erred in interpreting the meaning of his observation. What he saw was not a transit of Mercury, but a remarkably large sun spot. It did not take long for him to recognize his error. A few years later, Johannes Fabricius, the son of the well-known astronomer David Fabricius, made public the first information on the spots which could be seen on the sun with the newly discovered telescope. "Lucky I," Kepler later exclaimed, "who was the first in this century to have observed the spots." Because of his erroneous interpretation of the phenomenon he took shelter behind the utterance, still valid today: "How very changeable is the fortune of war in astronomy too, since the movable army of conjectures, with vacillating assurance, turns now here now there."

His many court connections proved useful to Kepler for another little script[1] which he published during his stay in Prague. It deals with the total solar eclipse of October 12, 1605. We know how valuable eclipses were for Kepler's researches. Whereas in the solar eclipse which he had been able to observe in 1601 the moon was at apogee, the one in 1605 was to take place near perigee. By the calculations, the zone of

---

[1] ED. NOTE....*De Solis Deliquio...Epistola.* Prague: Schuman [1605]; also available in *Johannes Kepler Gesammelte Werke*, IV, 37–53.

totality was to fall on the southern part of Europe. That offered favorable prospects. By comparing the results of the two eclipses Kepler hoped to draw conclusions about the eccentricity of the moon's orbit and thereby gain a surer basis for the discussion of the physical causes. In his *Optics*, whose publication he had accelerated for this reason, he broadcast to the world the summons that all astronomers must be ready. After the eclipse was over, in the above-mentioned special tract, in which he communicated his own observations, he appealed to the friends of astronomy, astronomers and laymen, to send him their observations. He introduced himself as imperial mathematician, appointed by the highest prince of Christendom to renovate astronomy. Gladly would he himself travel to the south of France and Spain in order to collect the reports. That, however, would not be possible. On the other hand, the foreign ambassadors at the imperial court are prepared to act as intermediaries for the reports. Any communication, however trifling, would be welcome. He wants to know everything: circumference of the eclipse and boundary of the zone of totality, weather conditions, color of the moon, light around the sun, visibility of stars. He only wishes this present eclipse to be so constituted as to show the human race what is to be sought by astronomers today. With his own observations Kepler was certainly not satisfied. He had performed them in the imperial pleasure garden. The craftsmen whom he needed to produce his instrument had put him off with empty promises up to the last moment. A crowd of courtiers surrounded him during the observation, and the gardener had not exercised sufficient care in preventing disturbances. With the assistance of the ambassadors and agents at the imperial court and other influential men, he got his script distributed. To the Jesuits, also, he had turned successfully because their far-flung connections promised him an advantage. In this manner, besides a series of German cities, his call reached to Italy, Spain, France and into Switzerland and the Netherlands. Even though the answers did not arrive in exactly large numbers, still Kepler did receive valuable supplements to his own observations and calculations. At any rate, the imperial mathematician's procedure, as described, presented a very noteworthy attempt to win the astronomers over to a communal work of solving a problem which interested them all. And it makes a happy contrast to the attitude of most of his astronomical colleagues of that time, who jealously held back their observations so that no one else at all would be able to use them for his own purpose.

In addition to the men with whom Kepler maintained personal contacts and carried on an oral exchange of thoughts, there are a greater number of others with whom he associated by letter. The spoken word disappears, the written is preserved, and so from his letters one gets to know the coloring which belongs to the expressions of his thoughts and feelings. It is so original and variegated that it frequently gives the impression of seeing the writer in person. The wealth of content extends to all domains of his scientific activity and to the multitudinous incidents and entanglements of his personal life. It corresponds to the wide gamut of emotions, which ranges from humor and merriment to sadness and despondency. It is not Kepler's fault that the Tübingen teachers and friends recede into the background in the letters from the Prague period. It has already been shown how Maestlin was more and more wrapt in silence and little by little ceased the earlier active sympathy for his former pupil. However, the old Maecenas, Herwart von Hohenburg, remained faithful. Indeed, the correspondence with him is even more ardent and more rich in content than previously in Graz. It is a pleasure to see how along with his concern for the circumstances of Kepler's life, he interested himself in all Kepler's scientific efforts and was able, by a sympathetic question, to stimulate him to present his thoughts about any subject whatsoever in detail. Kepler could rely on Herwart. Besides, letters arrived from foreign countries, for example, from Italy and England. From the former, Edmund Bruce sent him reports. Bruce was an Englishman who associated much with Galileo. The well-known English mathematician, Thomas Harriot, conversed with Kepler about optical questions. With several members of the Society of Jesus there arose a correspondence about the solar eclipse, and especially about chronological investigations.

Two more men stand out especially in the correspondence, as regards range and content. One is David Fabricius, the previously mentioned Frisian clergyman, the well-known discoverer[1] of Mira Ceti, the variable star in the whale, a zealous investigator of the proceedings in the sky, whom Kepler praised as the best observer after Brahe's death. Fabricius, to be sure, was also an arrant astrologer, so that he drew some reproofs from his critical friend. In letter writing he showed an uncommon alertness; indeed, he bombarded Kepler with letters. As soon as something occurred to him, he grabbed his pen and in his hasty, barely legible handwriting set down an epistle. Nearly forty of his letters to Kepler are preserved. The latter, however, always let several collect before he gave in to the other's pressure and answered his thousand questions. The two men never became personally acquainted; in 1601 when Fabricius stayed in Prague for a short visit

[1] ED. NOTE. 1596.

he met Brahe, but at that time Kepler had just gone back to Graz. Kepler knew so very little about his friend as a person, although he sincerely admired him, that he was exceedingly astonished when all of a sudden he received from Wittenberg a letter from Fabricius' son. Kepler had, he writes in his answer, up to then spoken with the father as with a comrade of the same age and almost as with a pupil. Now all of a sudden Fabricius appears with a very discerning son. Therefore Kepler would have to bestow greater veneration on the father. He would also like to know that the all too childish jokes were weeded out of his letters, since he would fear the son as a censor. Kepler's correspondence with Fabricius is of special importance because it tells us a great deal about the discovery of the first two planet laws and the origin of the *Astronomia Nova*. To this friend Kepler poured out his new thoughts. He told him of his efforts and reflections and reported to him in detail the step by step progress of his work.

But it should not be supposed that Fabricius accepted everything from his friend. On the contrary, he contradicted him considerably on all points. However, since he was an independent thinker and tried to support his contradiction with reasons, Kepler went into his objections patiently and thoroughly. It is in a long letter of forty folio sheets, which Kepler wrote to him late in the summer of 1605, that the first communication of the discovery of the law of ellipses is to be found.[1] This, too, was not clear to Fabricius. He sought rather to "improve" Kepler's theory by a system of circular motions and to do this made so much trouble for himself with calculations that, as he wrote to Kepler, his health suffered. This provoked Kepler to the witty rejoinder: "I should like to recommend a remedy to you. Refrain from the effort of trying to establish a new theory for Mars. For it is already established. I have spent so much pains on it that I could have died ten times. But with God's help I have held out and have come so far that I can be satisfied with my discovery and rest assured. Out of my peace you might draw an end for these discoveries." Thus this correspondence stimulated Kepler because he was compelled to defend his assertions and consequently to struggle through to greater clarity. To the reader of today, however, it becomes obvious how unprecedentedly new and for that time downright incomprehensible was his interpretation of the motions of the planets. Not only the ellipse law formed a stumbling block but still more the law of areas with its physical bases. Against

[1] ED. NOTE. *Johannes Kepler Gesammelte Werke*, XV, 240–80 (letter number 358), especially "Itaque omninò Martis via est Ellipsis,..." (p. 247) and "Viam planetae verissimam esse Ellipsin, quam Durerus itidem oualem dixit, aut certè insensibili aliquo ab Ellipsi differentem" (p. 249). Although the letter is dated Oct. 11, 1605, the part quoted here was probably written shortly after Easter.

these, Fabricius' special contradiction turned. He was very proud of his own purely geometrical theory of the motions of Mars but had to take a good-humored criticism from Kepler: "You say, that a daughter has been born to you by geometry. I looked at her; she is lovely, but she will become a very bad wench who will carry off the men of the many daughters which the mother, physics, has given me. Your theory will draw teachers and philosophers to it, it will offer loopholes to the enemies of celestial physics, to the patrons of ignorance, the master builders of solid orbits, the rough mechanics, loopholes by which they can escape the bands of my physical proofs and reach freedom in order to manufacture their own Gods. Simple nature has guided me, free of all the attire of a hypothesis. In the place of this highly repectable girl, nature, your young whore with her wench's apparel and her conduct conducive to lust not to virtue is supposed to appear—that is in the speech of Fabricius—one should listen to strict science not to crazy philosophers."

Of an entirely different type from Fabricius was Johannes Brengger, a physician, who practiced in Kaufbeuren. With him Kepler corresponded earnestly for a few years. In particular, all possible questions from the broad domain of optics, which Kepler had broached or treated in his book about this field of knowledge, formed the subject of the discussions. The difference between Kepler's two friends is even apparent in their handwriting. In contrast to Fabricius' hasty script, Brengger's writing is strikingly clear and neat. Corresponding to this is the difference in style and in manner of thinking of the two. Kepler was so much the more impressed by Brengger's clear diction and sharp thinking because he himself frequently lacked, not sharpness of thought, to be sure, but clearness of expression. "Both in consequence of a deficiency in my talents and because of the novelty of my material, I usually lack the wonderful clearness which I admire in your style." Besides, Brengger was much more often provoked to contradict Kepler. But the manner in which he expressed his objections was so very welcome to Kepler that he even expressed the wish that it might be possible for Brengger to read through the work on Mars before printing. Kepler was convinced that the other could stimulate him greatly by his shrewd manner, so that answering would have given rise to a clearer wording of the text. Kepler was just as unable to win over Brengger as Fabricius to the Copernican theory. To put the earth in the middle point was an integral constituent of Brengger's view of the universe, however freely he otherwise felt and expressed himself as opposed to scientific authority, especially Aristotle's. And it was this freedom in thinking on Kepler's part which found particular favor with

Brengger. In this the two men agreed best. When Brengger at the very beginning of the correspondence complimented his partner on this, Kepler was very happy because he believed one should not have anything at all to do with the opponents of this freedom. To be sure, the restriction in the use of freedom which Brengger mentioned is necessary in order that those, whom nature has created as slaves, do not long to be free and in order that the freedom, justly earned, should not be transferred to rebellion.

Not alone with high lords and scholars could Kepler associate. He also loved to engage in a dispute with simple people. Just as he shared in the sorrows and needs of the little men and without learned haughtiness paid attention to each one according to his attainment and type, so in his German writings, especially in his calendars, he did not turn first to the men of science and higher education, not to the "potentates," but rather to the "common people" whom he tried to lay hold of with his teachings and exhortations. From occasional remarks something is learned about how things went in conversation. Thus he once reported on the distrust which his astrological views and weather prophecies encountered: "Here in Prague I have a tougher job and meet those of the lower classes with straight forward and active minds, of whom there are always a good number here, who do not beat about the bush much but very dryly say what they mean and answer word for word and keep it up until one or the other conquers. When I dispute with such super-astrologers I have a difficult position and I get such a workout that I might well call them my teachers." The weather prognoses were no less important to the people of that time than to those of today. When once a few sextile aspects between the planets in the sky were in prospect, Kepler swore fourteen days before, in front of doubters, that there would be wind and rain on that day. In due course, on the day in question, came a fierce gale, driving black clouds, so that at noon it was as dark as a half hour after sunset. Amazed, the people asked themselves what was happening. Then the cry grew loud, "Keppler comes."

So Kepler stood in the stream of the surging life which surrounded him in Prague. Here he felt the pulse beat of his time. He experienced the struggle of the scientific, artistic, political and religious forces which pushed toward a new state of affairs and was himself a part of the force at work here. The strong sensation of life noticeable in his surroundings permeated him also and made him a panegyrist of his time and its achievements. It seemed, he said, that the world had slept for a thousand years and was only awake since 1450. Unprecedented advances were visible in the whole range of cultural life, in legislation, business, military affairs. The discovery of America and the sea route to the East Indies had developed trade to the highest peak. Innumerable downright incomprehensible mechanical apparatuses had been constructed. The invention of the art of book printing bears witness to the indescribable industry of the people of his time. It replaced the class of monks by that of scholars. Christianity would be carried into foreign lands. The freedom to argue, the abundance of writings, their easy dissemination by printing, the greater universal education, the universal unrest had brought about the separation of many lands from the Roman See. A new theology, a new science of law had come into existence. The followers of Paracelsus had renovated medicine, those of Copernicus, astronomy. "It seems to me as though the world now first would live or rather rush madly about." To advance beyond the present time, it was even necessary to think about the discovery of a new world or the invention of the art of flying by which people might fly to the moon or to another celestial globe, because, forsooth, the earthly globe had already become too narrow for them.

10. Home life

Opposed to this great world stood the little world of Kepler's family circle. The great intellects who by their ideas often alter the face of an epoch more fundamentally than do the politicians for the most part do not appear as principal actors in the world theater. They act in the quiet of their rooms. Kepler's domestic life was simply run, as he himself says. But it was not at all mean and it conformed to his position as well as to the pretensions of his wife who was pampered from birth. At the time of his move to Prague he had sold a portion of his household furnishings in Graz for 200 gulden, but in Prague had provided new ones instead for 100 gulden. In the dedication of a copy of his book about the new star to King James I of England he calls himself a Diogenes, who "out of the shunned tub recommends his philosophical writings to the Emperor-Philosopher." This is a joking comparison, in which he found pleasure. In the comfortable "tub" of his home, he certainly did not lead the life of a cynic. By his own account, in the beginning he spent 400–500 gulden annually for his household, later when his wife's sickliness required special consideration and children came along, 600 to 1,000 gulden. He changed his dwelling place several times. From the spring of 1602 on he lived, as we have already heard, opposite the Emaus monastery in the New Town. In the fall of 1604 he moved to Wenzel college in the Old Town, where

Bachazek also dwelt. When he had to quit this house in November, 1607, he found lodgings in "aedibus olim Cramerianis" [the former Cramer buildings] opposite the Jesuits by the bridge. His landlord received him reluctantly and at the outset prepared all sorts of annoyances. Here Kepler seems to have remained until the end of his stay in Prague.

At Kepler's arrival in Prague, his stepdaughter Regina had been the only child in the family. But soon the family increased. In Prague, Frau Barbara presented her husband with three children, on July 9, 1602, a daughter Susanna, on December 3, 1604, a son Friedrich, and on December 21, 1607, a second son Ludwig. All three were baptized by Utraquist clergymen. The rank of the godfathers permits us to draw conclusions concerning the father's social connections. At Susanna's baptism two wives of imperial guards acted as godmothers. Present also were some noble gentlemen whom Kepler knew from Graz, Baron Ludwig von Dietrichstein and the three barons, Weikhard, Herwart and Dietrich von Auersperg; evidently those gentlemen had taken over the honorable godfathership. For the second child Stephan Schmid, the imperial treasurer, Matthäus Wackher, the imperial court barrister, and Joseph Hettler, the Baden ambassador, appeared as godfathers. For the third child the baptismal register records the Counts Palatine Philip Ludwig and his son Wolfgang Wilhelm von Pfalz-Neuburg, and as their representatives the ambassador Johannes Tzeschlinus as well as Georg Gödelmann, the Saxon ambassador. The children brought activity and life into the house. The father's sigh is understandable if, after the birth of his little son Friedrich, he writes to Herwart that he is too distracted to be able to write correctly. "For what a business, what an activity, does it not make to invite fifteen or sixteen women to visit my wife, who lies in childbed, to receive them hospitably, to see them out!" For an understanding of Kepler's attitude to the various creeds, it is significant that when his stepdaughter was fourteen years old, he sent her to Dresden to receive communion.

His relatives in his native land were proud of their Johannes and wanted to see him in the splendor of his office. In the year 1602 his mother came to Prague on a visit, in 1604 his sister, Margarete, who solicitously clung to her brother. In 1608 she married Georg Binder. He was a teacher in her native Leonberg, and subsequently became pastor. Even the bad and restless brother, Heinrich, interested himself in the imperial mathematician. He turned up in Prague as imperial guard, remained for eight years in this position, and found a wife who subsequently bore him two daughters.

In the spring of 1608 "Venus made a noise in the house." Regina,

Kepler's stepdaughter, who had not lacked suitors, was married. She made a good match. She wed Philip Ehem, agent, that is representative, at the imperial court of Elector Frederick IV of the Palatinate. The bridegroom was descended from a distinguished Augsburg family and was the son of Christoph Ehem, a Palatine privy councilor and high chancellor at Heidelberg.[1] In 1610 the couple moved to Pfaffenhofen in the Upper Palatinate and in 1617 to Walderbach near Regensburg, where the husband became administrator of the convent which had been suspended as a result of the introduction of the reformation. However, Regina died there on October 4 of that year at the age of twenty-seven. Her burial place can still be seen in the stone floor of the presbytery of the parish church in Walderbach.

Kepler's married life with Barbara was not very happy. Details are gleaned from his papers. In spite of all that he asserted about or against his wife, these still show the tenderness of his feeling, because he excused his wife's weaknesses and did not make her solely responsible for the discord. Consider especially two documents. One is a long letter, in which he justifies himself when, after his wife's death, evil tongues circulated foolish chatter about his behavior toward her. The other is an unpublished writing concerned with the settlement of her estate; it is now preserved at Oxford. Public opinion, as Kepler reports, had presented his wife the palms of respectability, uprightness and modesty, with which she joined, in a most remarkable way, loveliness in appearance and politeness in her associations, as well as piety and benevolence toward the poor. He heard glory and praise of her wherever she was known.

However, within the four walls of her house she showed far less of the charming manner with which she made such a good impression in public. The amiability of behavior was a gift of her wifely nature with which she favored outsiders more than she lighted her own home, since she took few pains to govern the outbursts of her irascibility and her cross temper. The fine shape of the exterior did not conform to the training of her mind. For the work of her husband, she had no understanding; she did not divine the genius which was in him. Often she came to him with household matters at the wrong time, when he was absorbed in his studies, and she was offended if he was brusque with her or gave no answer. She cast furtive glances at the imperial councilors, with whom her husband associated, or much more at their

[1] ED. NOTE. According to Zedler, *Universal Lexicon*, VIII, 411, Christoph von Ehem died in Heidelberg in 1592 leaving only two sons, Eberhard and Siegmund. However, Miss Martha List, Prof. Caspar's collaborator, informs us that Prof. Caspar based his statements on a document preserved in the Bayer. Staatsarchiv, Amberg, in which Philip Ehem, himself, gives information concerning his ancestry.

wives, and was bitter at the difference of her own position. When people spoke mockingly about the "Mr. and Mrs. Stargazer," the husband parried the foolish rudeness by calling himself that; but the wife did not get over the taunt, remaining bound by the limitations of her nature. Yes, if she had had the heart and the means to make herself better known, Kepler imagined, much would have been better. But both were lacking. She had, as he says, "a weak, annoying, solitary, melancholic temperament."

Dissatisfied in surroundings not congenial to her nature, she more and more gave way to her impulses to melancholy. The money worries, about which we have already heard, she met with an absurd economy. She would not permit her husband to handle her property or even pawn so much as one single beaker; she would not touch her trifling spare money from fear of becoming a pauper on that account. She economized on herself, neglected her own clothing too much but, like a "woman entirely captured by love for her children," spent everything on them. This being so, there were often violent altercations which, however, ended quickly. When Kepler saw that what he said went to his wife's heart, he would rather have bitten his finger than have uttered further words which hurt her.

The deeper reason, by which he excused his wife's weaknesses, lay in her constant sickness of body and soul, which became continually worse with the years. The melancholy in which she was embroiled was repeatedly talked about. Over-pious people who doubted Kepler's orthodoxy and knew that he took exception to some matters with the preachers, blamed him for having brought his wife to her distressing thoughts by the Calvinistic error about predestination. Indignantly he protested against such aspersions. It never occurred to him to discuss such questions with his wife, and when he had expressed himself about them with others at home he had not used the German language. Moreover, Frau Barbara did not want to know anything about this. She read nothing, not even stories, but was always involved, day and night, in prayer-books, which she would read not only for devotional exercise but by inclination. To preserve peace at home required patience. And Kepler learned to bear with patience the inconveniences caused by his wife's abnormal condition, although "not much love befell him." "To be sure, there was much biting and getting angry, but it never came to any hostility, neither ever sued the other, both of us well knew how our hearts felt toward each other." Caring for the property of wife and stepdaughter made a few trips to Styria necessary. The division of the effects left by Frau Barbara's father, who had died two years before, took place on May 1, 1603. Barbara inherited about

3,000 gulden which however consisted of landed property. Understandably, Kepler wished to convert the claims into ready cash, although net proceeds from a sale of real estate were slight there at that time. Moreover, he wanted to remove from Styria at least part of his stepdaughter's property, which consisted principally of ready cash, no doubt more easily to get possession of the maintenance allotment which belonged to him. For this he needed the consent of the guardians. Besides, after Jobst Müller's death, the guardianship was supposed to be rearranged. To take care of these transactions, Frau Barbara, with her nursling, traveled to Graz in the spring of 1603; she remained away for the entire summer. In these transactions, Herwart turned for support to Peter Casal, Archduke Ferdinand's secretary. How matters developed is not precisely known. In the summer of 1605 Kepler traveled in the same cause to that place, so rich in memories, the scene of his first public activity. Contrary to expectation, he succeeded in arranging his stepdaughter's property affair satisfactorily. He himself was appointed her guardian.

In October, 1606, an entirely different reason took Kepler and his family away from Prague for some time. The plague had broken out there. Whoever could, left the city. The emperor remained in Brandeis. Kepler and his household went to Kunstadt in Moravia to visit acquaintances and then to Kolin on the Elbe. In the middle of November he returned alone to Prague; the emperor had applied for his services in Brandeis. At the beginning of 1607 the whole family returned to Prague. In April of that same year Kepler took a trip to Lusatia; on April 27, there is mention of an observation he made on the way to Görlitz.

In the spring of 1609 Kepler's most important journey led him to the Frankfort Book Fair, and to Heidelberg to complete the printing of the *Astronomia Nova*. The previous March 29, in Prague, he had composed its long dedication to Emperor Rudolph. On the way home he visited his Swabian homeland. In the second half of June he presented himself once more to his Tübingen friends. It was his first visit to the university city, so very familiar to him, since he had become imperial mathematician. Thirteen years had passed since his last visit, when as scholarly beginner he had to worry there about the printing of his first-born work. In the meantime he had changed from the beginner into the master of the science of the heavens. In Stuttgart, for a special reason, he presented a petition to his duke. It had to do with his future which little by little threatened to become uncertain in Prague. But before we enter into greater detail concerning this transaction and can follow Kepler's life's vicissitudes further, we must turn again to the scientific researches, which he carried out after completing the manuscript of the *Astronomia Nova* in the year 1605.

11. Tabulae Rudolphinae; Ephemerides; Antwort auf Röslini Discurs; Tertius Interveniens

The great task which lay ahead of him, the great commission which he had to carry out, was to produce the astronomical tables planned by Tycho Brahe. But because the difficulties which were encountered grew in proportion to the requirements set by Kepler, a speedy discharge of the task was not to be thought of. Working up the rich observational material left by Brahe and still further increased after his death necessitated the expenditure of great effort. Kepler could no longer depend on the theoretical hypotheses previously made in the calculation of the planet orbits. By the discovery of the planet laws he had created an entirely new foundation for this calculation. But since he had derived these laws solely from the observations of Mars, it was now necessary to demonstrate their validity for the other planets also and to calculate the elements of the orbits of these on the new basis. Here special difficulties were encountered with Mercury, whose orbit is very eccentric and the determinations of whose positions in the neighborhood of the horizon are always considerably influenced by refraction. To do justice to the observations, all previous astronomers, even Brahe, had been forced to make positively prodigious assumptions regarding the form of the motion. The moon, with its irregular motion, caused no less trouble. Nor could Kepler in any way be satisfied with the theoretical hypotheses of his predecessors concerning it. With ceaseless efforts he worried over the physical foundation for the peculiarities in the moon's motion. Kepler's papers dating from the time of his stay in Prague contain many hundreds of folio sheets filled with calculations which apparently are preparatory work for the *Tables* and which bear witness to the prodigious industry and the unbounded conscientiousness with which he attacked his problem. Yet two more decades were to pass before the work was published. The emperor, whose name it was to bear and who had commissioned it, was long dead by the time it appeared.

Associated with these astronomical researches is another work which likewise did not pass beyond the preparatory phase in the Prague period. The astronomical tables form the basis for calculating the ephemerides, that is the year books which give the positions of the sun, moon and planets for each day of a year. Such ephemerides were very much in demand by astronomers who used them in their scientific researches, by seafarers who needed them for place determination, and last but not least also by calendar makers and astrologers to whom they were indispensable as supports for their prophecies. Kepler now planned to issue such an ephemerides and, moreover, for no less than eighty years. The year 1582 was to be the starting point because that was the year in which Brahe had begun his observations.[1] From the year 1593 on, Kepler wanted to add his weather observations which dated from then, to get a sure basis for investigating the dependence of atmospheric conditions on the constellations in the sky. He wanted to complete this big work before publishing the tables. A double purpose guided him. In the first place, he wanted to be the one to reap the harvest which the treasure of Tycho's observations and his own improved planet calculations promised; after the publication of the tables someone else could easily precede him in this. Secondly, the ephemerides, which he proposed to calculate in advance for the years to come, were to form a test of the accuracy of this new mode of reckoning and consequently for the trustworthiness of his theoretical bases for these calculations which were to be published in the tables.

Foremost among the astronomers who at that time were prominent in the preparation of new ephemerides was Giovanni Antonio Magini in Bologna. He was a rival of Tycho Brahe and thereafter of Galileo. To avoid a competition injurious both to science and to his own profit, Kepler now considered a work in collaboration with the prominent Italian astronomer. For this purpose, in March, 1610, he invited him to come to Prague for a while. He had once, as early as 1601, requested Magini to send him observations, because his Mars study was in its first stage. At that time the Italian had not answered. Now, to be sure, he replied to the imperial mathematician, but by a refusal. Age, health, and consideration for his teaching position did not permit such a long, difficult journey; besides, he was not accustomed to warm stoves, nor to partaking of beer, nor to the excess of the Germans in food and drink. Apart from these expressed reasons, in the last analysis a certain mistrust on both sides also hindered the collaboration. Besides, since the political situation in Kepler's environs soon became much worse and there was pressure from other tasks, the preparation for this plan likewise came to a standstill. Later, as we shall see, it was realized, but only to a limited extent.

A few of the shorter writings, which were in progress side by side with these great plans, have already been mentioned above. Information concerning other works, with which the active and indefatigable

[1] ED. NOTE. 1582 also marks the beginning of reckoning by the Gregorian calendar.

Kepler busied himself, can be found in his letters and extant papers. His *Hipparchus* haunted and continually pursued him. The appearance of a comet in September, 1607, induced him to compose a little work in German.[1] Further details concerning it, such as what Kepler considered to be the nature and significance of such a heavenly phenomenon, will be given later. The chronological researches were earnestly expressed in a voluminous correspondence which started after he first wrote about Christ's birth year.[2] Sethus Calvisius,[3] the Leipzig cantor, Johannes Deckers, the Jesuit father from Olmütz (previously in Graz where Kepler met him), and Herwart von Hohenburg were the principal ones to participate in it. However, the correspondence deals not only with the earlier thesis about the birth year of Christ but also with other dates in the life of Jesus and with the duration of the public activity of the Saviour which he set as two and a quarter years, as well as with other events in the Jewish history of the period. It shows how deeply Kepler had plunged into the historical sources. Finally, and how could it be otherwise, the harmonic speculations emerge again. They had been closest to his heart when he came to Prague; indeed they had been substantially responsible for his accepting Tycho Brahe's summons to go there. He wanted to obtain a more accurate knowledge of the structure of the planet system and the planet motions for the very purpose of gaining from experience a surer foundation for those speculations. For years he had worried about Mars. He had met with success. So at the very time when he had rolled away the stumbling block of his Mars researches, a letter to the Englishman Heydon dated 1605 moans: "Would but that God free me from astronomy so that I can concern myself with my work regarding the harmony of the world." But God did not fulfill his request and His faithful servant followed the "divine summons to teach people astronomy." He applied himself to his calculations and with the greatest patience and perseverance took all imaginable pains to mold the observations into accurate knowledge of the heavenly phenomena. The day for the completion of his *World Harmony* had not yet come.

Joy in scientific disputes was in Kepler's blood. It gave him pleasure to use his forcible wit and to cross swords with an opponent. An introductory *Captatio benevolentiae* was not a matter of necessity for

[1] ED. NOTE. *Aussführlicher Bericht von dem newlich im Monat Septembri vnd Octobri diss 1607. Jahrs erschienenen Haarstern oder Cometen vnd seinen Bedeutungen.* Hall in Sachsen: Hynitzsch, 1608.

[2] ED. NOTE. References to Kepler's published writings which deal with the subject as well as to the published correspondence, can be found *passim* in Max Caspar, *Bibliographia Kepleriana.* Munich: Beck, 1936.

[3] ED. NOTE. Sethus Calvisius (1556–1615), a German musician and chronologer, was director of the school of music at Leipzig, where he died.

this. "This consolation is not required, that someone who has something against me should first want to praise me so that I could so much more patiently endure the sweat bath to follow: for such *collationes Philosophicae* is one of the most pleasing recreations which my daily life could experience if they are worth the trouble." In two writings of the latter part of 1609 Kepler gave himself this pleasure. To reach a wider circle they are composed in German although interspersed with numerous scraps of Latin.

In the first work Kepler takes the stage opposite his countryman Helisaeus Roeslin,[1] known to him since his student days, who was physician-in-ordinary to the count palatine of Veldenz and the count of Hanau-Lichtenberg in Buchsweiler in Alsace. Roeslin, a man of alert mind and not without scientific merit, was devoted body and soul to astrology. His prophecies took on grotesque forms. He had previously upheld the view that the famous new star of 1572 and a comet which appeared in 1580 would only exercise their full effect from 1604 on, and that in this year something would occur "which would surpass all wonders." Now when in 1604 a new star really appeared, he saw in it a splendid corroboration of his old prediction and announced anew a universal catastrophe in the Christian world as the consequence of this heavenly phenomenon. These fancies Kepler had handled roughly in his book about the new star. Roeslin energetically repelled Kepler's astrology-hostile criticism in a *Discurs von heutiger Zeit Beschaffenheit*. This effusion now provoked Kepler to the composition of a written reply *Antwort auff Röslini Discurs* in which he takes Roeslin's work page by page, wherever he finds his own name, and disposes of his opponent's absurd assertions, not without fun and humor. Kepler called it "holding a friendly though frank German talk." For him it was a matter "only of exploring the truth / and introducing philosophy to the reader with some cheerfulness / so as to avoid getting involved with other argumentative brawling cats." Thus he defends the Copernican picture of the world which Roeslin rejects. Kepler expounds the earth's motion, trying to explain it and that of the other planets in a remarkable manner, by whirling. He speaks of the meaning and the path of comets, of the aspects and the influence of the heaven on earthly happenings. In the face of Roeslin's astrological fantasies he summarizes his opinion in the sentence: "That the heaven does something in people one sees clearly enough: but what it does specifically remains hidden." Strangely enough, however, he wanted to concede to Roeslin an "instinctus divinus," a special illumination in the interpretation of heavenly phenomena which has nothing to do with astrological rules. He

[1] Ed. Note. See C. Doris Hellman, *op cit.*, pp. 159–66.

believes that Roeslin's tongue and pen are ruled from elsewhere and that he to some extent is a prophet, just as Kepler does not reject the belief that it could come about that God might let "insane and pure simpletons" announce strange and unusual things, even astronomical phenomena, as warnings.

Almost at the same time as Roeslin dedicated his *Discurs* to Margrave Georg Friedrich von Baden, this prince's physician-in-ordinary, Philip Feselius, submitted to his ruler a *Discurs von der Astrologia iudiciaria* which he had composed. In contrast to Roeslin, Feselius inveighed sharply against the whole of astrology which he repudiated altogether. He had been led to write his work by a fight in which he engaged with the clergyman, Melchior Schärer, who in his *practicas* handled the belittlers of astrology in the sharpest terms and chose to call them disgraceful creatures, coarse fools, absurd dunces, grunters, cyclopses, who look at nature with calves' eyes. The standpoint of complete rejection, on the other hand, was not in keeping with Kepler's opinion. He wanted to distinguish between Chaldean stargazing superstition and "physics," the pure science founded on experience which he was convinced established a definite relationship between heavenly phenomena and earthly events. In the fight against superstition he did not wish the child to be poured out with the bath, the good fruit rooted out along with the weeds. So, shortly, out of the wealth of thoughts surging within him, he wrote a book to which he gave the title *Tertius Interveniens, das ist Warnung an etliche Theologos, Medicos vnd Philosophos, sonderlich D. Philippum Feselium, dass sie bey billicher Verwerffung der Sternguckerischen Aberglauben nicht das Kindt mit dem Badt aussschütten vnd hiermit jhrer Profession vnwissendt zuwider handlen.*[1] With this book he appeared before Margrave Georg Friedrich as a third party in order to mediate between the two extremes as champion of "physics or psychology" because according to his "philosophical profession a watchful surveillance" in all ways befitted him.

Writings like that of Feselius are extremely rich in information about the manner of thinking of that period. They show in which world the thoughts of an educated man then moved and give livelier colors and sharper contours to our universal picture of that time, with its many individual arguments. In order to lay a firm foundation for his rejection of astrology, Feselius especially introduced a great number of citations and referred to many authorities. There could not be any influence by the heavenly bodies, so he argues, because there is no direct contact

[1] ED. TRANS. *Tertius Interveniens,* that is warning to some theologians, medics and philosophers, especially D. Philip Feselius, that they in cheap condemnation of the stargazer's superstition do not throw out the child with the bath and hereby unknowingly act contrary to their profession.

between heaven and earth. To accept a motion of the heavenly bodies in open space would be wrong. Then how could the planets adhere to their orbits, if these were not solid? All stargazing is, furthermore, uncertain because astronomers cannot set forth exactly the motions of the heavenly bodies. It is impossible to rely completely even on Brahe's observations. How was one supposed, after all, to be able to reach the heights of heaven with little instruments? Copernicus' theory was contrary to reason and to Holy Scriptures. How is one supposed to be able to distinguish between the influence of the innumerable stars? The heavenly luminaries are put up by God only as signs for determining time. Because they were created by God, the stars can cause nothing bad. The Holy Scriptures, and many theologians, too, repudiate the art of prophecy and also signifying from the stars. To refer with Paracelsus to the *signatura rerum*, to the color and form of creatures and then to try to draw conclusions about their effect is a pastime of fantastic heads. Let experience punish the lies of astrological prophesying. The arguments just stated are a selection from those expressed by Feselius in his booklet, which also gives a picture, though only a very superficial one, of the thoughts with which Kepler's book deals. For while disagreeing with Feselius' theses, Kepler reports an extraordinary profusion of his own physical, astronomical, meteorological, philosophical, mathematical and ethical thoughts in motley succession so that anyone who once has read the book keeps taking it up to enjoy Kepler's pithy speech and original thoughts. This author is permeated with a lively belief in a meaningful world order according to which all parts of the cosmos are connected with each other and directed toward a higher aim. For this reason it is impossible to sketch the contents briefly and it must suffice to present a few main thoughts.

His thesis, that astrology is not to be completely rejected, Kepler presents right at the beginning in the stimulating words: "No one should consider unbelievable / that there could come out of astrological foolishness and godlessness / also useful cleverness and holiness / out of unclean slimy substance / also a snail, mussel, oyster or eel useful for eating / out of the great heap of caterpillar dirt / also a silk spinner / and finally that out of evil-smelling dung / also perhaps a good little grain / yes a pearly or golden corn / could be scraped for and found by an industrious hen." As part of the dung in which the industrious hen scrapes, Kepler counts the majority of astrological rules, the acceptance of a qualitative distinction for the twelve signs of the zodiac and the distribution of the latter among the planets, the relationship of the twelve houses, reckoned from the vernal equinox, to a person's varied circumstances and ties, the differentiation between good and bad constellations, good and bad aspects, the division of lands, of human limbs and of the four elements among the signs of the zodiac, the theory about the ascendancy and annihilation of the planets, and whatever other similar stuff became applicable. However, Kepler insists on the special significance of individual points of the ecliptic, such as of the ascendant, that is the point which is just rising at the instant of a birth, and of the center of the sky, that is the point in which the ecliptic cuts the meridian at that instant. He also stresses the significance of the configuration, which the planets form among themselves and with those points mentioned. In the heaven it is not a question of good and bad; here rather only the categories harmonic, rhythmic, lovely, strong, weak, and unarranged are valid. The stars do not compel, they do not do away with free will, they do not decide the particular fate of an individual, but they impress on the soul a special character. That is an oft-repeated favorite thought of his which he pronounces clearly here: "Then firstly may I truthfully boast of this knowledge / that the person in the first igniting of his life / when he now lives for himself / and can no longer remain in the maternal body / receives a character and pattern *totius constellationis coelestis seu formae confluxus radiorum in terra* [of all the constellations of the heaven or of the form of the rays flowing onto the earth], which he retains until he enters his grave: This character afterwards leaves noticeable traces in the formation of the countenance and of the remaining shape of the body / as well as in a man's business affairs / his manners and gestures / so that he also produces through the form of his body similar inclinations and sympathies with his person in other people / and by his behavior produces corresponding fortune. By this then (caused by the mother's imagination before the birth as well as by rearing after birth) a very great difference between people can be made / so that one becomes good, lively, gay, trusting[1]; another sleepy, indolent, careless, obscurantist, forgetful, timid / and so on / qualities which are comparable to the lovely and exact, or the extensive and unsightly configurations, and to the colors and movements of the planets."

The horoscope which thus forms a person's character, he can faithfully hold fast, so Kepler supposes, for his whole life, so whenever a constellation out of it is repeated or the planets pass through the points indicated, the soul reacts instinctively and feels driven to action. This is possible since the soul is still, so to speak, a point, but potentially a circle, so that it is able by virtue of these powers to distinguish tendencies and hold them fast. But now, which are indicated tendencies

---

[1] ED. NOTE. The German word is "trauwsam" or "trausam" and Grimm's *Deutsches Wörterbuch*, XI[1] (1935), 1543, uses this passage from Kepler to illustrate its meaning.

or on what are the categories lovely and unlovely, strong and weak, based? Now once more appear those harmonic primordial relationships already in evidence in the Graz period, the divisions of circles, made by the knowable, that is constructable, regular polygons foreshadowed in the divine being. A constellation is more or less beautiful, more or less strong, in accordance with the quality and quantity of the corresponding aspects. Thus all animated beings, human, animal and vegetable, are influenced from heaven by the appropriate geometric instinct pertaining to them. It is true that "all their activities are affected, individually shaped and guided by the light rays present here below on earth and sensed by these creatures, as well as by the geometry and harmony such as occurs between them by virtue of their motion, in the same way as the flock is affected by the voice of the shepherd, the horses on a wagon by the shout of the driver, and the dance of the peasant by the sound of the bagpipe." Thus for Kepler the influence of the stars is a matter not of a coarse physical nor of a magical but of a psychic effect, and it is understandable why psychology appears as counsel in his venture.

But still more, not only is innate instinct thus excited by the heavens but also human intellect, in search of knowledge, everywhere comes up against geometrical relations in nature, which God, while creating the world, has laid out from His own resources, so to speak. To inquire into nature is to trace geometrical relationships. Since God, so our mystic imagines, in His very highest goodness was not able to rest from His labors, He played with the *signaturis rerum*, the characteristics of things, and copied himself in the world: "Thus that it is one of my thoughts / whether all of nature and all heavenly elegance is not symbolized in geometry." But in its activity, unknowingly or knowingly, instinctively or thinkingly, the created imitates the Creator, the earth in making crystals, plants with their ability to shape in building and in arranging their leaves and blossoms, man in his creative activity. And all this doing is like the play of a child, without plan, without purpose, out of an inner impulse, out of joy in configurations, so that the eye enjoys that which arises and the contemplating spirit finds and recognizes itself again in that which it created: "As God the Creator played / so He also taught nature, as His image, to play / the very game / which He played before her."[1] Is it not as though he, who spoke these wonderful words, stood in Paradise and walked before the eyes of God with the innocence of the first human?

[1] ED. NOTE. This passage is from section 126 of *Tertius Interveniens* and is discussed and partially translated on pages 171–2 of W. Pauli, *The Influence of Archetypal Ideas on the Scientific Theories of Kepler*, translated from the German by Priscilla Silz and included in *The Interpretation of Nature and the Psyche*, Bollingen Series, LI, Pantheon Books, 1955.

12. *Political disorders; Kepler on the watch for a new place to work*

Unfortunately, the rough, hard reality did not permit Kepler to linger over his paradisian observation. The happy scholarly work which had borne such rich fruits did not remain untouched by the critical political circumstances, which came ever more to a head in the later years of his stay in Prague and whose focus lay at the court of the emperor whom he served. Increasing tension was created by the disunity of the German princes. Since the central power was weak, they used the universal unrest to procure special benefits for themselves and, to some extent, did not fear joining with a foreign country to strengthen their own power or to safeguard the party of their own confessional creed. Incidents like that in Donauwörth, where the confessional dissension finally led to outlawry and the occupation of the city by Duke Maximilian of Bavaria, produced growing agitation. Instead of bridging the gap created by the antagonisms, the religious parties fighting among themselves let the cleavage become ever deeper. Claims and counterclaims opposed each other bluntly. So it was a serious and menacing gathering of the forces on both sides when, in May, 1608, a majority of the Protestant princes under the leadership of Elector Frederick IV of the Palatinate formed a union[1] and in July of the following year Duke Maximilian of Bavaria initiated a Catholic alliance, the Catholic League. The fight which broke out in the same year over the inheritance of the duchies of Jülich and Cleves already threatened to lead to a settlement by weapons. By this time it would have been exceedingly hard for a clever and energetic emperor to overcome all these difficulties and to solve them peacefully. It was absolutely impossible for a ruler like Rudolph II to master the confusion. Rudolph lost himself ever deeper in his shyness of people and in his whims, but in his willfulness was in no way inclined to accept the conclusions which the other members of the house of Hapsburg prepared to draw from his intellectual position. Both the weakness of the imperial regime and the question of the succession to the throne, which before long must become acute, caused them severe uneasiness. Thus, to increase the confusion, internal antagonism was also added to the confessional cleavage. Naturally, this benefited the Protestants in the Austrian lands. Archduke Matthias, who led the opposition against the emperor, had to make concessions to them in order to secure his position in opposition to the emperor. As a result of this settlement, as early as

[1] ED. NOTE. The Protestant Union.

June, 1608, Rudolph was forced to transfer the government in Austria, Hungary and Moravia to his brother Matthias. Only Bohemia and Silesia, which joined it, remained in his hand. In order to hold on at least to the Bohemian crown, the greatly distressed emperor, under strong pressure from the Protestant representatives in Prague, who wanted to take advantage of a situation favorable to them, took a step which was inconsistent with all his earlier conduct in religious matters. By the so-called royal charter,[1] on July 9, 1609, he granted religious liberty and protection for their creed to all the followers of the new doctrine in Bohemia and to the knights, lords and imperial cities he granted the right to erect churches and schools free of charge in all places.

In the midst of the exciting days of negotiation over the royal charter, Kepler returned from the journey mentioned above. He had gone to Frankfort and Heidelberg because of the completion of the *Astronomia Nova* and to his native Swabia. He had still been in Tübingen in the second half of June. Shortly thereafter he presented his imperial lord with the immortal work composed in his service. Could the emperor, entangled in tribulations, foresee the shining triumph achieved by his mathematician, some share of which also fell to him? Kepler was happy about the freedom which the royal charter granted his coreligionists. On July 18, he wrote to Stephen Gerlach, professor of theology at Tübingen: "We have been victorious by the grace of God. German sermons are publicly held in churches and dwelling places." Yet this joy could not obscure his view. Since through the years he had been able to observe the development of affairs at first hand, no special political farsightedness on his part was needed for him to comprehend the untenability of the emperor's position and consequently also the uncertainty of his own situation. As early as 1605 he had expressed apprehension in a noteworthy manner: "I do not know how long war will continue to shun us in Bohemia." At that time fighting around Hungary was in process. There Stephan Bocskay successfully resisted Rudolph's regulations regarding the Counter Reformation. Hereupon, Matthias, who supported the emperor's side, had to make concessions in the peace of Vienna. These, however, Rudolph would not recognize. This situation led to dissension between the two brothers. Meanwhile, circumstances had become worse, even though war had not yet broken out in Bohemia.

Under these conditions it was only prudent for Kepler in due time

---

[1] Ed. Note. *The New Schaff-Herzog Encyclopedia of Religious Knowledge*, IV (1950), p. 300, calls this "the imperial brief in solemn acknowledgment of religious freedom and the ecclesiastical organization of the Protestants."

to look for other spheres of action if he did not want to be left high and dry some day. Moreover, by this time, the payment of his salary was so far in arrears that to a large extent he really had to live on his own resources; in his petitions he could justly point out that he would be able to care for himself and his family much better by another occupation in another location. Just as at the time of his departure from Graz, he again thought first of his Swabian homeland to which his heart belonged. Now when he knocked as the highly esteemed imperial mathematician might he not hope more easily to find an open door? Of course, it was clear to him that the professorship which would have been under consideration for him in Tübingen was well filled by Maestlin. Yet perhaps his services could be used differently there. Since, besides, as a former holder of a stipend from the duke of Württemberg, he needed the latter's consent to accept a position in another land, he used the opportunity offered him by the aforementioned journey to present a written petition to his sovereign in Stuttgart in May, 1609. He asked the duke to consider "what great troubles might alarmingly grow up for me, under such blistering treatment, in the short time which I would be without a post without any assurance of finding shelter in my distant homeland." He did not directly request employment in the duke's service, but rather permission, should the opportunity arise, to accept a position with another ruler. Since the ducal advisers recognized that the petitioner was a "distinguished mathematician," this right was conceded him with the restriction that he would have to be ready to enter the duke's employ at any time on demand if the duke ever needed his service. For all that, this decision yet held the prospect of sooner or later getting to Württemberg.

Now, in order to clarify the state of affairs from the very beginning and to preclude eventual trouble Kepler, in his frank way, considered it advisable to inform the duke immediately in a second written communication that, should the occasion present itself, he would only conditionally be able to sign the Formula of Concord which the Württemberg theologians esteem as the strictest pattern of belief. He would give assurance that he would not fight against it and that he would work for peace. He also hinted that he leaned toward the Calvinists' communion doctrine and explained that from youth on he had never been able to find that someone who follows the Calvinists in this article should not anyhow, in spite of this dissimilar view, be called and considered Brother in Christ by a follower of Luther. At great length he impressed on his sovereign that now was precisely the correct time to become reconciled with the Calvinists, because certain changes in their doctrine were perceptible and their leaders in Germany, namely the elector of the Palatinate and his advisers, were not very happy with the "cruelty" of the theory of predestination itself. In the face of Kepler's letter addressed to the duke from an anxious heart, the question must be asked: did he know the prince's clerical advisers and their point of view or did he not? If he knew them, then it must be assumed that a position in Württemberg was not of very much consequence to him and that he cared more about obtaining complete freedom by securing a situation in another land. Since, however, the tenor of the letter contradicts this and Kepler's sincerity cannot be doubted, it must be concluded that he did not know those gentlemen and underestimated their intolerance. There was no immediate answer; the duke approved a document which stipulated that no change should be made in the future duties of the applicant. But it was not to be too long before Kepler really got to know those guardians of the Württemberg orthodoxy.

It cannot be said what steps he took after returning to Prague toward finding the kind of place he needed. Since the matter was handled in oral discussions within the circle of his court acquaintances, reports are lacking. But it is known that, already at that time, he kept up a friendly correspondence with some nobles from Upper Austria who hoped to get him to go to Linz. Nor is it out of the way to assume that he directed his glances toward the Electorate of Saxony with whose politics he openly sympathized. So passed the year 1609, without any one plan for changing his position being worked out. In the first months of the following year the uncertainty of his situation acted like a tether on his creative power. After the great achievements of the past years he was obliged to be inactive; his mind was "grown stiff in a distressing frost."

13. Galileo's first discoveries with the telescope

Then occurred a scientific event which, with one jerk, alarmed his spirit, alerted his powers and kindled anew the lust for work. On a March day in 1610, his friend Wackher von Wackenfels drove up to his door, called him out and, without getting out of his carriage, informed him that a report was received at court that in Padua Galileo, using a two-lensed "perspicillum," or telescope, as it is now called, had discovered four new planets. The information greatly excited and amazed Kepler. A lively dispute immediately ensued between the two men. Was the report true? Of what type were the newly discovered heavenly bodies? Were they companions of fixed stars or did they belong to the solar system? Neither man doubted the reality of the phenomenon. They differed in their interpretation. Previously they had ardently debated Giordano Bruno's cosmological views. In a daring flight of thoughts Bruno had assumed that the stars are suns, similar to our sun, and that, infinite in number, they fill the infinite space. Wackher agreed with this view and believed that the new discovery corroborated it, in as much as he construed the four planets as companions of fixed stars. Kepler completely disagreed with Bruno's view. He was of the opinion that the phenomena could only be moons circling other planets as our moon does the earth. He rejected the idea that the new heavenly bodies would circle the sun like the planets, because he considered it definitely established by his *World Secret* that the number of these planets is six.

Both men eagerly awaited conclusive reports. These came soon. In the very same month Galileo came forward with a book informing the world of his discoveries with the marvelous new instrument. *Sidereus nuncius, Starry message* or *Starry messenger*, is its title.[1] The little work forms a milestone in the history of astronomy. Since it offered the first observations of the heavens made with the recently invented telescope, it introduced the epoch in which this science celebrated prodigious triumphs with this instrument and brought about a complete reversal in the exploration of heavenly phenomena. The armed eye drew the furthest regions of the universe within its reach. Since then, an unpredicted abundance of new objects were discovered and the methods of measurement were refined in an unprecedented manner. The credit for having laid the foundation of this brilliant development is divided between Galileo and Kepler. The former was the first practical observer, the latter, however, appeared right next to him as the gifted theoretician who, unlike anyone of his time, clearly recognized the significance of the new instrument and by impressively establishing its physical effectiveness created the sure basis on which future research built. From the very beginning, theory and practice joined hands and found it necessary to work together to achieve abundant success.

What kind of phenomena did the new instrument with its thirty-

[1] ED. NOTE. Edward Rosen in "The Title of Galileo's *Sidereus nuncius*," *Isis*, XLI (1950), 287–91, contends that Galileo intended the title to mean message on the grounds that a message could contain observations whereas a messenger could not. Most frequently one finds the title interpreted as meaning messenger. This is the interpretation of Kepler, as the title of his answer indicates, and, more recently, of Stillman Drake in his *Discoveries and Opinions of Galileo*, Garden City, New York: Doubleday, Anchor, 1957, *passim*. For other comment on Galileo's and Kepler's interpretations of the title, see Franz Hammer in *Johannes Kepler Gesammelte Werke*, IV, 442, 450.

fold magnification[1] reveal to the astonished Galileo in the winter months of 1609–1610? The moon, naturally, was the first object to present itself to him. Galileo was the first to behold the splendid picture that ever since then has enchanted all those who look at the complex moonscape with its mountains, ramparts and craters, even though they use only a small telescope. He was amazed when he saw that the boundary between the dark and light part of the lunar disc was not a sharp line but resolved itself into projecting and receding sections. With astonishment he noticed bright points near the light boundary in the dark portion and realized that these are mountain tops already being lighted by the rising sun while the lower sections still lie in the shadow. He even made a skillful attempt to use these phenomena to get an approximate measurement of the height of the lunar mountains. In observing the stars, he experienced a further surprise. Whereas the planets appeared as little discs, no magnification of the fixed stars was shown; only the brightness of their twinkling light was notably increased. Correspondingly, his instrument showed him an abundance of stars whose light made so feeble an impression as to be beyond the range of visibility by the naked eye. For example, in the constellation of the seven stars or Pleiades, he could now count more than forty stars. More remarkable was the picture offered by the Milky Way. He realized that its weak gleam arose from the accumulation of a countless multitude of tiny stars. The discovery most important to him as well as to his contemporaries was, however, that of those four new satellites, notice of which had so greatly excited the imperial mathematician. The latter had been right in his surmise. These stars were companions of Jupiter, caught by Galileo in careful observations from January 7 to March 2, 1610. In honor of the ruling house of his native city, Florence, he called them the Medicean stars. This discovery particularly aroused great attention because it provided the first actual proof that, contrary to the assumption of the Ptolemaic system of the world, there are celestial bodies which circle around a central point other than the earth. Galileo himself considered this discovery so significant that now, for the first time, he ventured to adhere openly to the Copernican world system.

[1] ED. NOTE. In the *Sidereus Nuncius* (*Le Opere di Galileo Galilei, ristampa della edizione nazionale sotto l'alto Patronato di S.M. Il Re d'Italia e di S.E. Benito Mussolini*, III[1], Florence: Barbèra, 1930, pp. 60–1) Galileo speaks of the magnification. This information is repeated by Kepler in his *Dissertatio Cum Nuncio Sidereo* (*Le Opere di Galileo Galilei, op. cit.*, p. 108, or in *Johannes Kepler Gesammelte Werke*, IV, 291). Galileo described his method for determining the magnifying power, by comparison of two similar geometric figures on the same wall, the larger one being viewed with the naked eye, the smaller with the telescope (Galileo, *op. cit.*, pp. 61–2; Kepler, *op. cit.*, p. 443, notes of the editor).

14. Dissertatio cum Nuncio Sidereo

On April 8, through the agency of Julian de Medici, the Tuscan ambassador in Prague, Kepler received a copy of the anxiously awaited book with Galileo's request to render an opinion of it. Obviously the judgment of the imperial mathematician was of great importance to Galileo. However, he would not have needed to make a special request for a written opinion. Kepler was so moved and inspired by the new things he learned that he could not restrain himself. "Whom does knowledge of such important things allow to be silent? Who is not filled by a wealth of divine love so that he pours himself forth, overflowing in word and writing?" On April 19, the courier was supposed to return to Italy. In the intervening eleven days Kepler composed his answer. It is in the form of a letter to Galileo, but as early as May reached publication under the title *Dissertatio cum Nuncio Sidereo*, conversation with the starry messenger, a little, slender work which attracted much attention in the rising conflict of opinions and was reprinted a few times, including once in Italy in the same year without authorization.

The significance of this tract can be correctly evaluated only by surveying the general opinions at the time when Galileo published his booklet. Anyone proceeding from the sensation usually called forth nowadays by scientific discoveries might believe that the new insight into nature's wonders not only would have aroused surprise everywhere but also would have been received with favorable astonishment if not with enthusiasm. Yet the opposite was the case. Everywhere, especially in Galileo's native land, doubt, mistrust, challenge and contradiction sprang up. Nowadays the method of research into nature based on observation and experience is everywhere taken for granted and almost daily offers astonishing new discoveries, so that at length the uncritical layman considers the impossible as possible, absorbs reports so much the more eagerly the more fantastic they are and is ready to believe the blatant follies with which pseudoscientific writers tickle his curiosity. In Galileo's time, on the other hand, the contemplation of nature was still so entangled in the rigid conceptual constructions of the Aristotelian cosmology that it was extremely difficult for anyone to free himself from these patterns of thought and to accept facts irreconcilable with them. So, from the very beginning, the contention proved by Galileo's observations, that the moon is a body exactly like the earth, contradicted the conventional conception, taken over from the ancients. They supposed that the heavenly bodies in contrast to the earth, to which the familiar four elements belong, are composed of another substance, *quinta essentia* [the fifth essence], the heavenly aether. In addition, men were completely in the dark about the laws of optics which formed the basis of the telescope. Even Galileo could say nothing worth mentioning about them; it is true that he promised to inquire into them, but, for obvious reasons, he never redeemed his promise. How was it possible, with such an ingenious instrument, to perceive things which were entirely hidden from the eyes? Was this not something ghostly, which the "spirits of the telescope" practice with humans? To these real considerations were also added personal motives which acted as a hindrance to the recognition of the new discoveries. Galileo had a special talent for making himself disliked by his colleagues because of his demeanor. He made others feel all too severely the intellectual superiority which distinguished him and in his desire for fame and recognition handled the lesser intellects unpleasantly when they dared to contradict him. Thus the publication of his *Starry Messenger*, with observations which the majority could not believe, gave his opponents the welcome opportunity to appear against him. They wanted at least once to tear the mask from his face and straightways brand him as an impostor. Even his contention that he had independently rediscovered the telescope after he had been informed about it was considered insolence, since it was known that such instruments had at that time reached Italy from France. Should credence be given to a man who from the very outset appeared with doubtful claims? Galileo, visiting the home of Magini, the widely respected astronomer in Bologna, wanted to convince the latter of his credibility by showing him Jupiter's moons with his telescope. The attempt was made during the night of April 24 to 25, in the presence of numerous invited guests. Since none of those present was able to distinguish the moons, Galileo's opponents thought their triumph certain. The reason that those men did not see the moons was certainly due to their lack of practice in telescopic observation and to the imperfection of the instrument, which still possessed all the defects necessarily attached to a telescope composed of two simple lenses. But that was not yet known and consequently it was believed that there was cause for accusing Galileo of deception. While Magini with his opposition remained in the background, Martin Horky, the son of a Bohemian pastor, a young man who had belonged to that company, was allowed to publish a work entitled *Peregrinatio contra Nuncium Sydereum*. In this pamphlet, dedicated to the Bologna teachers' college, Horky declared Galileo's reports to be fables. The hazy points of light visible in the telescope are nothing but false reflex pictures, and the resolution of the Milky Way into stars is an ancient tale. Truly a discordant echo of the rousing message learned from the heavens and announced by Galileo!

What a contrast between Kepler's position and this narrow-minded and miserable point of view! "Here highly significant and exceedingly admirable show pieces will be offered to philosophers and astronomers. Here all friends of a true philosophy will be called together for the beginning of lofty reflection." These are the words in his written answer with which he introduces the vigorous acceptance of Galileo's message. He takes up his pen so that Galileo "should be armed against the sour-tempered critics of everything new, who consider unbelievable that which is unknown to them, and regard as terrible wickedness whatever lies beyond the customary bounds of Aristotelian philosophy." For him it is a question, not of the individual but of the subject. "I do not believe," says he, with good cause, in his preface to the reader in the edition of his *Dissertatio*, "that Galileo, the Italian, has deserved so well of me, the German, that I ought to flatter him by patterning the truth and my innermost conviction after his ideas." Yes, Galileo had really lacked regard for and appreciation of Kepler. How much had Kepler, according to his own words, desired notice by Galileo for his *Astronomia Nova*! But the Italian was wrapped in silence. However, Kepler did not let him suffer for this attitude by joining the ranks of the malicious critics. The objective manner with which Galileo presented his discovery, his scholarly earnestness and his love for truth vouched for his credibility as far as Kepler was concerned. Kepler was not easily talked out of anything.

Joined with this feeling for the truthfulness of the individual was also his sense for the truth of the subject. Galileo had seen new things in the heaven. What he could report about moon, Milky Way, and satellites had anyhow previously been more or less clearly foreseen or conjectured by those with great intuition. Moreover, Kepler could also point to thoughts of his own expressed in earlier works. His fantasy and keenness at divining had put these into his mind. Thus his intellect was prepared for the acceptance of the new. While now, with admiring words, he approved Galileo's discoveries and commented on details, he still could not refrain from adding information about those facts to his approval in order thus to arrange the new observations historically. Naturally, this acted like a damper on the Italian, who touched lightly on that which others had done and was jealously bent on having his own claims appreciated. The form in which Kepler reduced these claims to their true size, however, left nothing to wish for in courtesy, since he sugared over with praise every word which could have a bitter taste for Galileo.

Kepler also looked ahead. What might still be expected after these first promising beginnings? The whole question of optics stimulated him powerfully. What he could already say right then far surpasses Galileo's paltry suggestions. Above all, he expected the new instrument to bring a refinement in the methods of astronomical measurement. As he reports, he had previously believed it absolutely impossible to surpass Tycho Brahe in this respect. This view he upheld against the vehement disagreement of his friend Pistorius. Now he has learned better and sees new paths before him. His fantasy is given wings. Why should Jupiter be the only planet to have satellites? If it has four, does not the symmetry then demand that Mars be accompanied by two, Saturn by six to eight? And to what end do those four satellites serve Jupiter? Is it not necessary that creatures live on that planet and contemplate the marvelous changing spectacle? Indeed this cannot have been made for us dwellers on the earth, where we do not even see these moons. His fancy flies yet higher: "Ships and sails proper for the heavenly air should be fashioned. Then there will also be people, who do not shrink from the dreary vastness of space." But the new facts also give rise to reflection. If there are bodies similar to the earth, how can human beings save their claim that everything was created for their sake, that they are masters over the works of God, a claim on which, moreover, Kepler's whole view of the world depends? Now the speculating inquirer comes forth to speak on the basic ideas of his mystery of the universe, which were not shattered by the new discoveries, as he had at first feared. Indeed, the earth occupies the most distinguished place in the universe, since it circles the sun in the middle between the planets. The sun in turn represents the middle place at rest in a spherical-shaped space enclosed by the fixed stars. Everything is regulated according to the eternal laws of geometry. And here he sets down the motto already mentioned once: "Geometry is one and eternal, a reflection out of the mind of God. That mankind shares in it is one of the reasons to call man an image of God."

The best way to illustrate the contrast between Kepler's and Galileo's manner of thinking is by comparing the *Sidereus Nuncius* and the *Dissertatio*. Whereas the Italian adheres strictly to an exact investigation and representation of facts, the German goes further. Despite his admiring interest in all the phenomena of nature, his genius drives him ever further into the domain of metaphysical speculation, where he attempts to organize the facts into a single universal system determined by purpose and focused upon man.

The echo of Kepler's *Dissertatio* was overwhelming. Each person read from it what pleased him, and showed, according to the rules of psychic resonance, the words on which his thinking apparatus was adjusted. The imperial mathematician had to hear from all the people around him that he should have dealt somewhat more sparingly with his praise for Galileo, so that there would still have been space for the opinion of very important men who differed with him. To this he gave the excellent answer: "I have always adhered to the habit of praising what in my opinion others have done well, of rejecting what they have done badly. Never do I scorn or conceal other people's knowledge when I lack my own. Never do I feel servile to others or forget myself when I have done something better or discovered it sooner with my own power." Galileo himself was apparently well satisfied. Of course, he took four months to answer Kepler's prompt tract defining his attitude. In a letter of August 19 he thanked him for having been the first and almost the only one with frankness and intellectual superiority to have given him full credence without having seen the thing himself. To Vinta, the Tuscan minister, Galileo reported that Kepler approved everything contained in the *Starry Messenger* without having expressed even the slightest disagreement or doubt in any detail. Magini, on the other hand, expressed the opinion that Kepler's method of answering would not exactly please Galileo because Kepler had cleverly and amicably called his attention to what he owed to others. George Fugger, imperial ambassador in Venice, even expressed the opinion that Galileo would soon notice that his mask had been torn from his face. And Maestlin himself acted in consort with them by writing to his former pupil: "You have pulled out Galileo's feathers exceedingly well in your little work, which I read with the greatest pleasure." Also Martin Horky in his ill-willed attack in the pamphlet mentioned above had referred to Kepler, in whose house he previously had visited, and had embarrassed the latter by utilizing isolated critical remarks in the *Dissertatio* for his lowly purposes. On that account he must have submitted to a sharp reprimand from the angered master, who excitedly expressed his indignation to Galileo. Later, in an oral argument with Horky, Kepler nevertheless arrived at a milder interpretation of the matter. From all this it is apparent that in the scholarly republic, too, judgments are not always expressed *sine ira et studio*.

Now, above all, Kepler wished and had to wish for the possibility of verifying the new discoveries in the sky. He would like to get a serviceable telescope, so that he himself might observe the wonderful things. The kind obtainable in Prague was wholly inadequate and did not make it possible to perceive the moons of Jupiter. As early as April, Julian de Medici advised Galileo to place a tube at the disposal of the great astronomer at the imperial court. But Galileo turned a deaf ear. To be sure, during the following months he presented a series of instruments to high lords; however, he had none left over for his valiant brother-in-arms in the fight for truth, who, moreover, had the first claim and would have made the best use of it. Probably jealousy was back of this attitude. Meanwhile, the voices of the doubters increased, so that at the beginning of August Kepler was provoked into asking Galileo for the names of witnesses, who had observed the new bodies. Not that he himself had begun to doubt; he only wanted to have proofs in order to be able to confront the envious gossip. The answer which arrived from Padua was not exactly elevating. Galileo did not name an expert as witness, but only the grand duke of Tuscany and Julius de Medici, the brother of the Tuscan ambassador. He added the information that the grand duke had even summoned him to his court at Florence, made him a present of over 1,000 ducats for his achievement and settled on him an annual salary of about 1,000 ducats. What may this boasting have made the poor imperial mathematician feel? Indeed, he served a higher lord but had not once been paid his entire 500 gulden.

In the same month of August, Kepler's anxious wish, to be able to observe the moons of Jupiter himself with a suitable telescope, was to be fulfilled, thanks to the kindness of Elector Ernst of Cologne, duke of Bavaria, a great friend of optical science. At that time this prince was spending a few months in Prague on the occasion of the meeting of princes then deliberating on the settlement of the fight between Emperor Rudolph and his brother Matthias. From a journey to Vienna to negotiate with Matthias, he brought along one of the telescopes presented by Galileo, and placed it at the disposal of the imperial mathematician for a little while. Thus the latter was finally in a position to observe with his own eyes what he had so longed to see. In the presence of Benjamin Ursinus, the young mathematician, and several other guests, he observed Jupiter from August 30 to September 9. To preclude any error, each one individually, without the knowledge of the others, had to draw in chalk on a tablet what he had seen in the telescope; only afterwards were the observations from time to time compared with one another. Kepler published the results of these observations in a booklet entitled *Narratio de Jovis Satellitibus*. This was reprinted in Florence in the very same year so that, in Galileo's native land too, it served as a strong witness for the credibility of the new discoveries.

Without detracting from Galileo's merit it can still be said that, after the telescope was invented and had immediately been further circulated, there was only a short step to directing that instrument to the objects in the heaven, where the phenomena observed by him must have been visible to anyone who went to work with care and knowledge. In fact Galileo immediately became entangled in a priority fight with Simon Marius of Gunzenhausen regarding the discovery of the moons of Jupiter. Marius was court mathematician to the Margrave of Brandenburg in Ansbach. In his *Mundus Jovialis* which appeared in 1614 he maintained he had discovered those moons at the same time or indeed even a few days before Galileo. Although there is no convincing reason for doubting this assertion, nevertheless, in after-days, the continuation of Galileo's attack did injustice to the astronomer from Ansbach who, for one, loyally acknowledged the achievement of the Italian in the observation of the moons of Jupiter.

15. Dioptrice

Thus, whereas the discovery of the satellites of Jupiter was one which many could make, the pressing demand of the theoretical investigation of the new instrument was an entirely different matter. In order that the enormous possibilities harbored in this instrument could develop it was necessary to clear up the theoretical laws by which it worked. And this achievement was reserved solely for Kepler. With the energy peculiar to him, inside of a few weeks, in the months of August and September of the same year, 1601, he composed a book tracing basically once and for all the laws governing the passage of light through lenses and systems of lenses. It is called *Dioptrice*, a word which Kepler himself coined and introduced into optics. The substance of this book has since become the common property of science and the possession, as a matter of course, of everyone who has examined optical questions to any extent. All the schematic figures, well known from physics text books, are found there. The work is divided into 141 theorems which are distinguished as definitions, axioms (theorems needing no proof), problems (theorems to be proved by experiments), and propositions (theorems which follow out of definitions and axioms by logical conclusions). The author begins with the law of refraction which, indeed, he was here as little able to express exactly as in his earlier work about optics. Since in the work under consideration, however, only small angles of incidence are dealt with, he managed well by assuming the proportionality between the angle of incidence and that of refraction. He himself determined the ratio by measurements. By investigating the path of a ray in a glass cube and three-sided prism he discovered total reflection. Next in his exposition comes the treatment of the double-convex converging lens. He sets to work with great thoroughness. There appear the ideas well known to us of the real and virtual, the upright and inverted image, the distance of the image and the object, the magnification or reduction of the image. From the path of the ray for a simple lens he proceeds to two and three lens systems. In problem eighty-six in which he shows "how with the help of two convex lenses visible objects can be made larger and distinct but inverted"[1] he develops the principle on which the so-called astronomical telescope is based, the discovery of which is thus tied up with his name for all time. Further on follows the research into the double concave[2] diverging lens and the Galilean telescope in which a converging lens is used as objective and a diverging lens as eyepiece. By the suitable combination of a converging lens with a diverging lens in place of a simple object lens he discovers the principle of today's so-called telescopic lens by which an inverted real image of an object can be produced, an image which is in fact larger than that formed by a converging lens alone. Even this scanty account of the main content shows the epoch-making significance of the work. It is not an overstatement to call Kepler the father of modern optics because of it.

"I offer you, friendly reader, a mathematical book, that is a book that is not so easy to understand and that assumes not only a clever head but also a particularly intellectual alertness and an unbelievable desire to learn the causes of things." With these words Kepler introduced the preface of his work. With it he exquisitely distinguished the work and at the same time himself. Study immediately shows that, after his previous depression, the author plunged to some extent into the complex of questions which excited him. Driven by that unbelievable desire he brought forth out of the deep recesses of his mind the precious information which he presents to us. Prodigious had been the trouble to trace the causes, he wrote in a letter to Galileo, but he had been no less delighted thereby than had Galileo by his discoveries with the telescope. The wealth and the singular dualism of his genius is continually astonishing. The author was the same man out of whose mind, scarcely a year before, there bubbled forth the *Tertius Interveniens* with the confusion of its series of thoughts running this way and that, of its murmuring prattle, crammed full of far-reaching and fantastic speculations, rich in cross-references, and allusions. Yet here he sets forth in concise language a work of such concentrated objectivity and convincing logic as to please even those who hold that a work in the

[1] ED. NOTE. "Duobus convexis majora et distincta praestare visibilia, sed everso situ." (*Johannes Kepler Gesammelte Werke*, IV, 387.)

[2] ED. NOTE. The word "biconvex" in the German is merely a slip.

field of natural science is truly "scientific" only if it suppresses feeling, does not digress into metaphysics and does not try to find meanings beyond the visible phenomena.

In September, 1610, Kepler presented the finished manuscript to his patron, Elector Ernst of Cologne. However, the printing which followed in that city, probably through the good offices of Markus Welser, the Augsburg patrician, was delayed. In the meantime Galileo successfully continued his observations of the heavens. In the summer he had directed his instrument to Saturn and found that this planet does not appear as a simple disc. He thought he perceived a three-membered system, a main body with two smaller spheres clinging tightly to it. The imperfection of his instrument kept him from the correct interpretation of the picture; it was nearly a half century before Christian Huygens, the Dutch astronomer and mathematician, saw and recognized what today enchants every owner of a small telescope: the unusual ring system surrounding Saturn. However, of greater significance for the astronomy of that time was a second discovery which came to that zealous Italian observer a few months later, namely the discovery of the fact that Venus does not always appear as a circular illuminated disc like the upper planets, but shows phases like the moon, depending on her position relative to the sun. Consequently he correctly concluded that this body does not shine in its own light but only reflects the light of the sun, which it circles. This furnished an important, effective argument against the Ptolemaic system of the world. Galileo did not publish these new discoveries in a printed work. He gave information about them by letter to his friends, including Julian de Medici, the ambassador, and through his good offices Kepler. The report about the first discovery took place in the first half of August; that about the last one is dated December 11. For reasons which one must explain for oneself, the cautious man, however, did not express his discoveries clearly, but hid them in puzzles of the letters of the alphabet, which must have been absolutely ununderstandable and insolvable for everyone who received the information. He wanted publicly to prevent anyone else from passing himself off out of jealousy as first discoverer and to urge others to say openly what they had seen. What he knew about Saturn, he expressed through the hodgepodge of letters: "Smaismrmilmepoetaleumibunenugttauiras." He hid his observations of Venus in the anagram "Haec immatura a me jam frustra leguntur oy" (This was already tried by me in vain too early). Naturally Kepler's curiosity was excited to the utmost by this game of hide and seek. He worried excessively over deciphering the secret messages. Since he guessed at there being two moons of Mars, in an astonishing trial of patience he glued together out of the heap of letters the "half-barbaric" verse, which conforms within one letter: "Salve umbistineum geminatum Martia proles" (Be greeted, double knob, children of Mars). In November Galileo condescended to unveil the secret, because he had learned that His Majesty the Emperor wished to become acquainted with the meaning of the letters. The statement is worded: "Altissimum planetam tergeminum observavi" (I have observed the highest of the planets three-formed). The emperor's wish was like a command for him. He let the imperial mathematician fidget. The latter then also wrote him after receipt of the second puzzle: "I adjure you not to leave us long in doubt of the meaning. For you see that you are dealing with real Germans. Think in what distress you place me by your silence." On this occasion the Florentine scholar communicated the solution promptly on January 1, 1611, as follows: "Cynthiae figuras aemulatur mater amorum" (Venus imitates the phases of the moon). The postponement in the completion of the *Dioptrice* gave Kepler the opportunity of adding a long preface to the book, apprising the world of Galileo's discoveries and publishing the letters about them directed by the latter to Julian de Medici. But in this preface Kepler gave free rein to fantasy in his usual manner and let his heart speak out. So in his enthusiasm he bursts forth with the lovely words: "O you much knowing tube, more precious than any sceptre. He who holds you in his right hand, is he not appointed king or master over the work of God!"

With the *Dioptrice* Kepler took leave of optics; he worked no further at this science. Moreover, after the publication of this very important work, remarkably little is heard about it. Was it beyond the comprehension of his contemporaries? It did, however, influence the further development of optics more strongly than some accounts indicate.[1] The greatest attention was paid the work in England where, later, optical studies were carried on with special zeal. There also it was twice reprinted in the seventeenth century.

It is surprising that Galileo said not a word about the book, which, indeed, concerned him above all, especially after Kepler had shown himself to be a faithful co-fighter in the dispute about the *Nuncius Sidereus*. Yet this is typical of the attitude which the haughty Italian exhibited toward him. It was Galileo, also, who now again, as once in 1597, broke off the exchange of letters with his German colleague, and this time it was forever, except that there is still preserved a brief note which Galileo wrote in 1627 recommending to Kepler a young scholar traveling to Germany. Whereas Kepler in 1610–1611 directed at

[1] ED. NOTE. See Ronchi, *op. cit.* (1956), pp. 189–202, and *op. cit.* (1957), pp. 47–8.

least six letters to Galileo, the latter wrote to him directly only once; the other communications about the new discoveries went to the ambassador, Julian de Medici, who transmitted them to the "Signor Glepero" for his information, a procedure which did not exactly testify to great respect for Kepler. To contend that the exchange of letters between the two men had contributed toward the German astronomer's achieving a freer interpretation of nature and that only with this had he made his great discoveries, is to be entirely on the wrong track. A closer knowledge of the sources completely contradicts such an astonishing contention.

Later Kepler seized the opportunity to set Galileo's claims straight as regards the importance of his discoveries, particularly the observation of the phases of Venus, in showing the truth of the Copernican world system. He did this in a tract in which he undertook to defend Tycho Brahe and his correct view about comets against the Italian. Indeed, the observation of the phases of Venus can have but one meaning, namely that Venus circles around the sun. But only the Ptolemaic system is refuted by this, not the Tychonic, according to which the phenomena must exhibit the very picture which was seen through the telescope. As close as the proof of the Copernican system lay to Kepler's heart, he nevertheless shows himself here as the stronger logician.

16. The unfortunate year, 1611

Upon the upheaval and tension, which the scientific events of 1610 had brought about, followed a sudden collapse, the great unlucky year of Kepler's life, 1611. The political development led to the tumult of war and bloodshed in Prague and to the abdication of his imperial master. His Swabian homeland conclusively rejected her great son. Need and death visited his own house. Sorrows engulfed him, fate pursued him. The fruitful and, despite everything, happy period which the astronomer had passed in Prague came to an end.

The trouble began at the end of 1610. Frau Barbara became very seriously ill with Hungarian fever and epilepsy and showed traces of mental disorder. Barely had she somewhat recovered when the three children were seized with smallpox in January, 1611. Whereas the oldest and the youngest recovered, Frederick, the six-year-old darling son, died on February 19. Great was the sorrow of the parents. The mother especially became crazed with love for her children because of the painful loss which she suffered. The lad had shown great talents. "To look at the bloom of his body or the charm of his behavior, or listen to the prophecies of promised happiness made by friends, gave good cause to call him a hyacinth of the morning in the first days of spring, which, tenderly fragrant, filled the room with ambrosian good odors." Such was the picture of his darling son which the grief stricken father carried in his heart.

While the child was lying on his deathbed, Prague became the scene of martial events. The quarrel between the emperor and his brother had progressed. In 1610 a congress of princes, lasting the whole summer and into the autumn, had been held. The emperor was urged to come to agreement with his brother. Yet the existing difficulties were of the sort that could not be abolished by negotiation; actions alone could effect change. The main hindrance lay in the person of the emperor, whose whimsicality by this time bordered on mental disturbance. Distrust and shyness more and more assumed abnormal sizes and shapes. Stubborn and tenacious, he insisted throughout on his authority. To get backing, he plotted with his young relative Leopold, bishop of Passau, a brother of Archduke Ferdinand in Graz. Leopold raised troops ostensibly ready for the fight near Jülich, however really intended for another purpose. Did Leopold want to assure himself of the succession to Rudolph? Did the emperor, with his help, want once more to set aside the freedoms in Bohemia which had been wrested from him in the royal charter? Or did he only want to insure himself of a support for his throne against his brother? At any rate, it came as a menacing surprise, when in the winter of 1610–1611 Leopold set his troops on the march, invaded Austria and turned from there to Bohemia, which the soldiery, predatory and unpaid, devastated severely. In February he pushed into Prague and occupied the Small Town. At this act of force, armed Bohemian troops banded together in another part of the city and there, amid vile excesses, attacked cloisters and churches. The representatives interfered; however, they too felt menaced by the foreign troops and longed for their departure. Thus in addition to the antagonism between Rudolph and Matthias the split between the emperor and the representatives was also opened. As a consequence, the representatives and Matthias were obliged to act together even though their aims regarding the domain of the confessional ran counter to each other. Matthias with an Austrian army hurried on to answer the representatives' call for help. But before there was further bloodshed, the Passau forces which the emperor had paid off out of his own resources suddenly marched out of Prague. Matthias was ruler of the situation. He forced Rudolph to abdicate and himself assumed the crown of Bohemia. This happened May 23, 1611. On June 13 of the following year Matthias was chosen emperor in Frankfort.

Kepler took an active part in these events. At heart he remained faithful to the emperor, although his religious attitude placed him closer to the other side. He was asked for his expert astrological opinion, a serious matter, by followers of both parties. Yet he tried determinedly to leave astrology out of the picture in these events. At Easter of that year he composed a letter expressing himself on this subject. Very prudently it bears neither address nor signature, but it is at any rate directed to a close confidant of the emperor.[1] He knows that, indeed, the position of the stars is not referred to in negotiations. "But this little fox lies in ambush at home so much the more concealed, in the bedroom, on the couch, inside in the soul and meanwhile gives thoughts which, guided by him, one then puts forth in a session, without saying whence they come." He is aware that astrology can cause prodigious harm to monarchs if a crafty astrologer wants to play with the gullibility of people; and the emperor was gullible. Kepler sought to further the emperor's cause diplomatically by giving the emperor's opponents upon request, but contrary to his own astrological discernment, the information that the stars stood favorably for the monarch. However, he was on his guard not to say any such thing to the emperor himself. He urgently requested that not only must astrology be kept entirely out of the emperor's sight but also completely out of the heads of those who wanted to give him the best advice.

17. Rejected by the Württemberg theologians; successful negotiations with Linz; death of Frau Kepler; departure from Prague

Under the circumstances it was now high time for Kepler, too, to make a decision about a new sphere of action, a question which had already been occupying him for two years. He dared not hesitate further. It was not advisable for him to trust his luck in Prague any longer. Since, as he says, it was now his sole desire to serve his fatherland sometime, on March 19, he turned once more, as he had done

---

[1] ED. NOTE. In Kepler's *Opera* (Frisch, ed.), VIII, 343–5, this is one of the letters in the section entitled *Wallensteinii Nativitas*. It was preserved in a Pulkova manuscript and first printed by Otto Struve: *Beitrage zur Feststellung des Verhältnisses von Keppler zu Wallenstein*, St. Petersburg, 1860, pp. 11–12, from where Frisch took it. It explains Kepler's general attitude regarding prognostication. It can be found in *Johannes Kepler Gesammelte Werke*, XVI, 373–5 (letter number 612), dated April 3, 1611, and there it is clearly stated that the letter is addressed to an anonymous nobleman, that the original is not known, and that it was first printed by Struve. A note by Dr. Caspar (*Gesammelte Werke*, XVI, 465) adds that the letter is no longer in the manuscript from which Struve took it.

two years before, to the duke of Württemberg. In the past year the emperor had granted him a draft of 2,000 talers on the Silesian House. Since, naturally, Kepler had not yet received the money and Duke Johann Friedrich's sister was married to a prominent member of that House, Kepler asked his prince to intervene. He wrote that, with the aid of these 2,000 talers, provided he obtained possession of them, he would like to go to his fatherland to wait on the duke. And now he inquires, begging, if the duke could not use him as a professor of philosophy or for a political service, thus allowing him a little peace to complete and publish the studies he had begun and which were generally well known. It is not easy to conceive what kind of a political service he here had in mind. But he also tendered his services to the duchess, the duke's mother, whom he likewise asked to intervene in his financial affair. The request goes to the ducal chancellor's office. Since Kepler would discharge a function at the university *summa cum laude* and would be very suitable for this and since Maestlin, the incumbent of the mathematics professorship was very old, the chancellor imagines that the applicant might be put in line for the position. But the application still goes to the Consistory. There the petitioner is scrutinized more sharply. The old deed is pulled out and here is uncovered Kepler's admission that he could not think that someone, tending toward Calvinism, because of his dissimilar opinion should not be called or considered a Brother in Christ. Could a man with such a criminal view be used in Württemberg? From this sentence, so thought the clerical advisers, it is apparent that Kepler is a sly Calvinist. He would certainly inspire the youth with the poison of Calvinism and, since he is "an opinionist in philosophy," he would otherwise also arouse much unrest at the university. For these "highly moving" reasons and because the statutes of the university require its teachers to be followers of the Augsburg Confession and to subscribe without reservation to the Formula of Concord, the Consistory considers it advisable to deny Kepler's request. The duke, who had obviously agreed with the chancellor's decision, complies and simply gives his approval to the act. The document is dated April 25, 1611. It closed with finality the door to Kepler's permanent return to his native land. He must have found it hard to bury the hope with which he had so long been preoccupied.

Now it was a question of grabbing elsewhere. A slight prospect came from Italy where Kepler had a good offer.[1] After the departure

[1] ED. NOTE. The next sentence has been omitted. Its translation is: "He [Kepler] had been accepted in the newly founded Academy of Lynxes in Rome as early as 1604." Dr. Caspar also made this erroneous statement in a note in *Johannes Kepler Gesammelte Werke*, XV, 536. This note is to the close of a letter of April 1, 1606, from Johannes Heck

of Galileo from Padua for Rome, Kepler was considered for the vacated professorship in Padua. He indicated a desire to go there and Galileo had publicly recommended him to the council in Venice to whose sovereign rule Padua belonged. But the plan came to nothing and Kepler must have been intent on finding a position in German lands. He had been invited earlier by the Upper Austrian lords, for example by Helmhard Jörger in the previous December, to move to Linz. At the end of May, immediately after the abdication of Rudolph, he went there to tender to the representatives his services as district mathematician. His application was accepted and papers dated June 11, 1611, were drawn up for him. The question which had troubled him for so long was decided. Now he knew where he could find refuge after the collapse in Prague. The Wittenbergians, who a few weeks later considered calling him to the mathematics-professorship, which had been vacated by the death of Melchior Jöstelius, came too late. Moreover, this plan could not have been realized without friction since to some circles Kepler's friend, Ambrosius Rhodius, who was the other candidate proposed and who actually obtained the position, was more pleasing for various reasons.

In chosing Linz, Kepler, as he himself says, had given his wife much thought. She had never been able to feel at home in Prague. He hoped that Linz would please her better; there she would be nearer to Graz and would meet people of her own type. Besides, the money worries which had so often occasioned marital quarrels would disappear there. He must have been happy to be able to bring his wife good news from Linz. Accordingly, he must have been horrified when on his return to Prague on June 23 he found her sick again! The Austrian troops had brought along contagious diseases, and his wife had been infected. By July 3, death tore away his wife from this severely tried man. He recounts her suffering and her death in these words: "Stunned by the deeds of horror of the soldiers and the sight of the bloody fighting in the city, consumed by despair of a better future and by the unquenched

---

to Kepler (XV, 316), which ends "salve et vale, valere te iubent Lincei." Prof. Erwin Panofsky repeated the information about Kepler's membership in the Lynxes in his *Galileo as a Critic of the Arts*, The Hague: Nijhoff, 1954, p. 20. Prof. Edward Rosen, who has carefully examined the records of the Accademia dei Lincei, says that Kepler was never a member (*Isis*, XLVII (1956), 78). Prof. Panofsky says (*Isis*, XLVII (1956), 182) that his source was Dr. Caspar's note. A letter from Miss Martha List, Dr. Caspar's collaborator, gives his source of information as *Isis*, XXIV, 87. However, the article by Guiseppe Gabrieli entitled "Per la storia della prima romana accademia dei Lincei" (*Isis*, XXIV, (1), December, 1935, 80–9) clearly indicates that Dr. Gabrieli did not count Kepler as a member of the Lynxes. It must be concluded that he never was one. An accurate scholar like Dr. Caspar would never have wanted this slip to be perpetuated. The 1958 edition of the biography merely states that from 1604 Kepler had most friendly relations with the Academy of Lynxes.

yearning for her darling lost son, to bring an end to her troubles she was infected by the Hungarian spotted typhus,[1] her mercy taking revenge on her, since she would not be kept from visiting the sick. In melancholy despondency, the saddest frame of mind under the sun, she finally expired." As a clean shirt was put on this dying woman, her last words were the question: "Is that the dress of salvation?" Thereupon she fell silent, as though overwhelmed by sleep. The burial was performed by Matthias Hoe, the well known Lutheran theologian, who afterwards became First Chaplain in Ordinary in Dresden, and also played a political role as a sharp opponent of the Calvinists. Kepler would have liked Romans 8, 26ff. for the text of the funeral oration, had not another, by the established order, been next in turn. Later he dedicated a memorial to his wife and his little son. Entitled *Funera domestica duo luctuosissima*, it went unnoticed until very recently and was preserved in only a very few copies. He put in print a poem which he had composed in Latin and German. His wife had always found pleasure and solace in it. Let it be reprinted here as witness of his religious inclination and as proof of his poetic speech.

Sic nunc inanes cernis imagines
Si functo aevo ipsissima lumina
   Cernes: quid haec amittere horres
   O Ocule, et meliora apisci?

Oh physical eye so weakly made /
Your seeing is only dissembling /
In these dark fields /
When however your shine is extinguished here /
From countenance to countenance /
You will behold the eternal light /
Make the change with confidence /
Do not fear /
Do not be childishly afraid.

Si mutilâ tam suavè scientiâ
Mulceris, ut laetaberis integrâ?
   Fidenter obliviscere illa,
   O Anime, ut citò noris ista.

Oh mind, your knowledge is piece work /
And though it may bring you some pleasure /
And sweet fantasy /
When, however, the final perfection /
Will have destroyed this vanity /
How greatly will this please you /
Hurry and forget /
That which is uncertain /
So that this may turn out well for you.

---

[1] ED. NOTE. Miss Martha List, Prof. Caspar's collaborator, has been kind enough to supply the information that "Fleckfieber," the word in the German text, is equivalent to "Flecktyphus." This can be translated as "spotted typhus."

Si vivere hic, est, perpetuum mori;	*O man you now live a perpetual death /*
Semelque, vitae principium, mori:	*For true life death's agony /*
Quid quaeso differs interire	*Is but the beginning /*
O Homule, et moriens renasci?	*All at once like grain you will /*
	Be reborn to eternal life /
	Through Christ you will succeed /
	Do not wish for delay /
	Hurry to die /
	In order to achieve life.

Frau Barbara had not made a will. Kepler writes that she left him
nothing. So, in addition to the care of his children, he was caused much
trouble by the evaluation and the division of the estate between the
two children and his stepdaughter Regina. The settlement with his
son-in-law brought friction; a legal decision was even invoked. Kepler
had put together the expenditures for rearing his stepdaughter and for
her dowry. Her husband considered them too high and would not
recognize them in entirety.[1]

At Rudolph's request Kepler put off his move to Linz and continued
with the deposed monarch. The research work was untouched; the
astronomical study was, as it were, knocked from his hand. But in
order to furnish proof that in spite of all untoward seeking of a home
he had not yet neglected all study, he prepared still another publication
in which he assembled his correspondence with Seth Calvisius, Markus
Gerstenberger, Johannes Deckers S.J. and Herwart von Hohenburg
about various chronological questions relating to the time of Christ.
The book was later printed under the title *Eclogae Chronicae*. Kepler
wrote that this activity, indeed, required no more work than a bowed-
down man was strong enough to achieve. He dedicated the book to
Tobias Scultetus, an imperial adviser, who had been appointed fiscal
procurator in Silesia and to whom he wanted to appeal urgently by it
in the trouble over his 2,000 talers. Of course, this friendly gesture led
to no more success than had the duke's intercession.

The unfortunate Rudolph passed away on January 20, 1612. Now
nothing further could hold his former mathematician in Prague.
Kepler left the city in the middle of April and brought his motherless
children to Kunstadt in Moravia to a widow named Pauritsch. He
himself traveled via Brünn on to Linz, where he arrived immediately
after the middle of May. A lonely man, he entered the Upper Austrian
capital.

[1] By an unknown fate the deeds concerned have arrived at Oxford. (Transferred from
text.)

IV

DISTRICT MATHEMATICIAN
IN LINZ
1612–1626

---

1. *Kepler in a new intellectual situation and official position*

It was an important event for the charming Danube city when Johannes Kepler, the imperial mathematician, took up his residence there. That he remained for fourteen years, longer than at any other halting-place in his earthly wanderings, must be entered on a separate page in the city's history. He had come to Prague as a sick refugee. Favorable circumstances, his emperor's understanding open-heartedness, the pliant strength of his own personality won him a distinct place by which he was buoyed up and firmly established. From his environment he had received great stimulation and encouragement, first from Tycho Brahe, then also from others. Numerous friends, scholars and lovers of knowledge met his need for discussion. He had learned to mingle with people of importance. He even had had insight into the political machinery from close up. It was different in Linz. Here his situation was in every respect more restricted and pettier. He was the same as he had been in Prague. But since the circumstances of one's manner of living have much more influence than one realizes, his attitude changed into a reaction to his narrow surroundings. The more restricted these were, the greater was his contrast to them. He towered above the little world in which he lived. However, a little world takes this amiss and will not tolerate it. He towered above it by means of his erudition, which was not understood. He towered over his environment by means of the spiritual freedom in which he breathed and which lifted him far above the range of the self-sufficient herds. As he had to acknowledge, he was a "thorn in the eye" for many. Instead of finding stimulus and understanding he met resistance and misunderstanding. More than in Prague he was dependent on his own inner world.

Loyalty to himself and to his calling kept the contrast to the others alive. Under this differentiation he suffered so much the more because he knew that he excelled the others. He sought fellowship with them and wanted to be connected with them. He needed that. But the multitude rejected him. He could not surmount the difficulties which grew out of this situation just by striding past them. His nature did not make it that easy for him. To his surprise right at the beginning of his stay in Linz he felt that certain tendencies of his character bestirred themselves strongly, "the blind confidence, the display of piety and compassion, the grabbing for fame with surprising new plans and unusual deeds, the restless searching, interpreting, analyzing of the most varied causes, his agonized doubts about his own salvation." He knew no arrogance toward simple men.

He came to Linz as district mathematician and teacher in the district school. We know that the teaching profession was not highly regarded at that time. Again he had to subordinate himself to superiors, the representatives and their commissioners. He had devotedly offered them his "humble services in mathematical, philosophical and historical studies." He had promised that he was willing to "recognize gratefully all the favors and benefits which he might receive and to absolve obediently, with faithful diligence and sincere Teuton honesty, all the tasks demanded of him, as far as his humble powers permitted." The priest, as inspector of the school, was also a superior. As spiritual shepherd he held his little sheep closely together. So from the very outset there existed a disagreement between Kepler's official position and his personal stature. Even though he had been accustomed to complete dependence from his early years, he had outgrown this in Prague. Now the situation was again as it had been in Graz. Tensions were inevitable. He was not a man to be ignored. He either had to be spurned or respected although it was never his intention to pose such a decision for others.

With matters as they were, Kepler is to be believed when he says he had chosen Linz as his dwelling place principally out of consideration for his wife. In fact the circumstances there were more suitable for her than for him. For him a position at a university would have been desirable where, in lively intercourse with scholars and students, he could have developed a rich activity. For full success in education the wealth of his thoughts, as well as the difficulty of his style of lecturing, needed a small group of young men who were particularly gifted and eager to learn. Why was he willing to spend fourteen of his best adult years in Linz? (At his arrival he was in his forty-first year.) There are various reasons for this in addition to his gradually getting accustomed to less favorable conditions. Even if Kepler was exposed to great opposition and open hostility in Linz, he also found friends and patrons there in the prominent circles whom he already knew in part. They granted him complete protection and good will and, in their love for science and in the admiration which they paid to the unusual genius, they gave him the happiness of intimate association. Foremost among these were Barons Erasmus von Starhemberg and Georg Erasmus von Tschernembl, the Protestant leaders in Upper Austria, as well as Lords von Polheim, Maximilian von Liechtenstein and Helmhard Jörger. In choosing Linz he had thought of these men when he took into consideration "that especially [in] these places there are many noble minds which, after the very praiseworthy example of their district princes and lords of the Austrian House and disregarding all other entertainment, are sensibly devoted to the mathematical skills and contemplation of the wisest and finest works of God in creating heaven and earth."

In addition, immediately after Rudolph's decease on March 18, 1612, his successor, Matthias, had confirmed Kepler as court mathematician and granted him an annual salary of 300 gulden to which were added an additional 60 gulden for dwelling and wood costs. In this position, which was of great importance to him, Kepler was dependent on the emperor in the choice of his dwelling place. Matthias did not hold his mathematician to the court but called on him only occasionally for counsel and gave his express consent to the move to Linz. Thus a change of dwelling place did not rest solely with Kepler. Linz, so he also hoped, would be the right place where he could have peace to continue and complete the works he had begun "under the protection and to the glory of the Austrian House" within its territories and rule, as was fitting. We know what work he particularly meant here and which had long stood before him, the *Rudolphine Tables*. The longer their completion was put off, the longer he felt himself tied to the Austrian House and consequently Linz. Until completed, the tables determined his dwelling place. That work became his fate.

With the years, moreover, a change came in the Linz situation. The above description of Kepler's position in this city holds good chiefly for the first period of his stay there. When six years after his arrival the great war began, when the Counter Reformation struck all Austria, those who were his opponents had other worries. The most ardent delight in creating came over Kepler. A series of great works, among them his master work, the *World Harmony*, took form. His mother's witch trial kept him away for a considerable time. Political and martial events changed the picture and pushed personal desires in the background. And when after fourteen years Kepler left Linz and two years later withdrew from the service of the district, this occurred on friendly terms with the representatives.

At the appointment of Kepler to Linz, the prospects were different from those in Graz. Whereas in Graz a suitable man had been sought for a vacated place, in Linz a position was created for a man whom a few influential, magnanimous leaders wanted to have in their city, in order to free him from the confused conditions in Prague and to provide him with the possibility of carrying on his scientific works. So Kepler's written application was certainly agreed upon ahead of time with these men who had decisive influence with the representatives. It was formal in nature, since the answer actually followed the petition through the diet one day after it was addressed. At the district school Kepler had no precedents. This school was smaller and less important than the one in Graz. After a drawn out start, it had received its constitution at about the same time as the Graz school. This was accomplished through Johannes Memhard, a Württembergian, who acted as rector for twenty-two years and was discharged in 1598. Two years later as a result of Rudolph II's Counter Reformation regulations the school had to be closed but it was revived again in 1609 when Matthias, in the dispute with his brother, was forced to make concessions to the Protestant representatives in Upper Austria. In the year 1611 the entire teaching body consisted of the rector, the co-rector, and four professors; in the following years it was enlarged by the appointment of a few additional teachers, including Kepler. In addition to his teaching activity the district mathematician was given the further obligation of completing a map of Upper Austria. Baron von Tschernembl appears to have been particularly interested in this. It was he who as early as October, 1595, during a visit to Graz, had wanted to interest the young mathematician there, Kepler, in renovating a map of Lower Austria.

That Kepler's appointment was to a position created especially for him is also apparent from the decree of installation made out on June 14, 1611, in which his duties were described. The representatives take him in their service "because of his famed ability and praiseworthy virtues" and charge him first of all to complete the astronomical tables in honor of the emperor and the worshipful Austrian House, for the profit of the commendable representatives and of the entire land, as well as also for his own fame and praise. After that he is exhorted to neglect nothing "which he can produce, not only in mathematical but also in philosophical and historical studies, which is useful and suitable for the praiseworthy representatives in general as well as for each one in private and also for noble youths." The general style of this instruction is surprising. Nothing further is said about work at the school. Four hundred gulden were granted to Kepler as annual salary. The appointment followed a half year's notice to both sides. As regards the completing of the map, all that was said was that the costs of necessary travel should be fairly compensated.

2. Exclusion from communion

The beginning in Linz prophesied trouble. In the first weeks there came a blow which struck the newcomer at the very core and gave the signal for all the chicanery and persecution to which he was exposed in the following years. Kepler was excluded from communion by the pastor of the Lutheran congregation in Linz. Since this sacred rite served as symbol of the fellowship between the members of the church, this exclusion meant expulsion from the congregation. Act two of his tragic conflict with the Church in which he had grown up had begun. We have seen the first act, in which a position in his native land was denied him because of his creed; the third and final act will be discussed later. The schism, here torn open, furnished the foundation for the outer discords in Kepler's life. The attitude which he here showed gives deep insight into his inner life. At a time which least wanted to understand it, and against the current moral pressure, he demonstrated freedom of conscience and overcame intolerance in a typical manner by a new example of religious and moral conduct. Consequently, there developed concern about the meaning of a purely private experience.

In Upper Austria at that time the situation of the confession was as it had been in Styria when Kepler arrived there. The majority of the population had accepted the new creed and left the Roman Church. The congregation in Linz was being cared for by Württemberg clerics; Kepler described it as a Württemberg colony. In 1610 Daniel Hitzler, who was born in Heidenheim, and eventually had acted as deacon in Waiblingen, was sent forth as chief pastor in the Upper Austrian capital. He was five years younger than the new district mathematician and had gone through the same course of instruction in his native land. When Kepler moved to Linz, he at once requested communion of his minister. In his candor he immediately communicated his religious beliefs. This Hitzler would have scarcely required. Previously he had received from Württemberg, where everyone was informed about everyone else, suspicious information about the imperial mathematician who had caused much talk about himself. He knew that in the eyes of the custodians of orthodoxy in Stuttgart and Tübingen this man was an "unhealthy sheep" and leaned toward the hated Calvinist doctrine. He consequently asked the new teacher at the district school to give his signed consent to the doctrines stipulated in the Formula of Concord before receiving communion. Kepler made a reservation regarding the doctrine of ubiquity. Hitzler asked for unconditional subscription. Kepler declined. The breach was there. The matter was not confidentially handled and Kepler was talked of by high and low. Whether the masses of the believers understood the subtle dogmatic distinction in question is more than doubtful. But the less they understood about it, the more heresy could they attach to the man whom the minister felt compelled to exclude. "The first servant of the church and at the same time inspector of the school has branded me as the public picture of a heretic." Kepler with his delicacy of feeling and conscientiousness suffered severely under the public scandal which thus began. He decided to appeal[1] to the Stuttgart consistory, to see whether it was not possible, by virtue of its authoritative remonstrance with Hitzler, to procure admission to communion and therewith to eliminate the scandal. This last desire was especially close to his heart. Had he still not become acquainted with the Stuttgart lords? The answer of the consistory arrived on September 25. It agreed with Hitzler in everything and rejected Kepler's request.

The consistory's lengthy communication deals in detail with the points of view which the petitioner had pleaded. In the memorable conflict it represents a document that clearly illuminates the position of the evangelical church of that time with regard to dogma and discipline. There were two opposing parties which were distinguished in various ways. On the one side were the Württemberg church authorities which had great prestige and far-reaching influence in all Protestant Germany. On the other side stood the great and famed astronomer who pleaded his point of view so energetically not because he was stubborn and dogmatic, but because he struggled conscientiously to find common ground with his coreligionists and to arrive at a true faith.

Kepler had asked the consistory to decide whether Hitzler really could exclude him from communion on the grounds of conscience and which was the lesser of two evils: admission to communion in spite of his protestation or being compelled to receive the sacrament

---

[1] Unfortunately, in spite of zealous search, Kepler's petition of August 20, 1612, clearly pleading his standpoint very urgently, has not been brought to light. (Transferred from text.)

elsewhere. The answer to the first question was unequivocal. No minister of the church, who wanted to be a true caretaker of God's secrets, could admit a person to communion who outwardly boasted of the true evangelical religion but in articles of faith was not exact in all things; who deviated from the sound doctrine, obscuring it with dubious meanings and absurd speculations; who was confused and confused others; who was carried away according to his own judgment in matters of faith and divine secrets; who did not wish to be committed to any definite form of the pure doctrine, shrinking from subscribing to the Formula of Concord as the symbol of the orthodox church founded in the Holy Scriptures and contradicting it in one article or another, unless he dropped his erroneous opinion and harmonized his teachings with those of our church. That Kepler had such a point of view was sufficiently known from his own utterances. In particular he denied the omnipresence of the body of Christ and admitted that he agreed with the Calvinists, if not in everything, at least in some articles. Therefore Master Hitzler had acted properly and well in not admitting him to communion. It would not do and would be bad for the worthy enjoyment of the sacrament if Kepler wanted to receive it during an undecided argument. Because Kepler's nature compelled him to scrutinize minutely the writings of all orthodox teachers and defend the opposite doctrines, as he conceived them, it was to be feared that on his part the argument would long remain undecided, no theologian would clear away his scruples and he would have to die in the middle of the controversy with a troubled mind and wavering conscience. As for the rest, Kepler shut himself out from the community by designating the communion as a sign for that creed which was set down in the Formula of Concord, while at the same time contradicting this sign and defending its opposite. "You don't want to have anything to do with our confession: how then could you ask for *confessionis notam ex animo?*" As regards the second question, the consistory gave the evasive answer: both evils could be avoided if the little lost sheep were willing to listen to and obediently follow the voice of the arch-shepherd. The blame for any scandal would fall on Kepler. He would have to answer for it if, as he wrote, all who saw him go to communion outside of Linz would have to believe that not only he but also the pastor, who admitted him, and his parish were true Calvinists. So he should keep his hands off theological speculations, pursue mathematical studies all the more, not step across the bounds of his profession, nor give further annoyance by unnecessary disputations. "Don't trust your own inspiration too much, and see to it that your faith rests, not on the wisdom of man, but on the strength of God."

One cannot deny that this argument of the consistory was logical and moderate. A church congregation will and must pledge its members to a well-defined doctrine if, from the beginning, it is not to bear within itself the seed of dissolution. It is not consistent, however, if that church authority enforces this assumption with all its conclusions but also supports the doctrine of a universal priesthood, imbues its theology students, to whom at one time Kepler, too, had belonged, with pride in the freedom in the explanation of the Holy Scriptures promulgated by Luther, because the Holy Spirit and the inner anointment teaches everything to each one and permits the content and extent of the binding doctrine to be determined by nobles and theologians, whose only advantage over the laity was that they had studied the sacred sciences. Here there is an obvious contradiction and it is this which must be blamed for Kepler's conflict with his church. To the extent that the contradiction was unbridgeable, to that extent was the conflict insoluble.

Kepler's spiritual conflict stemmed from that contradiction. For reasons of conscience he could not accept the doctrine of the fellow members of his community, but he still felt a strong urge to be united with them. He did not approve the assumption of a universally binding doctrine for the church community and took seriously the principle of freedom of creed, yet he also believed that such a union was possible. But the schism consisted in his opposing his subjective freedom to the community which felt bound by ecclesiastical authority. In his vindication, indeed, he did not refer to this freedom, as he had once done against the Catholic Pistorius with such great self-assurance, but without reserve he held fast to this freedom. The manner in which he makes use of it and tries to solve his conflict of conscience reveals the high moral earnestness which distinguishes him. In order to clarify the disputed points he draws on all available theological writers and especially buries himself in the works of Christian antiquity; at one point he explicitly names Gregory Nazianzen, Fulgentius, Origen, Vigilius, John of Damascus, and Cyril. What he finds here is his standard. "I felt the burden of antiquity." Since there and in the statutes of the old councils he did not find the ubiquity doctrine of the Formula of Concord, he felt himself strengthened in his rejection of the doctrine, in which he perceived an inadmissible innovation.

Naturally, the fight was not over with the decree of the consistory. Kepler complained that one only reproached him and warned him to cease his captiousness although he had not yet wished to advocate captiousness but to ask for communion by an open avowal. For reasons of conscience he takes a divergent position and does not share in the poisonous sentence of doom in regard to the one article of creed. He answered the consistory that he would behave quietly and make no further difficulties for Hitzler; only he would have to insist on the request for admission to communion, at a better time or more suitable place.

In spite of his forgiving nature the hostilities increased. One did not understand him or else did not want to understand him. He himself reported the many things imputed to him. "I have been denounced as unprincipled, agreeing with all, prompted not by a sincere heart, but by a desire to obtain the good will of all parties, whatever may happen, today or tomorrow. I have been called a godless scorner of the word of God and of the Holy Communion, who does not care whether or not it was granted, and who, far from being eager to receive it agreed that it be kept from him. I have been denounced as a doubter, who at his advanced age has not yet found a basis for his faith. I have been denounced as unstable, now siding with this one, now with another, according as something new and unusual is brought into the arena." He was called a weathercock who turned according to the wind. He was accused of giving in to the Catholics in individual points in order to profit personally. He was reproached for being neither cold nor warm. Since he inclined to the Calvinist conception in the doctrine of communion, followed the Catholics or, as he said, the Jesuits, as regards the doctrine of ubiquity, and on the other hand rejected Calvin's "barbarous" teaching of predestination and could not agree with Luther's book about the captive will, he was declared a newcomer, who wanted to add to the others a Keplerian creed of his own. His friend, Georg von Schallenberg, a noble, who admired his intellectual freedom and peace of mind, wrote to him, that many called him an atheist, a heretic, a flatterer, a self-seeker. He justified himself against all these reproofs and strove with troubled heart for understanding and respect. "It is indeed an annoying and, for the common uninformed man, a very strange notion that someone should be so bold, proud and puffed up, as not to agree with any party. But I bear witness before God that I am not pleased nor satisfied in this role, nor do I like to be considered as a man apart. It hurts my heart that the three great factions have torn the truth so miserably that I must collect it piece by piece, wherever I find a piece. I have no regrets, however. Rather I busy myself reconciling the sides, where I can do it with the truth, so that I can indeed agree with several. One result is that some consider me a mocking bird, when I say in an argument that I frequently agree with two parties against the third. See, I am pleased with either all three parties or with two against the third, always hoping for an agreement; on the other hand each of my antagonists is satisfied by only one party, imagining one eternal irreconcilable disunity and quarrel. My hope is Christian, God willing; what the imagination of the others is, I do not know. God already has rewarded the quarreling Germany with tribulations." Does he not trust his insight too much? Are those who accuse him of too great confidence completely wrong, when he believes himself able to collect the truth piece by piece according to his own judgment? Still, he supposes, should one not be allowed to think independently, to speak openly and frankly? "In these times the theologians only want to have good German hirelings in matters of creed, because one takes money from a single lord, and for him risks life and limb, and does not worry much whether he is right or wrong."

However much Kepler troubled to bring clarity into the disputed questions of dogma, however much he clung to that which he had recognized as truth, he was not spurred by a mere wish to be right. His deepest desire is not strife but peace; he wants to bring into prominence not that which disunites, rather he wants to cultivate what serves unity. He prays daily, as he once admits, for the reuniting of the separated faiths. "I tie myself to all simple Christians, whatever they are called, with the Christian bond of love; I am an enemy of all misinterpretation, and speak well wherever I can." "My conscience tells me that one should not wrong the enemy, but love him, and should not help to increase the causes for further separation; it tells me, I should set my enemy a good example of all moderation and mildness; perhaps I would cause him to act the same, and may God then, at last, send the dear desired peace." He tests the ugly forms which the confessional struggle in his time had frequently assumed. "My argument in matters of religion is only that the preachers in the pulpit are becoming too haughty and do not abide by the old simplicity; that they arouse much dispute, bring up new matters which hinder devotion, frequently accuse one another falsely, stir up the nobles and lords against each other, interpret many actions of the Papists too maliciously, and thus cause many again to drop out when once a persecution commences."

It is hard to resist quoting from the many passages in which he reveals his sincere and ardent love for peace. It is permissible to introduce one more instance which shows the deep agitation of his heart when reference is made to the religious question. In the autumn of 1616 Maestlin informed him of the death of Helisaeus Roeslin. Since the latter had sympathized with the Schwenkfeldians,[1] Maestlin expressed

---

[1] ED. NOTE. Followers of Kaspar von Schwenkfeld (1490-1561), German religious reformer who held that the Last Supper nourishes the spirit but produces no change in the elements.

his apprehension for his eternal salvation; in religion Roeslin had always been an individualist and followed no particular one entirely. Kepler understood the sign and certainly noticed the warnings aimed at him. He writes back, "Since this was God's will, let us wish good fortune to Doctor Roeslin on having completed his life and arrived at peace. I do not doubt that he adhered to the foundation of the faith. May God, who forgives those who repent the wickedness of their hearts, also have mercy on the errors of our mind. May He be merciful, however, not only toward those who go their own way but also to the rabbis on the earth who strive for the favor of the people and are full of self-confidence." Then he sets forth with emphatic words his point of view in the religious dispute. What he finally confesses shows this harassed man's magnanimity and his greatness of soul: "I could quell the whole fight, by subscribing unreservedly to the Formula of Concord. Yet I have no right to be hypocritical in matters of conscience. I am ready to sign, if the reservations I have already presented are accepted. I want no share in the anger of the theologians. I shall not judge brothers; for whether they stand or fall, they are my brothers and those of the Lord. Since I am not a teacher of the Church, it is better for me to pardon, report good and interpret favorably, than to indict, vilify and distort." If only Kepler's way of thinking had been general and alive in his time! Then the Germans would not have been forced through a Thirty Years' War to settle their differences in mutual toleration. They would not have had to learn first in this hard school that differences of faith should be arbitrated with intellectual weapons and in Christian love.

But Kepler's example and words had little effect. Men did not stop at the making of insinuations, they warned people against him, undermined the public confidence in him, planned ways and means to remove him, in such a way that in other evangelical places also the door would be shut to him; they persecuted him. Not just once was his personal safety threatened (*non una via vel vice de incolumitate mea periclitatus fui*). He himself tells all this for the very period of 1612–1617, which we are considering here. He does not, to be sure, tell us the detailed incidents of these threats. We only discover from a letter written in the spring of 1617, that difficulties were put in the way of his teaching profession. Moreover nothing is learned about this from the extant documents. "The pupils are kept together in the farmhouse and overburdened with lessons. The hours are so filled up that it is impossible for anyone to come out to me or to anyone else, in order to learn geometry or languages." And yet it was, as he says elsewhere, his earnest endeavor "to imbue" the young "with admiration for the works of God and to ignite them with love for God, their Creator."

In the autumn of 1616 his position was severely endangered. The commissioners found the disbursements for their mathematician too high in comparison to what he had so far accomplished in the service of the district. They were of the opinion that, "These expenses could well be saved by dismissing him." The matter came before the body of representatives, which was divided into four groups: prelates, lords, knights and cities. Obviously, there were rivalries between the two middle groups. Kepler elucidates this situation in a letter. "My salary was the subject of argument at the session of the district; a great many from the ranks of the knights were against me, the barons for me. I was victorious (without knowing about the matter), winning more votes. Indeed, an honorary gift was even granted me on January 1, as consolation for the insulting opposition. I fear this occurred at the instigation of the triumphant party so as to arouse the anger of the losing side. For the envy of my opponents appears entirely unconcealed. Everyone knows that my view of the creed is the reason which will always be used as pretext." It was, in any case, the above-mentioned Barons von Starhemberg, von Tschernembl and so forth who stood for him, whereas the lesser nobility opposed that group of lords.

At almost the same time Kepler received a letter from Giovanni Antonio Roffeni, professor of philosophy at Bologna, informing him of the death of Magini, Roffeni's teacher, and inquiring of Kepler if he would care to take over the vacant chair. Kepler declined, not the least of his reasons being that that city belonged to the Pontifical State. In his elegantly styled reply he referred again to his freedom. "From my youth until my present age, as a German among Germans, I have enjoyed a freedom in deportment and speech, the practice of which, should I go to Bologna, while not putting me in actual danger, could easily expose me to abuse, distrust, and the denunciations of meddling persons." Here Kepler certainly was too arrogant. What has been told above amply proved that in Germany, too, an attempt had been made to clip his wings.

3. Second marriage and household

After a year, Kepler's private life in Linz took a significant turn. He entered on a second marriage and built his home anew. He had thought about it late in the autumn of 1611, while still in Prague, a few months after the death of his first wife. He first considered a widow, whom his wife had introduced to him and who was so placed that she would be pleased to be free of the situation in which she found herself and drawn into society. Yet the plan was not carried out. A suitor should not let himself be guided in his choice by sympathy for a woman whom he wants to marry and reverence toward his dead wife, even if these feelings are good in themselves. Similarly a second match in Prague and another in Kunstadt, whither he had brought his two children when he moved to Linz, came to nothing. There, too, his heart was touched and he won confidence, since a marriageable maiden took loving care of the children. In Linz he immediately became entangled in further plans. There is extant a very remarkable letter revealing his character. In it Kepler invites an anonymous nobleman, quite likely Baron Peter Heinrich von Stralendorf, to his wedding, and recounts in detail the history of his second marriage. No fewer than eleven candidates appear successively, the three mentioned above and eight from Linz, on whom the vacillating suitor cast his eye one after the other. Although this tale brings a smile, still this letter does not in the least reveal the humor to which it lies so close; it is the most serious letter imaginable. He intends "to reflect on divine providence in general and, as it behooves pious faith, its particular application to his own marriage." He searches his soul and in almost self-tormenting conscientiousness asks himself the grave question: "Was it divine providence or my own moral fault that in the last two years and more my heart was pulled in so many directions, that I lay in wait for so many matches and took into consideration many more who, I admit, were entirely different from one another? If it was divine providence, what were its intentions with regard to the individuals and their actions? If, however, it was my own fault, in what did it consist? In sensual desire, in lack of judgment or in ignorance?" In this connection, the oft-cited lovely sentence flows from his pen, expressing in a few pertinent words his innermost being and aspiration: "There is nothing which I desired more to investigate thoroughly and to know than this: can I also find God within myself, God, whom I readily grasp when contemplating the universe?" This man who takes such sure forward strides in his science, who is never at a loss in a dispute, who knows how to strike the right tone with princes and commoners and otherwise shows himself master of every situation, exhibits such helplessness, indecision and lack of self-assurance in his search for a suitable wife, that it is almost pitiful. He does not know what he wants. He is influenced by advisers of both sexes. Some push him toward this, others toward that match. He considers marrying a well-to-do woman, because in Prague he had known financial worries. Yet he fears the expenditure she would make. If the woman is poor and unassuming, he is afraid to be burdened with her family. He imagines that a widow would be most suitable for a philosopher, who had passed the peak of manhood, since with her the passions are already extinguished and the body is by nature dried out and relaxed. On the other hand, the youthful freshness of a maiden attracts him. The bad health of one deters him, the corpulence of another would not have suited his slight stature. He recoils before people's gossip and fears their mockery. He tries to be smart and test the party by telling her what there is against the match. Now, however, things go even worse. He is hurt by the doubts about his orthodoxy. He suffers from lack of self-assurance and power of judging people who think unfavorably of him. Sick at heart, he goes from one failure to another courtship. "Why has God permitted me to occupy myself with a plan that could not reach a successful conclusion?" He spoke thus after one failure; after another he felt that divine mercy did him a favor. He tells himself that other people, too, encounter complications in marriage affairs. "The difference, however, is that, unlike me, the others do not brood over it, they forget and make light of something more easily than I, or they have better self-control and settle their unfortunate experiences by themselves; or if their experiences are less complicated, they owe it to the fact that they are less gullible than I."

Tossed restlessly hither and thither, he finally returned to the one who chronologically had been the fifth to strike him and who had commended herself by her love and the promise of unpretentiousness, economy, industry, and love for the stepchildren. "You see," he writes in his letter, "how I was driven by divine decree in this distress so that I should learn to despise high society, riches, relations, none of which is to be found in the present case, and to aspire calmly to the other simple traits." The choice was Susanna Reuttinger, a twenty-four-year-old maiden from neighboring Eferding. She was an orphan; her father, Johannes Reuttinger, had been a cabinetmaker. Baroness Elizabeth von Starhemberg had adopted her and remained her guardian for twelve whole years. Evidently Kepler had become acquainted with her during his customary visits at the home of his patron. For him her good upbringing took the place of a larger dowry. The wedding took place on October 30, 1613, in Eferding. The festival meal was served at the Sign of the Lion, an inn still standing today. As court mathematician the groom had also invited Matthias, his imperial master, to the fête. The representatives, who likewise had to be invited, presented him with a goblet valued at 40–50 gulden as a wedding gift.

Kepler is full of praise for his bride: she suits him in form and manners; instead of being proud or extravagant, she is industrious, has some knowledge of conducting a household and is ready to learn the rest. Kepler himself admits that public opinion did not approve this union. He asks himself why he loses or appears to lose respect, and in his letter implores the noble addressee to attend his wedding, in order to give strength by his presence to Kepler's disdain of public opinion. Regina, his stepdaughter, also wrote him firmly cautioning him against this union and recommending a more aristocratic match. She thought "it would be a marriage if my Herr father had no child." Evidently she considered the wife too young to bring up her two stepchildren, Susanna, the daughter, who at that time was eleven, and Ludwig, the son, who was six. The marriage, nonetheless, turned out happily, more happily than the first. In that security of his new home which he needed, Kepler found a peaceful place, where he could feel sufficiently free from worries to complete his scholarly works. To be sure, in his letters and in the other documents of his lifetime, there is barely any further talk of his wife. This may, however, certainly be deemed a good sign, if it is true that those wives are the best who are least spoken of. Now Kepler could again have his two children with him. Several months before he had brought them from Kunstadt to nearby Wels, to the family of Johannes Seidenthaler, a relative of the later Linz schoolmaster, Moriz Seidenthaler. Now he fetched them home.

During the stay in Linz Frau Susanna presented her husband with six children. The first three, however, died very early. Margareta Regina was born January 7, 1615, and died September 8, 1617, of a cough, consumption and epilepsy; Katharina, born July 31, 1617, died February 9, 1618, from similar symptoms; and Sebald, born on January 28, 1619, died on Corpus Christi day, June 15, 1623, from small-pox. Cordula was born on January 22, 1621, in Regensburg, where her mother paid a short visit. Fridmar was born January 24, 1623, and Hildebert on April 6, 1625. Kepler chose the last child's name because he recalled an eleventh century ecclesiastical author of that name who had written "exceedingly well about the correct punctual observance of the ceremony of the Eucharist."[1] The sponsors include, for the first two children, Eva Regina von Starhemberg, the baron's daughter, and Katharina von Herberstein, née von Polheim. The others belong to the upper middle class. They are the men and women whose names are met in later documents: Dr. Abraham Schwarz, the jurist, a native of Württemberg, legal adviser at Neuberg and counsel to the Upper Austrian representatives; Dr. Johannes Oberdorfer, the physician, who

[1] This is the theologian who first introduced the term "Transubstantiation" into the dogmatic doctrine of communion. (Transferred from text.)

previously had lived in Graz where Kepler had become acquainted with him, had had to leave there because of the Counter Reformation, and had later lived in Regensburg; Dr. Stephen Marchtrenker of Wels, the jurist, who, after Kepler's death, took care of the latter's family in Regensburg; Sebastian Baumeister, Helmhard Jörger's steward; Balthasar Gurald, the district secretary, who subsequently, under the pressure of the Counter Reformation measures, also repaired to Regensburg; and Gurald's wife.

There are various documents showing how the thoughtful father cared for the education and instruction of his older children. He eagerly taught Latin to his son Ludwig. A handwritten fragment of a translation of the first book of Caesar's *Gallic War*, which certainly resulted from this instruction, is extant. Likewise, there are still a few printed copies of a translation of the whole first book of Tacitus' history in which Kepler tried "whether it is possible to bring out in good German translation, be it brief or more elaborate, the whole far-reaching meaning which the author generally had couched and as it were hidden in short, rapturous quite majestic words." It is charming to read how Kepler discharged this task and took the trouble to render the Latin phrase in his pithy German and also to put into German the occupational and other designations.[1] For three years in weekly practice, as reported by Ludwig Kepler, the growing boy had to render sections of this translation into Latin; then if the father had corrected the version, the youth had to write the Tacitus text in the margin. Especially close to the heart of the pious head of the family was his children's religious instruction. As appears from a letter by Gregor Eichler, the Görlitz pastor, Kepler had, already in Prague, celebrated family prayers on every Sunday and feast day. When his two oldest children reached a suitable age, he composed for his "children, domestic servants and relatives" a "lesson from the Holy Sacrament of the body and blood of Jesus Christ our Saviour," divided into questions and answers, which he had printed. A single copy is still preserved at the library of the University of Tübingen. He had the children learn the answers by heart "in the hope that if they learned it by heart and had it in their memories, the admonition itself read aloud in church would be so much the more understandable to them, and by means of the power of the Holy Ghost would be more fruitful in continuing right true Christianity with them."

Kepler changed his Linz home several times. But it is difficult to

[1] For example: consul [consul] = Ratmeister [master adviser], augur [soothsayer] = Lösseler [solver], praetor [leader] = Schultheiss [village mayor], tribunus [tribune] = Zunftmeister [master of a guild], procurator [administrator] = Kammerpfleger [administrator of the court]. (Transferred from text.)

make exact and complete statements about this. For the beginning of his stay the deputy canon assigned him a room in a house described only as "in the suburb called Weingarten." When he married again in 1613, he took over a house in Court Street (Hofengasse). But it has not yet been possible to determine which house in this short street is to be regarded as Kepler's dwelling. Nor is it known how long he lived there and why he left. It can only be said that he did not move out before 1619. According to a precise assertion by Kepler, the house which today is No. 5 Town Hall Street (Rathausgasse) was designated as his dwelling in 1625. It is quite certain that in the course of that same year, by a unanimous decision of the commissioners, a dwelling in the country house at the city wall with a view of the graves and the outskirts of the town was put in order for him.[1]

Money worries were less pressing in Linz than in Prague, because the representatives regularly paid their district mathematician his salary and besides gave him handsome gifts for the writings he presented. The salary which was to run simultaneously and which had been granted to him as court mathematician, however, reached him only spasmodically, and was sometimes skipped altogether. The insolvency of the imperial treasury was transmitted from one incumbent of the throne to another. Thus the sum owing Kepler, from his salary claims under Emperor Matthias until the latter's death in the year 1619, amounted to 985 gulden. As long as he lived, he was as little able to recover this sum as that still owing him from Rudolph, because of which he had undertaken a fruitless trip to Prague in the fall of 1612. Also the revenues which he had previously drawn from his first wife's property ceased. In 1615 out of the share of the inheritance which the latter had left to her two children, he invested a sum of 2,000 gulden with the treasury of the Upper Austrian representatives. In return, a promissory note to him on his account was drawn up.

Here, too, the reasons which prevented Kepler from keeping a permanent assistant were, above all, financial. We know of two helpers who worked for him in Linz. Both joined in his family life and gained his trust and praise by their conduct and efficiency. One was Benjamin Ursinus, who had already helped him in the closing period of the stay in Prague and later followed him to Linz, remaining until the

[1] Au. Note. It is due to Rudolf Reicherstorfer, the Linz regional investigator, that the question of Kepler's Linz dwellings was clarified on the strength of thorough research into the archives as far as the extant documents permitted. In doing this he cleared up various erroneous conjectures. He is also responsible for the fact that the commemoration tablet which was put up several decades ago on the house at No. 10 Kepler Street (Keplergasse) was removed and that instead a similar one was put on the house at No. 5 Town Hall Street (Rathausgasse). Kepler never lived in the old Leatherdresser Street (Lederergasse), which in 1869 was renamed after him.

autumn of 1614; he afterwards became professor of mathematics at Frankfort-on-the-Oder. The other, indeed the most efficient of all the helpers, was Janus Gringalletus, a native of Geneva. He came to Kepler at Linz in 1617 through Florian Crusius, the physician from Strasburg. Gringalletus left his honored master in 1620 when the latter, because of his mother's trial, repaired to Württemberg; this assistant died at an early age in his native city.

From this it is not difficult to draw a picture of Kepler's life in Linz. It was, at least in the first years of his stay, after he once had gained a solid footing and had established himself domestically, more peaceful than in Prague, in spite of the hostility caused by his religious principles. Yet, of course, it was also narrower. His correspondence was less far-reaching than before. He felt somewhat apart. He complains incidentally to a correspondent in Antwerp: "We in Linz miss the benefit of a post. The class of messengers is, however, hostile to scholars, because they are paid meager little sums. They bother us ceaselessly with their charges, although we make no profit from our writing, live poorly, and draw a miserable salary. If I am not mistaken they charged me $\frac{1}{3}$ gulden for transmitting your letter." He advised his correspondent to make use of the rectors of the Jesuit colleges in sending letters; the Jesuits in Linz would certainly be happy to render assistance. Apparently, in this city, also, he was on good terms with the members of this society.

A fortunate meeting on July 17, 1612, laid the foundation for a bond of friendship which was to last until Kepler's death and which brought him help and consolation in many situations. That day, traveling past, Matthias Bernegger, the noble, famed Strasburg humanist, who later played a big role in the intellectual life of that city, had sought him out. Bernegger was about to undertake the professorship in history there. He was a native of Hallstatt in Upper Austria and ten years younger than Kepler. In the domain of astronomy he had the credit of having translated Galileo's Italian *Dialogue* about the systems of the world into Latin, thereby assuring him of a wider circulation. A continuous exchange of letters lasting nearly two decades began between these two men of similar aspirations. Their friendship continued although they never met again and there developed a confidential relationship which gave rise to Kepler's communicating his personal desires and cares and absorbing his friend's services. Bernegger was the best and most faithful friend that he ever found. Their exchange of letters therefore forms an especially important source of information about Kepler's life from then on. To be sure, the originals of the letters are no longer extant. In 1672 Bernegger's son published the correspondence. Because, however, it was the custom or rather the bad custom of that time when issuing such publications to make omissions and changes for various considerations, it must unfortunately be presumed that the texts of Kepler's letters were not rendered completely and verbatim.

4. The year of Christ's birth; the Gregorian calendar

Because as yet there was no press where he lived and Strasburg offered good possibilities, Kepler engaged his new friend's services for the very first scholarly work which he wanted to publish in Linz. The work dealt with chronology once more. He had not yet secured the peace and equanimity necessary for his astronomical studies. The trials and tribulations of his last Prague years, as well as the moving about caused by the change of residence to Linz, had, as he himself says, knocked the science of the heavens out of his hands; he had nearly forgotten it even though he had a deep-felt intellectual love for it. Everyone knows how difficult it is to refasten the severed thread in an intellectual work which, because of its difficulties and because of the extent of the material to be conquered, requires exerting all one's ability. Upon his move from Graz to Prague, Kepler was immediately put into a research project at Tycho Brahe's, which engrossed him and monopolized his time. In Linz, on the contrary, he found no one with whom he could discuss the astronomical questions which filled his mind. He had to erect his own world and in this world he was lonesome and alone. Time was needed to erect and develop this world and to materialize his plans. The harvest ripened slowly, but was to become rich and blessed. In a stately series of works he poured forth the knowledge which ripened in him. In Linz he presented mankind with that very valuable fruit of his mind, his *World Harmony*.

The work on chronology which he had printed in Strasburg in 1613 dealt with an old question, the birth year of Christ. In the appendix to his book about the new star he had already been concerned with it and had stated that the starting point of our time reckoning was a few years too late. In the intervening time, Helisaeus Roeslin had written against this thesis in a work dedicated to Emperor Matthias, attempting to prove that the difference only amounted to five quarter years, not four to five years, as Kepler claimed. Since other contemporary writers, such as Joseph Justus Scaliger in Leyden and Seth Calvisius, the cantor of the Thomas church in Leipzig, also differed from him, Kepler felt compelled to defend his thesis in an enlarged and somewhat improved draft. The new work was supposed to serve the reader so that he "could silence the Godless Epicurean scoffers / and scorners of Christianity / of whom to this day there is not an inconsiderable number among Jews and Christians in large towns / at court and in commercial cities / if they reproach [him] with the incorrectness of our numbering of the years; that furthermore this particular example would help him to have fewer doubts in other points / too / to which they fasten their hatred / even though it may be harder than in this case to get at the root with this /." At the basis of our chronology, which was set up in 532 by Dionysius Exiguus, the abbot, lies the assumption that Christ was born in the 754th year of the city of Rome, that is in the year 46 of the Julian era. Opposing this, Kepler, by penetrating research and comparison of the reports of Flavius Josephus, the Jewish writer of history, of the Roman historians, and also of the Evangelicans, furnishes the information that Herod died in the 42nd year of the Julian era, so that the birth of Christ, which took place before Herod's death, should be fixed in the year 41, thus fully five years earlier than Dionysius assumed. Since a lunar eclipse is mentioned in the account of the last part of Herod's lifetime, Kepler does not fail to make use of this astronomical event, too, as an argument for his thesis. So that his book might circulate abroad also, he published it in the following year, with additions, in a Latin translation under the title: *De vero Anno, quo aeternus Dei Filius humanam naturam in Utero benedictae Virginis Mariae assumpsit*. He set the start of Jesus' public activity as occurring in his thirty-third year. Today Kepler's thesis about the birth year of Christ is universally accepted.

In the second half of 1613 a higher order made Kepler interest himself in another question dealing with chronology. The emperor had had him ordered to proceed with the princely household to the Reichstag which had been summoned to Regensburg, because there, too, the introduction of the new calendar was to be deliberated. Kepler went into action as court mathematician. In his application to the representatives for a leave of absence he did not neglect to point out that this journey would serve as "adornment of his profession." Naturally, the representatives had to grant the leave, but they appended the comment that the applicant should return to their service as soon as possible.

For the question under discussion, Kepler was prepared in the best manner. The controversy over universal introduction of calendar reform in Germany, which Pope Gregory XIII, by the bull "Inter gravissimas" of February 24, 1582, had directed should be started, had already lasted for thirty years. Mathematicians and theologians had taken up positions in this controversy and advocated or disapproved the innovation with religious, political and scholarly arguments. To follow this controversy is to gain the impression that the Germans would take pleasure in finding something about which they could bicker and squabble.

The call for calendar reform had been raised for a few centuries. Popes and councils had undertaken the task, however, without solving it. Since it primarily concerned the celebration of church fêtes, because of the commanding position held by the Pope in united western Christendom, it was chiefly ecclesiastical courts which should and could carry through that reformation. After long controversies, the Council of Nicea in the year 325 had provided that Easter be celebrated on the Sunday after the first spring full moon. Carrying out this provision, to which the Christian church has held for more than fifteen hundred years, necessitates two astronomical determinations: first that of the spring equinox, namely, the exact measurement of the length of the tropical year, and secondly the computation of the length of the synodic lunar period and the fixing of the days in the solar year on which the full moon occurs. Now, with both calculations errors had crept in, which in the course of decades became increasingly noticeable, so that one could no longer do justice to the claims on which the celebration of Easter had really been based. In the Julian calendar, introduced by Julius Caesar in 46 B.C. and in force ever since, the length of the tropical year had been fixed at $365\frac{1}{4}$ days; accordingly, every fourth year an intercalary day was inserted. However, that amount is eleven minutes fourteen seconds too great, so that in 128 years one day too many was inserted. The result was that the beginning of spring, which at the introduction of this chronological reckoning fell on March 21, gradually moved up, and in the sixteenth century took place about ten days earlier, that is on March 11. For the determination of the time of full moon, the nineteen-year lunar cycle,[1] already known to the ancients, was used. At the basis of this observation lies the fact that in nineteen years the moon makes 235 circuits of its orbit, so that at the end of such a period the full moon occurs once more on the same day of the month. Yet, since nineteen Julian years are about one and a half hours longer than 235 months, the full moons occur in each following period this much earlier than in the preceding one. In 310 years this amounts to a whole day. Thus the cyclic calculation of the occurrence of new moon, also, leads to a discrepancy from the actual phenomena in the heaven.

How Pope Gregory tried to eliminate these flaws of the old calendar

[1] ED. NOTE. The Metonic cycle.

is known. In order to bring the beginning of spring to March 21 again, it was decreed that ten days should be omitted in October of 1582. October 15 should immediately follow October 4. To prevent a further disarrangement of the beginning of spring, in the future three intercalary days should be omitted in every four hundred years; consequently, in all full centuries not divisible by four hundred without a remainder the intercalary day should be left out. To calculate the full moon more closely, the so-called epacts were introduced in place of the nineteen-year lunar cycle. They specify the age of the moon (reckoned in days from the previous new moon) for January 1 of each year. Suitable tables for the coming century were set up.

This reform plan had been worked out by experts and submitted by Pope Gregory to all Christian princes and Catholic universities for their opinion before he instituted it. Most of the opinions received were affirmative, at least in principle. Then the reform was forthwith introduced in Italy, Spain, France, the Catholic Netherlands, the Catholic cantons of Switzerland, Poland and Hungary. In Germany the Catholic princes and cities likewise obeyed, as Emperor Rudolph II, in the year 1583, after deliberation had directed the acceptance of the calendar. The Protestant princes, however, principally led by their theologians, maintained a negative attitude. In this "papist innovation" there was seen threatening interference in the personal rights and freedom, which had been wrested in the face of the tyranny of the Pope. The Pope therewith reaches for the princely crowns of the princes of the empire. One should have nothing to do with the enemy of Jesus Christ so as not to become a party to his sins and infamous deeds. This is not the place to go into detail about the flood of polemical, libelous and sarcastic pamphlets in which the reform was being fought, nor to adduce the reasons brought forth in detail against it. The often ugly and somewhat monstrous attacks directed against the innovation which, even if not perfect, was still basically very sensible are mortifying; they show how deep and unbridgeable was the division between the faiths in Germany at that time. Among the harshest shouters in the fight were the Württembergians, the theologians Lukas Osiander and Jacob Heerbrand, as well as Maestlin. Helisaeus Roeslin likewise could be heard in this chorus, whereas Tycho Brahe, always objective, recognized the improvement of the calendar and condemned the resistance to it, while the opponents were motivated by blind passion, selfishness and hate for the Pope.

Now, at the Reichstag the troublesome question was to be dealt with again. Naturally the specialist in mathematics at the court of the Catholic Emperor Matthias was committed. But Kepler was not a supporter of calendar reform because he was in the household of the imperial prince; rather he was there because he was in favor of the new calendar. Previously in Prague he had written a stimulating German dialogue on this subject which, however, he had not had printed. In it he collects "at any rate, what the contending parties, clerical and lay, as well as the mathematicians, had to say rationally or what they thought about the matter." His dialogue is "presented and directed by a lover of the truth for the instruction of those on whom the existence of the calendar officially devolves and who yet do not have time to run through all writings in this matter and to distinguish therein the personal emotions from the truth." Moreover, already in 1597 as a young master from Graz he had written a letter giving his teacher Maestlin a long lesson on the subject of the calendar. The occasion was Maestlin's informing Kepler that, because of the pains he had taken with the printing of the *Mysterium Cosmographicum*, he had neglected the completion of his verdict about the new calendar and, as a consequence, been reproved by the senate of the university. Now, as his third document toward putting calendar reform into effect, Kepler stated his opinion: "what the Roman Imperial Majesty would like to accomplish fruitfully for the three electors of the Augsburgian Confession, concerning the matter of the calendar."

As one can believe from his oft outspoken and substantiated love for peace in religious things, in all three writings he indeed proved himself to be "lover of the truth," as he designated himself. Certainly, he also knows better than anyone the calendar's astronomical faults. He knows that the length of the year is not absolutely exact and that from the lunar calculation according to the epact cycle in certain rare marginal cases a different result is obtained for the appearance of the full moon than is found by an exact astronomical calculation. He concedes that one may not and cannot simply place the Protestant princes under a papal order and understands the apprehension of his coreligionists that a concession on the calendar might imply the recognition of more extensive demands by the Pope in ecclesiastical and religious matters. On the other hand he advocates the astronomical use of the new calendar reckoning. Not only is it much better than the old, but it is also sufficiently exact to satisfy all needs for many centuries. Indeed, the slight deviations had been well known to the Pope's advisers. The method chosen for computing the moon's position had, however, been introduced because it recommends itself by "its fine nimbleness" compared with exact astronomical calculation. The material evaluation of the reform entailed difficult matters. "If the sagacious, over-fiery or idle heads in Germany occupied themselves enough with *studiis* so that they would understand this matter and had to ask us mathematicians, who are generally poor negligent people, about it, they would find that the heat and smoke of quarrel vanished, and they would see that they had no cause to get so worked up over this project." The further considerations of his coreligionists are unfounded and their fears baseless if the emperor were to publish a purely political edict and the princes collected the opinions of their mathematicians. Then approval would be not of Gregory's bull but rather of the proposals of the mathematicians. In this sense, he proposed to the emperor that, in his expert opinion, the emperor might of his own accord turn to the electors and request that they come to an agreement with him on a general imperial decree, which would have its origin and authority not from the Pope but rather from the emperor and the electors. Because all parts of the Roman Empire had contact with each other in commerce, in public fairs and court-days, it was imperative that in these common matters there be the same sort of time regulation. Also, since the new calendar has been introduced in most of the lands of the rest of Europe, Germany cannot set itself apart. Should one perhaps wait for a *Deus ex machina* to appear and elucidate all those regulations by evangelical light? And, further, should this wonder have occurred, still these lands would not again repudiate the once accepted innovation. As is shown by the position of the Protestants in Bohemia, Austria, the Netherlands and a few imperial cities, which had accepted the innovation, nothing there had changed in religious matters. By introducing the new calendar the emperor does not want "to give regulation" to anyone in religion, divine service and conscience. Each individual prince and representative should, within his land or domain, arrange divine service as he feels he can answer for it to God "who takes no pleasure in the squabble among neighbors." "Thus it would be a good thing if the others in German countries would follow the large majority and accept the calendar reform in obedience to his majesty as the proper authority and in brotherly love toward all Christianity and finally for the sake of promoting orderly conditions within the German Empire."

It is not known if and to what extent Kepler with his logical and moderate thoughts spoke at the Reichstag. At any rate, he was not successful. As everybody knows, the Protestant princes in Germany first accepted the new calendar in the year 1700, after the schism in time reckoning had lasted for more than a century.

There is an episode from Kepler's visit in Regensburg which should be reported. In the lovely cathedral he showed a group of acquaintances the round sun pictures cast by beams falling through the cracks of windows, and pointed out the sunspots. These images are made in accordance with the principle of the pictures of small apertures. Sunspots had first been observed and investigated about two years before by Johannes Fabricius, the son of David Fabricius, then by Christoph Scheiner, Galileo, Simon Marius and Thomas Harriot. A vehement controversy over the priority of the discovery flared up between Scheiner and Galileo. Kepler was much interested in the sensational phenomenon and made observations himself, but published nothing on the subject because Galileo had dealt with it so well. Kepler noted with special satisfaction the rotation of the sun, which was proved by the spots. He had postulated it a few years before in his *Astronomia Nova* in connection with his attempt at a physical explanation of the motions of the planets. In contrast to others, who wanted to explain the spots as bodies circling the sun, Kepler agreed with Galileo that these spots are matter somewhat similar to the clouds in the outer layer of our terrestrial sphere. However, he left undecided the question whether those heavy sooty smoke clouds come out of the mighty firebrand of the sun's body; the analogy may not be extended that far with certainty.

5. Stereometria Doliorum *and other scholarly work*

Kepler remained in Regensburg from July to October with three interruptions. He returned to Linz October 21, a few days before his wedding. Shortly thereafter he resumed work, this time on a new mathematical question which led him to compose one of the most significant works in the history of mathematics.

This year the wine harvest, he relates, was especially good. Many barges went up the Danube and in Linz the entire bank of that river was filled with wine casks. This being so, as husband and good paterfamilias, he considered it his duty to provide his home with the necessary drink. Accordingly, he had some casks made and installed in his house. Four days later the dealer came with a measuring rod, which was the only instrument he used to gauge all casks without distinction, without taking into consideration their shape or making any calculations whatsoever. The dealer simply stuck his rod slanting through the opening, used for filling, up to the lower edge of the bottom of the cask and read off the contents on a marker of his rod. How is that possible? Can such a measurement of the contents be justified mathematically, Kepler asked himself. He recalled the troublesome manner and custom by which it was usual to measure the contents of casks on the Rhine. There he had seen how this problem was solved, either by pouring bucket after bucket into the cask and counting them or else undertaking all possible measurement and from these making difficult and unsatisfactory calculations.

The answer to this question resulted in a stately folio volume, the *Stereometria Doliorum Vinariorum*. Once captured by his task, Kepler reached out further; the main question led him to various related questions. Right at the start he went back to Archimedes, the master who in antiquity had successfully concerned himself with the calculation of curved surfaces and of bodies bounded by such surfaces. Whereas Archimedes dealt only with the geometrical figures formed by the rotation of a conic section about its axis, Kepler widened the sphere of his researches to include bodies produced by the rotation of a conic section about any straight line lying in its plane. In his pronounced logical need for organization and division of a group of thoughts (characteristic of his mathematical manner of thinking) he arrived at ninety-two types of figures. The bodies, whose volumes he tried to calculate, he frequently designated by names of fruits and thus spoke of an apple-, quince-, plum-, olive-, or lemon-globe. He demonstrated his main question as follows. He imagined a cask made up of two equal frustums of a cone. These, especially in the case of the Austrian cask which "among all others has the nicest shape," approach the form of a cylinder. He showed that the casks with the greatest relative contents are those in which the diameter of the bottom bears the same ratio to the length of the staves as do the sides of a square to the diagonal. For such casks, he further proves, measurement with a simple gauging rod as described above is valid. Since the Austrian casks have that ratio, his question is answered in the affirmative.

Kepler had here proposed a new type of problem and he had to find his own path to the solution. Kepler knew that his proofs and calculations lacked the exactness demanded by strict mathematicians, exactness to which Archimedes had already applied himself, although his problems were simpler. Kepler consciously renounced this rigor and wanted to take over from Archimedes only so much as "is sufficient for the pleasure of the lovers of geometry." Exact proofs can be gleaned from that man by oneself, "if one does not shrink from the thorny reading of his writings." Indeed, Kepler's problems which transcended Archimedes could not be rigorously mastered with the set of tools placed at his disposal by the mathematics of his time. It is therefore evident that, guided by conclusions by analogy and trusting his mathematical feeling (which like so many other feelings is sometimes deceptive), he reached faulty results a few times. He was variously credited with deficiency in method, as for example by his contemporary, the well-known Paul Guldin of the Company of Jesus, with whom he later corresponded. Guldin's criticism is, however, inconclusive since he also says that although Kepler possessed very distinguished talent he had not been able to brood over an individual discovery long and persistently like a hen over her eggs because he always devoted himself to many matters. Certainly, this man did not know the *Astronomia Nova*, in which its author had shown such unique perseverance for the attainment of his high goal that anyone who wants to keep step with him on his way almost loses his breath. What in the *Stereometria* also amazes everyone who likes mathematics is the boldness in putting the question and the inventiveness in the attempts at solution of the problems set. Just as Kepler had prescribed new ways and goals for mathematics by the integration problems for the models of motion which he attempted in the *Astronomia Nova*, so also with the infinitesimal considerations in his *Stereometria*, he put himself, as will be generally recognized, as creative mathematician in the forefront of the men who paved the way for the revolutionary method of the integral calculus, which was being ushered in. Thereby he had achieved what he himself had set as a goal for his book: he wanted to push open the gate to a vast geometrical domain, incite the zeal of contemporary mathematicians and show what remained in that domain to work over, what still to search out.

When Kepler began his activity in Linz, there was still no printer in this city. Consequently, in December, 1613, he sent his work, as much as was then finished, to Augsburg to Markus Welser, the well-known patrician, requesting him to take care of the printing there. No printer in Augsburg, however, was willing to print, at his own risk, a book written in Latin. As a consequence the printing was delayed, but this proved very advantageous for the work. For only now, captivated by the problem he expanded it to the size in which it now appears. To begin with he had only thought of a little work of few pages. He himself was therefore pleased with the delay. Since Markus Welser soon died, Kepler had to see how to recover his original manuscript. To make the publication of his planned works convenient and feasible Kepler, in 1615, brought Johannes Plank, the printer, from Erfurt to Linz. The same year Plank printed the *Stereometria*, which thus represents the first printing in the Upper Austrian capital. The author defrayed the printing costs out of his own resources. In order to reimburse himself, he had to see about increasing the sale. That he was disappointed is apparent from a letter from his friend Peter Crüger in Danzig, whom he had asked to offer the work to the book dealers of his land. Crüger was obliged to report that except for the two mathematicians in Danzig and Königsberg, the Königsberg library and a certain noble lord, in all Prussia there was no one who would buy such a book.

In order to make the issues of his *Stereometria*, which are important for theory, accessible to wider circles, Kepler in the same year edited a German edition of it under the title: *Ausszug auss der Vralten Messekunst Archimedis* (Excerpts from the ancient art of mensuration by Archimedes). The learned accessories of the Latin work are omitted here; the whole material is arranged in a different sequence; and an appendix, giving a large number of individual problems about mass and weight, as used in antiquity and in Kepler's time, is added. In a playful dedication to the mayors, judges and counsellors of the Upper Austrian cities, the author presents geometry as the "ancient little mother of all authorities, communities, good positions, sensible tradesmen, free artists and artisans," in whose service he stands and says he has taken a great liking for Upper Austria. He points toward the help that this little mother renders to the various craftsmen and professions and offers his work as a contribution ("Beutpfenning") which geometry has raked up from her treasure; she had found it once when she had "stirred around" with a gauging rod in a wine cask in which she had taken shelter. He himself, with much trouble and expense, had "polished well" this penny "in a German style."

Besides its importance for geometry, Kepler's *Art of Mensuration* also occupies a special place in the history of mathematical technical language, since Kepler deliberately and successfully troubled to translate the numerous technical expressions into German words. At the close of the book he listed the German designations used, and explained them by the corresponding Latin words. The book indicates the pleasure which the translation into German afforded him. He gladly seasoned the reading with humorous remarks. Even if most of the German technical terms were not carried over into mathematics, nevertheless a few found lasting acceptance. For example the German designation "Kegelschnitt" (conic section) goes back to him. He who has a taste for changes in language and can find enjoyment in the ideas of a head gifted with a strong faculty for form possesses in Kepler's book a rich source for his inquiries.

A damper was put on the pleasure which Kepler himself experienced over his *Stereometria*. He presented the work to the representatives in the "humble hope, the honorable representatives would graciously take pleasure in this my work, which I produced for the best of the land with great pains and at my own expense of up to 250 fl., and thus would consider well spent the three-quarters of a year which I applied; especially since I undertook this work with the approval of some highborn and noble gentlemen who have knowledge of these matters and since I have done my work in conformity with their opinion." The representatives, however, showed little understanding for this accomplishment. They replied to their mathematician that he should suspend work of that type and complete the more important matters for which he had been employed, the *Rudolphine Tables* and the map. For all that, by granting him 150 gulden they gave him a tangible consolation.

Up to the time of that decree (it was in the spring of 1616), Kepler had not yet progressed very far with the production of the map. The work did not suit his taste and his special abilities which, indeed, lay more in the theoretical domain. He had, to be sure, collected material and undertaken tours and journeys into the country so that, by means of the conventional primitive procedure, he might establish the positions of individual localities, mountains, boundaries and water courses more exactly than they were shown in Augustin Hirschvogel's 1543 map of Upper Austria. He is able to reproach the representatives animatedly with the fact that for him such journeys were no pleasure and connected with unpleasant incidents. "Everywhere, in precincts and villages where I carried on inquiries, in the fields and on the mountains where I followed my aims, when I traced the waters and came upon strange paths, I often had to suffer scolding and threats from inexperienced, coarse, and suspicious peasants." But he was not yet clear about what the representatives expected of a new map. Should the new map only be made "more in proportion" and the names improved? That he could do at home. But if, instead, a more exact drawing of the boundaries, a parceling out of the land by parishes and provincial courts, and a survey of the passes for the defense of the country was wanted, then that would entail high costs and would require much time. By cleverly bringing these difficulties to the fore and also pointing to the impossibility of working on the astronomical tables and the map at the same time, since he obviously discovers that he "confuses, entangles" himself "and wastes noble time" by simultaneous occupation with both tasks, he attained, what he obviously strove for, deliverance from the commission which was burdensome to him. The completion of the map was thereupon transferred to Abraham Holzwurm, the representatives' engineer. It is apparent that the severance occurred without discord since two years later the representatives submitted the draft worked out by Holzwurm to Kepler for his opinion.

It was an entirely different matter with the additional task assigned to him at his installation, the completion of the tables. Here Kepler felt in his element. To be sure, for this work, too, he still could present nothing in any way complete and had to urge the representatives to be patient. The lords had to know "that *in re literaria* the *Tabulae astronomicae* would have to be a carefully prepared major effort and could not be performed overnight like a comedy, or like a poem consist of mere flashes of inspiration, or like a commentary on Aristotle be shaken out of a sleeve; rather, many years would have to be employed in deliberating and being occupied with *observationibus* and *calculationibus*, if it is desired to make the reckoning so comprehensive that the tables would be valid for many hundred, yes thousand, years ahead and past." The work's progress was hampered by Kepler's being unable to keep assistants for the many calculations due to his lack of means and because of the negligence of the imperial treasury. So "everything rests on my shoulders alone, not only the speculation and invention, but also the deduction and calculation of the observations, furthermore not only the conceiving (concipirung) of the text but also the most tedious and lengthy calculation of the tables, yes even the copying, also the tracing of the figures on the wood, and finally the manifold corrections in the printing, together with the last correcting and changing of the text, which would otherwise be very pleasant for me."

Naturally Kepler could explain to the representatives only the external impediments to the completion of the tables. The inner difficulties, which he made for himself by the innovations in his calculations of the planets and by the great demands which, as distinguished master, he set for his work, he could not present to the eyes of laymen, if he could not even make such important astronomers as Maestlin and Fabricius understand his new thoughts and plans. Previously when describing Kepler's scientific activity in Prague, these special difficulties, as well as his keen and penetrating efforts to conquer them, were pointed out. He had brought along from Prague a wealth of material from which he could start. His work with mathematical ideas in his *Stereometria* had been a refreshing and strengthening mineral bath (which anyone who has ever followed such a treatment can appreciate). His old eagerness to work had come over him, and from 1614 on, with new strength, he had returned to his beloved astronomical studies, which were directed to the tables as the final goal.

There were two plans which he now wanted to realize; both had already engrossed his attention in Prague. The one concerned the ephemeris. The necessary information has already been given about the importance of this as preliminary work for the tables. Again, Kepler was occupied first in calculating the elements of the orbits of

Venus and Mercury, a difficult task considering the extent of the Tychonic observational material. For whole months in 1614–1615 he sat over these calculations. The moon, with the numerous inequalities in its motion, gave even greater trouble. He worried over attempts to explain these inequalities physically, especially the variation, and to follow through the calculation of its physical concept accordingly. Hitherto it has hardly been noticed that here, too, he ran up against innovations in the framing of mathematical questions which deserve the interest of the mathematician and which were difficult to overcome with his tools. The year 1616 was primarily devoted to this painstaking work. Attacking the calculation of the ephemerides, the zealous scholar felt as though he had emerged from an abysmal sea. Waiting from March to May, 1617, at the court in Prague on the emperor's order, he utilized his free time to complete the ephemeris for 1617; that for 1618 he calculated in the following months at home in Linz. Johannes Plank began the printing immediately. In 1616, to meet the printing costs, Kepler for the first time in a long while wrote a calendar for the following year; that would at least be somewhat more honorable than begging, and in this way the emperor's honor would be preserved. The emperor left him completely in the lurch so that in spite of all orders to the imperial treasury he might have starved. The calendar is noteworthy as the first publication of planet positions calculated according to the new laws, so that Kepler rightly could designate it as the "first fruits of the *Rudolphine Tables*." Unfortunately, at present no copy of the calendar has been located.

The second plan concerned a textbook, an outline of the Copernican astronomy, the *Epitome Astronomia Copernicanae*. Soon after the appearance of the *Astronomia Nova* in 1609 it had occurred to Kepler to present a comprehensive astronomical world picture, as he himself had built it, based on the Copernican theory with his own new discoveries. He wanted to make the shape he had given astronomy suitable for "school benches of the lower classes." "Since this science can be successfully learned only if each one, who at maturity wants to pick its fruit, has cultivated the seed as a boy, so I wanted to come to the aid of all by an easily understood presentation of low price and an appropriately large edition." Now the time had come to realize this plan, too, and by it to publish the "fundaments of the Rudolphine tables." Kepler had finished the composition of the first part of the work in the spring of 1615. Under his supervision this book also was to be printed by Johannes Plank. But there were difficulties since Kepler had previously signed an agreement with Hans Crüger, the Augsburg publisher, who would not agree to Plank's taking over the printing.

After some vexations Kepler, nevertheless, carried out his plan, and in the spring of 1617 the first part was in the press.

6. *Kepler's mother's witch trial*

So Kepler again had arrived in his element of happy creation. His plans ripened and he was pleased at the sprouting seed. Yet, as though Fate did not want to leave him any peace, when he was in full swing threatening clouds collected over him, throwing heavy shadows on his life's course in the next years. From his home arrived news of malicious machinations against his mother, which finally led to accusations of witchcraft and to a criminal legal process. Kepler received this bad news "with unutterable distress," as he writes, "nearly causing my heart to burst in my body." The events not only made it necessary for the dismayed son to intervene by letter but also compelled him, twice in the six years that the unpleasant affair lasted, to journey to Württemberg in order to assist his mother and by using his prestige as imperial mathematician to save her from certain death at the stake.

The time in which Kepler lived is better understood by recalling that witch hunting in Germany reached its peak precisely then. The belief in demons, which is as old as the human race and without which a good many horrible black pages in the life of men and in the history of humanity cannot be understood, had increased to a mass psychosis, a mania, which led to the most fearful pursuit of blameless persons. Behind the various unfortunate occurrences, even if harmless, which can overtake man and beast, one suspects the influence of evil spirits, of people who have an alliance with the devil and supposedly command secret powers for the harm of others. Clerical and secular princes, jurists and theologians, men of intellect and education labored no less than the common folk under the delusion, indiscriminately, as much in the Catholic as in the Protestant parts of Germany. Even Kepler never expressed himself against the belief in witches as such. In his enlightened manner of thinking he most likely recognized the folly and the untenability of the indictment which it was customary to bring and his feeling of righteousness must have passed sentence on many of the harsh and inhuman rulings of the approved judges. Yet the belief in demoniac influences and effects was part of his thinking. When he appeared at his mother's trial there was nothing for him to do but to attest that she was no witch, that is, that the allegations against her either were invented or were explicable in an entirely natural manner, but not that after all there are no witches. Moreover, he would have fared badly with this last thesis.

Inside of a few months during the year 1615–1616 six women were sentenced as witches in Leonberg, where Kepler's mother resided. That was just the time when old Katharina Kepler became involved with the tribunal. In nearby Weil der Stadt in the years 1615–1629 no fewer than thirty-eight witches were surrendered to a horrible death. Reflecting on the exciting and suggestive effect of such events in a small community, one can picture from the beginning the danger hovering over an old woman on whom evil-willed people managed to stamp the mark of a witch. Indeed, for that only a few grounds of suspicion were necessary. These could always easily be hunted up and made believable and for the most part they soon sufficed to provoke the judge to agonizing questioning under torture. Since the very voluminous acts of the trial of Katharina Kepler together with the many petitions, with which her great son intervened in the procedure, have been preserved, the course of the tragedy can be followed from its first beginnings on through all the phases. It is possible to trace how the hate and jealousy, malevolence, thirst for revenge and stupidity on the part of the alleged bewitched and injured, Frau Kepler's clumsiness, the unfairness and partiality of the public prosecutor, the weaknesses, readiness to be influenced and superstitious prejudice of the highest court in the state worked together to bring on the stage a troubled drama, in which only the entrance of the imperial mathematician prevented the worst end. It is desirable to present the course of events in sequence so far as is feasible.[1]

The wicked spirit who fanned the fire was Ursula Reinbold, the wife of Jakob Reinbold, the glazier, a female who was unbalanced and had a bad reputation, since she had been punished for public prostitution. Kepler's mother had often associated with her in the past. But through a dispute, an irreconcilable enmity had arisen between the two women. Christoph Kepler, who was approximately fifteen years younger than his brother, the astronomer, was a tinsmith in Leonberg and held an esteemed position in the city. On the occasion of a business negotiation he had in anger reproached the Reinbold woman for her misdeeds. His mother took part in the abuse and also reminded the other woman of her bad conduct. Thereafter the glazier's wife nursed a deep hatred for the Kepler family, expressing it at every opportunity. Since now, through the witch hunt in the neighborhood, feelings

[1] AU. NOTE. The acts of the trial, first extracted from the state archives in Stuttgart by J. L. C. v. Breitschwert, were published in their entirety in 1870 by Ch. Frisch in the eighth volume of his complete edition of Kepler's works.

were aroused and incited to scent out traces of evil spirits everywhere, it occurred to the wicked woman that once, after a drink at the home of Frau Kepler, she had become ill. She spread the rumor that only poisoning could have been the cause. The Kepler woman is a witch who had caused her serious illness by that magic drink. In reality her suffering arose from an abortion by which she had sought to eliminate the consequences of her promiscuity. What wonder if now others also were at once able to produce everything possible to strengthen and corroborate the suspicion of witchcraft. The story went that Kätherchen, as the Kepler woman was called, had grown up in the home of a kinswoman who had met her death as a witch at the stake. It was recalled that, many years before, the Kepler woman had asked the grave digger in her birthplace, Eltingen, on the occasion of a visit to her late father's grave, for his skull in order to mount it in silver and have a drinking vessel made out of it for her son Johannes; only the grave digger's explanation that he could not comply without the permission of authorities dissuaded her from her plan. It had supposedly been heard that the Kepler woman's son, Henry, who had meanwhile died, had said his mother's conduct was scandalous; she had ridden a calf to death and had wanted to prepare a roast from it for him; he, however, had answered, the devil should eat the roast. There was Beutelspacher, the lame schoolmaster, who could relate a moving story that, as a former schoolmate of Johannes, he had to read the son's letters out loud to the mother and on one such occasion (already ten years past!) had been invited to take a drink, which had brought about pains and finally his lameness. Bastian Meyer's wife, who had also drunk of it at the same time, had even fallen into a lingering illness and died. The wife of Christoph Frick, the butcher, was able to report that her husband once had suddenly felt pains in his thigh when the Kepler woman had passed by, without touching him; when, some time afterward, he had seen her before him in church, he had said to himself: Katharina, help me for God's sake. Thereupon she had looked around at him and the pains were as though blown away. Daniel Schmid, the tailor, blamed the Kepler woman for the death of his two children. She had kept coming to his house without cause, leaned over the children's cradle and said a blessing over them. The children, however, became ill. Indeed, the Kepler woman had taught his wife a speech, which she had to speak at full moon under the open sky in the church yard. But in spite of this saying the children died. Others complained that Kätherchen had bewitched their livestock, so that it had become mad and had kicked, or else had died. Indeed, there were people who maintained Frau Kepler had given them their biggest fright by coming to them through locked doors. Ursula Reinbold found an excellent supporter in the wife of Jörg Haller, a day laborer. Because of her poverty, this woman, who was called *Schinderburga*, was dependent on Ursula Reinbold but was also in debt to the Kepler woman. In her own way she undertook to furnish proof that the Reinbold woman's illness was due to witchcraft. With ceremonious words she measured the sick woman's head and naturally found a support for that suspicion in the result of the measurement.

That is a selection from the rumors which buzzed in the air and were embroidered and circulated. Everyone who heard anything about it shook his head: it did not look natural. It was repeated to the neighbor, male or female, who in return had heard something else, still worse. It is easy to picture this. An invisible net, as it were, was spun about the old woman, in which sooner or later she had to be caught. On hearing the silly charges, an image is formed of the nearly seventy-year-old little woman, small, lean, garrulous, feeble-minded, hot-tempered, inquisitive. Ever restless, she cannot bear to stay at home but runs into other homes, where she has no business, meddles in matters which do not concern her and forces her advice where it is not requested. She had concerned herself a great deal with the preparation of medicines and wants to introduce her recipes everywhere. In her little tin jug she always has a drink ready, and she gladly offers it, heedless whether it had spoiled from long standing in its container. As she heard about the accusations she was seized with anger and rage. The more she defended herself, however, so much further did the slander spread. Naturally, most of the babble was sheer nonsense, some in turn was twisted or fabricated. Yes, she had desired her father's skull from the grave digger, but not for some witchery, rather because she had heard in a sermon that among some ancient people the custom prevailed of using the skulls of dead relatives as drinking cups. She got into a fight with Henry, her ill-advised son, when he made demands in food which she could not satisfy; when his mother refused his demand for a roast, the hot-headed fellow in rage impaled a calf in its stall. The laming of Beutelspacher was not the result of a drink, but of a fall when, burdened with a heavy basket, he tried to jump over a grave, and thereby injured his spine. Yet stupidity is blind and does not want to see the truth, and malevolence is deaf and does not want to be taught about that which is right.

Complications arose when in August, 1615, the Reinbold relatives were carried away and insulted the Kepler woman. Ursula Reinbold had a fine brother, Urban Kräutlin, who practiced surgery and was also court barber for Duke Johann Friedrich's brothers, who were living in Tübingen. When Kräutlin came to Leonberg at that time, in the retinue of Prince Achilles, who intended to hunt, the Kepler woman was criticized at a meal in which the lower magistrate, Luther Einhorn, participated. In their cups the magistrate and the barber, who were boon companions, had the old woman come to the courthouse where, in the presence of the Reinbolds, they pressed her very hard. She was charged with having occasioned the Reinbold woman's suffering by witchcraft and was asked to restore her health by counter-sorcery. In the devilish procedure, what they were demanding was considered particularly witch intrigue, and the judges regarded it as an *indicium ad torturam* (sign for torture), when a person undertook to do this (*maleficium contrario maleficio solvere*—to remove evil by counter-evil). Frau Kepler stood up for herself and refused the unreasonable demand just as she also took no blame for the woman's suffering. The barber now carried his threats and curses so far that he touched the aged woman's breast with drawn sabre and swore she would fall by his hand unless she cured his sister. Ultimately, the magistrate, after watching for a long time, put an end to the ugly scene.

This incident was not to be left unatoned. Because of the insult, Frau Kepler, supported by her family, preferred a complaint with the city court in Leonberg. In addition to her son Christoph, she was helped by her daughter Margarete, and the latter's husband, Georg Binder, the pastor in Heumaden, in the district of Stuttgart, about five hours away from Leonberg. Naturally, the children were also defending their own interest when they stood at the side of their slandered and threatened mother. It is evident that the complaint must have been especially agonizing for the magistrate even if it did not point to him. He counted as qualified witness of the shocking preceeding against Frau Kepler which he had allowed to happen in the court building. Accordingly, it is understandable that this Luther Einhorn moved heaven and earth to quash this civil trial and promoted the intrigues of those who wanted to ruin the Kepler woman. Help for this came from his pretty friend Kräutlin, who had the princes' attention and was in their good graces, which was also the reason why the magistrate was nice to the barber. However, a secret channel led via the princes to Stuttgart to the duke himself and to his advisers. The course of the trial clearly shows that these are no empty conjectures.

Not until now did the Kepler family at home consider it necessary to inform the son and brother in Linz of the intrigues which had been going on for a long time. Until then he had been ignorant of all that transpired at home. On December 29, 1615, he received the letter with the bad news; it was an evil mail at the close of the year. Immediately on the following January 2, he sent off a sharp dispatch to the counsellor at Leonberg. Aroused to indignation, he protested against the wrong inflicted on his mother, and pilloried the practices of the persecutors, who had let themselves be blinded and guided by the miserable devil who is a God of all stupidity, all superstition and all darkness. Hot wrath rose against the "devilish people," who at his mother's arrest in the courthouse wanted to set a trap for the old woman and produce grounds for bringing her to the rack. To his surprise, he had to hear from his family that he, also, "had been accused of forbidden arts." Filled with indignation, he rejected this gossip, making reference to his many years of imperial service. He would only consider it as an intrigue by his insolent opponents, who "wanted to deprive" his mother "entirely of her courage." Finally he demanded of the court categorical transcripts to all acts up to then and "exhorted" the counsellor to pay proper attention to his interests. He made it plain that he was intending to fight for his claims regarding his deserving mother as was right and proper for him to do, by staking life and property and such favor and reputation as he had gained by services as well as by calling on his relatives, friends, and patrons; and he planned to pursue the matter until it should be closed in accordance with the written laws.

This document contained an especially severe reproach for the magistrate, even though his name was not mentioned in it. But a man of his sort was not intimidated; rather the sharp tone incited him all the more to carry his crafty plans into effect. He now knew, to be sure, that he was dealing with an opponent who saw through him and would cause him more trouble than all others. But he also knew that the other was far away, and that he himself held trump cards.

In any case, Kepler had accomplished one thing at the beginning: no one ventured any more to attack his name slanderously. The source of the wicked rumor so far as he himself was concerned with witchcraft is, as he later mentioned in his posthumous "dream," to be sought in Tübingen, so that the path to Leonberg is easily perceived after the above. That is, in Prague Kepler had already formed the plan for a description of the moon, and in his delight in telling a fable he gave it a fanciful expression. In it he saw himself as a youth, who grew up in Thule under the guidance of his mother who was in touch with spirits. We will have more to say about it in its place. A Baron Volckerstorff brought a copy of the first sketch of the exciting book to Tübingen where, as Kepler supposes, it was also spoken of in the barber shops and that rumor was started out of stupidity and superstition.

It is a mistake to believe now that the civil complaint preferred in

August, 1615, had been immediately decided. Even then judges took time, and in the case in question, besides, the official representing the ducal power in the city was interested in not having to give testimony under oath and admit his own breach of duty. The matter was postponed. In the meantime could not the grounds of suspicion against the Kepler woman become solid? Could not she herself, in her wrath, make thoughtless utterances if she was incited and provoked? So it happened, that the first evidence-taking in the trial was only fixed for October 21, 1616. But this date, too, was not met. The accused party managed to bring about an incident which was to create a new situation.

Eight days before this appointed time Frau Kepler, walking on a path through a field, met a bevy of young maidens carrying raw bricks to the kiln. In passing she grazed their clothing. The maidens made way for her, so that Frau Kepler naturally knew that she was decried as a witch. Afterwards the old woman turned around once more, looked back, and then proceeded on her way. One of the maidens was the twelve-year-old daughter of the aforementioned *Schinderburga*. This girl immediately asserted that on turning around, the Kepler woman had hit her on the arm. She pretended to feel pains, which increased hourly, until she finally could move neither hand nor finger. Now that was what was needed. Out of that a rope could be twisted for the hated woman. The magistrate was quickly approached. Even the court barber came from Tübingen. The maidens and the brick maker's family were summoned. Of course, only one particular girl confirmed the blow on the arm. None of the others knew anything about such a thing. The incident, suitably elaborated, was introduced among the people. When the Burga woman met the Kepler woman, she went after her with a knife and shouted she should make her daughter well again. Frau Kepler also ran to the magistrate. With the authoritative air of a witch judge, the latter scrutinized the girl's arm. He found: "It is a witch grip; is even the right impression." He rejected all of Frau Kepler's objections. No further notice was paid the fact that in a few days the young girl, whose pains obviously came from carrying the bricks, was completely recovered. The magistrate had the date for hearing testimony cancelled and explained that he had to inform the ducal chancery about the difficult and exceedingly serious incident. Then in her confusion and senility Frau Kepler did the stupidest thing she could do: she offered the magistrate a silver cup as a gift if he would omit the report to the chancery and keep the appointed date for the hearing. Naturally, he rejected the gift and triumphed at having in his hand the bad conscience of the persecuted. The report went off in suitable form to the chancery. The old woman immediately went to her daughter at Heumaden. Christoph, Margarete and her husband were alarmed at the turn of affairs. They conferred with one another and finally were able to persuade their mother to journey to Linz to her son Johannes. As matters stood, this was best. On December 13, 1616, accompanied by her son Christoph, the poor woman arrived in Linz where her great son had long been expecting her.

By such maneuvers, as Kepler writes in a later report, the magistrate avoided the necessity of having to swear an oath and give evidence about the state of affairs three days later. By such maneuvers this woman, feeble-minded because of her age, was pushed against the wall. And the ducal advisers? It might be believed that these gentlemen had seen through the intrigues. As follows from the further progress of the trial, members of the high counsel[1] were not in complete agreement; at any rate, the decrees promulgated give evidence of a certain indecision. In the present stage, however, the report was followed in turn by an unequivocal decision, which conformed entirely with the wishes of the magistrate. The latter is instructed to arrest the Kepler woman on sight and, after keeping her for one or two days, to examine her earnestly with reference to the case of Jörg Haller's young daughter and the theological articles in the presence of the special tribunal, and also to ask why she had moved away, to report on her confession, and to await further orders. Examining on the theological articles is the same as starting the witch trial. The situation was obviously very grave. For the present, however, since the woman disappeared from the duke's domain, the order was not carried out.

Although the poor accused's journey to her son was fitting, yet by making it she nevertheless gave her opponents further support for their dark plans. It was asserted that she had fled because she had a bad conscience. There were further reports by the magistrate. Kepler himself appealed to the duke and gave excellent reasons for his mother's journey. He wrote to Sebastian Faber, the vice-chancellor, who was kindly disposed to him, and disclosed all the baseness of the machinations against his mother. His request was that the pending civil suit be carried out as soon as possible, regardless of his mother's absence. With the same end in view, Christoph and the Binders appealed to the duke. The decrees accumulate. The intrigues expand. The date of the "production day" for taking testimony was arranged by the high adviser in March, 1617. With the magistrate's assistance,

---

[1] The competent board which together with the church counsel already known to us and the revenue office administered the land. (Transferred from text.)

the defendants were able to delay it a couple of times by various tricks and pretexts. The question arises, how was it possible for the magistrate to act contrary to the express orders of the chief adviser? He must have had secret support at the court to be able to achieve such a misdeed.

Whereas Johannes Kepler joined without reservation in all his demonstrations in favor of his mother, although he well knew her present faults, Christoph and Pastor Binder showed themselves to be weaker. They made written and oral statements in which it could be observed that they took into consideration their mother's blame, even though as a distant possibility. How could they, without harming her, write to the duke: if matters were found to accord with the magistrate's report, they would willingly desert their mother and let her pay the penalty! Both men harbored fears for their own position.

Not the least important part in the whole disgraceful transaction was played by money considerations. The allegedly injured parties (the general rabble and the scum of the people, as Kepler rightly calls them in a petition), especially Jörg Haller, demanded recompense. Frau Kepler's property, valued at about 1,000 gulden by the magistrate, was to be blocked or at least itemized. Christoph and Margarete agreed that none of it would leave the city. Johannes demanded the assignment out of it of the cost of maintaining his mother, who was staying with him. The longer the transaction took, the more did the costs of the trial accumulate. The prospect of a bountiful compensation in case of a verdict against Frau Kepler incited her persecutor to intensified efforts. There are few expressions of human baseness not manifested in this affair.

In October, 1617, after an absence of more than nine months, Frau Kepler returned home to her daughter Margarete, with whom she now took up permanent residence. Shortly thereafter Johannes followed her. He believed that by being present in person he could more quickly and more surely direct his mother's case to a satisfactory conclusion. His journey took him through Walderbach near Regensberg, where his stepdaughter Regina had just died; the widower had begged Kepler to leave his fifteen-year-old daughter, Susanna, to take care of the motherless children for a while. Very reluctantly had Kepler torn himself from his studies. As reading for the journey he had taken with him the *Dialogo della musica antica e moderna* (Dialogue concerning ancient and modern music) by Vincenzo Galilei, the physicist's father. The world of harmony lifted him above the depressing thoughts which forced themselves upon him from outside. He arrived in Leonberg on October 30, at the time of the Lutheran jubilee; the preceding night Vaihingen, a neighboring city, had been destroyed by fire.

The result of his pains fell short of his expectations. It is true, in November, at his request, a new term for the hearing was set in categorical form by the chief council. Yet this time, too, the order was not carried out quickly. As the sole positive result of the journey, Kepler obtained permission to take his mother back to Linz with him again. However, she did not want to go. She could not tear herself away from the surroundings familiar to her since childhood. So in December Kepler returned alone to Linz. After a short stay in Regensburg and Walderbach, he arrived home on December 22, 1617.

At last, nearly three years after the presentation of the complaint, in May of the following year—this memorable May of 1618—the magistrate placed the record of the first examination of witnesses before the chancery. The date of the term for the hearing is not known, since, curiously, that record is missing from the trial acts. Consequently, also, the exact result of the examination is not known. In a later petition to the duke, Kepler himself reports that his mother had succeeded, particularly on information given by four old honorable council members, in proving the immaculateness of her reputation, and in demonstrating that the accused Reinbold woman, because of her conduct, was in no way comparable to her and did not have the slightest basis for her slander.

Sufficient time had elapsed, Kepler fancied according to the petition, for the judge to have pronounced an impartial sentence, which would not have pleased the opposition. But such a pronouncement did not follow. In order to prevent it, the Reinbold woman and her gang made every possible effort to get the witch trial started. As she expressed it, she wanted to play the game into the hands of the ducal chancery, which was responsible for the initiation of the criminal procedure. She wanted to change from defendant in a civil suit to complainant in a criminal suit. The libel complaint should vanish in a devilish procedure, by which she and Frau Kepler exchanged the parts they had played up to then. The magistrate helped. He provided the side which hitherto had been the defendant with the opportunity of legally presenting its accusations. So a bill of indictment against the Kepler woman came into being, in which, in forty-nine articles, was collected all that was mentioned above as circulating rumor. It was a disgraceful piece of work. What was put forward there in these "most horrible and most shameful" articles, says Kepler, is nothing but "sheer old female babble," ferreted out and collected by people like the Schinderburga, as every sensible person can conceive without hearing testimony. One especially aggravating charge was to the effect that the Kepler woman had wanted to seduce to witchcraft a young girl, Barbara, the daughter of Bastian, the hunter. Now the acts swelled to thick volumes. Thirty to forty witnesses were heard in detail, all statements exactly recorded. Meanwhile, the poor woman, whose life was at stake, lived at her daughter's in Heumaden. Her son-in-law, however, had not allowed her to come in his sight. The proceeding dragged on for nearly two years. There is a gap in the acts from the fall of 1618 to the fall of 1619. It is questionable whether, as some believe, a pause occurred because of the political events of that time. It is more likely that the cause of the delay is to be sought in the process itself. The record finally submitted by Melchior Nördlinger von Merklingen, the court clerk, in charge of the taking of testimony, is no less than 280 pages long. And this document was just the beginning.

Finally in the spring of 1620 as the net about the accused woman drew tighter and barely a doubt remained that she would be criminally examined, that is tortured, Kepler, greatly concerned, appealed to the duke. He requested that the duke, before giving the order for criminal examination, should graciously await his, the oldest son's, well-founded report on declarations of the witnesses and for this purpose have him presented with a complete copy of the testimony, because hitherto, in spite of various written efforts, he had been unable to obtain an exact statement of the forty-nine "points of disgrace." The request was not granted. The duke himself directed the chief council to conclude the suit most speedily. On July 24, the chief council ordered that the Kepler woman be arrested, wherever found, diligently examined on the theological articles, and confronted with the witnesses and that, in case she should not plead guilty, in accord with the report in hand, she be tortured. During the night of August 7, she was apprehended while asleep in Heumaden and, so as not to cause a sensation, she was carried out of the parsonage in a closed chest, and later imprisoned in Leonberg. The hearing took place immediately. The magistrate was now in a great hurry. The accused steadfastly denied the charges against her and declared that she had never practiced sorcery or witchcraft, nor had anything to do with the evil spirit, nor ever caused harm to people or livestock. She stuck to her statement, even when confronted with the witnesses. The chief council ordered that she be examined a last time with all seriousness and under threat of the executioner; if she still would not confess anything amicably nor speak her mind freely, the magistrate should accuse her *ad torturam*, and should let this be carried out when the court had so decreed. Afterwards the confession on the rack should again be reported. This all took place in August, 1620. This was the time of greatest need. Margarete wrote to her brother in Linz. He alone could still help. But, alas, just at the very time when Kepler received the bad news from home, his residence was engulfed by the martial events, fraught with disaster, which for two years had been raging in the Hapsburg lands.

On May 23, 1618, the signal for the rebellion had been given by the defenestration in Prague, and the tension which had long existed in Bohemia had exploded. The horrible war which was to bring unspeakable sorrows to Germany for three decades, had begun. The imperial cause fared badly from the beginning. The Bohemians under the leadership of Count Heinrich Matthias Thurn made good progress and the rebellion was greeted with sympathy in the Protestant divisions of Germany, as well as by the numerous followers of the Augsburg Confession in the Hapsburg lands themselves. On March 20, 1619, when the weak-willed Emperor Matthias died, his cousin, Ferdinand, who had been chosen as his successor, faced a difficult situation. The Bohemians reached Vienna and threatened the capital of the Hapsburg lands. Hesitation by Thurn, their leader, saved the situation for Ferdinand. On August 28, 1619, the latter was elected emperor in Frankfort. Two days before, however, the Bohemians had transferred royal dignity, to which Ferdinand had claim, to Elector Frederick V of the Palatinate. The Upper Austrian representatives welcomed this choice and hoped the new situation would strengthen their position against Ferdinand's Counter Reformation plans. But now the new emperor was able to persuade the Bavarian Duke Maximilian to take an active part in the campaign against the Bohemians by pledging Upper Austria to him. In July, 1620, Maximilian's army marched into Linz on the way to Bohemia.

The occupation of their city created a completely new and very difficult situation for the Protestant representatives and the entire population. Kepler's situation, also, was uncertain. Would the Protestant school at which he was employed continue? Was it not to be expected that here, also, as once upon a time in Graz, the preachers and teachers who followed the Augsburg Confession would be expelled first of all? To whom should he cling? He had been court mathematician to two emperors and, even if Ferdinand had not yet also approved him as such, nevertheless, his fate was connected with the House of Hapsburg by the astronomical tables. Did not abandoning this house question the finishing of the tables? And still he was a convinced follower of the Augsburg Confession to which, in spite of the persecution of his coreligionists, he wanted to remain unconditionally loyal. The new king of Bohemia, by his Calvinist bent, also gained his sympathy. Many of the best friends he had made in Prague favored him. But there was also another reason in favor of taking the side of the counter king and consequently against the Hapsburgs. Frederick, a Palatine, was the son-in-law of James I of England. This monarch was Kepler's great hope in matters of creed. (The astronomer was more at home in the starry heaven than in the political heaven.) He had already given repeated public expression to his admiration for the important peacemaker on the throne, as he considered him. In this connection, Sir Henry Wotton, the English ambassador, had recently been at his home and invited him to come to England. Should he accept the invitation? Could and might he remain in Linz? He clearly summarized the result of his deliberation in a letter written in these critical days to his friend Bernegger: "I may not simply abandon my second fatherland because of the indignities which it suffers at present, if I do not want to appear ungrateful—in the event that I am not more ungrateful if I become still more of a burden to it. About that it must decide itself."

But now was no time for further deliberation. He had more to worry about than his position and his residence. That would come. Now it was a question of his good name, of the good reputation of the name Kepler, of the honor and life of his beloved mother, who was in the greatest danger. He wrote to the duke of Württemberg to let him know that he himself would take part in the legal dispute. He did this in solemn form: since by divine and natural right he was obliged at all times to render proper assistance to his mother in this her need and where possible to prevent further misfortune, he could not otherwise be sure in his conscience whether he had done his duty—because of his close blood relationship—unless he personally attended the impending trial. The chief council thereon decided to concede the prisoner a respite of five to six weeks to submit her written defense. The "feeding cost" she would, however, have to pay herself. In September Kepler set out for Württemberg. Since he did not know whether the changed circumstances would permit him to return to Linz, and since he foresaw a long absence, he took his whole family away with him and found them accommodation in Regensburg. Continuing the journey, he also stayed with the Jesuits in Ingolstadt. On September 26 (old style) he had his first meeting with his mother in the prison.

She, however, was no longer in Leonberg. For in the meantime her son, Christopher, had petitioned to conclude the trial elsewhere. It is obvious from the motivation for his request that he was cut from a different pattern than his brother, Johannes. He had, so he wrote, not the least objection to the manner in which the truth might be brought out. But according to him all previous transactions in Leonberg brought him and his family very great disrespect, together with daily pain and most extreme distress, so that in his trading as well as in his position as drill master for the militia, to which he had been appointed by the duke, sundry noticeable prejudices, scorn, indeed uttermost opprobrium would arise. The good fellow thinks of himself first. The duke granted the request and shifted the conclusion of the legal dispute to the little town of Güglingen. Johann Ulrich Aulber, the magistrate there, was a worthy counterpart of Luther Einhorn. With an understanding blink of the eyes, he accepted his Leonberg colleague's remark at the transfer of the deeds, that nothing further was necessary for ascertaining the truth than Master James, that is the hangman.

The first thing Kepler did for his mother was to petition the duke and get her a better shelter, because she complained greatly "about the cold and the cheerless lonesomeness." She was supposed to be brought from the tower in which she was first confined to the house of the town beadle and be guarded there by a keeper. But since the town beadle stated he had no free room, she was brought to a little room in the tower gate, which day and night served the gate keeper and his servants as the passage to their chamber. Since the little room in the gate was open, it was considered necessary to put the poor woman in chains and have her watched by two keepers. Since both keepers had to be paid out of the woman's own funds, a petition was made, which, to be sure, was in vain, to have her watched by a single guardian. The keepers, two good-for-nothings, liked their job. They were lavish with the firewood, since they did not have to pay for it themselves. They were deeply in debt and did their best to contribute to the lengthening of the imprisonment, listening to the old woman and repeating every clumsy word to the judge so as to provide him with new grounds for suspicion. When, as a result of this housekeeping, the expenses kept mounting the Reinbold woman, who was afraid there would be nothing left over for her, submitted the request that "for the sake of God's mercifulness" the funds of the defendant be spared. Kepler protests, he indignantly protests, the malicious misuse of the name of God's mercifulness for the unmerciful persecution of his mother. But he also has to defend himself against his brother Christopher, who severely reproaches him as having alone by his interference caused such huge expenses!

And now to the progress of the trial. All the moves and countermoves with which the two parties operated against one another cannot be described here in detail. Right at the beginning, on September 4, 1620, the bill of indictment was read aloud. This was followed a few weeks later by the written defense in which Kepler and Johannes

Rueff, the counsel for the defendant, refuted the points of the accusation. Since Aulber, the magistrate, as agent for the indictment, did not feel as though he were a match for his opponents, he asked the chief council for support by a lawyer, which was afforded him in the person of Hieronymous Gabelkofer, the prince's counsel. In December the latter, in a writ of acceptance, rejected the reasons presented by the defense. To prove his superior knowledge in this matter and domain, the eminent jurist enriched the indictment by a series of "additional articles," which had hitherto escaped the less educated co-indicter. So he established, what in a witch trial was certainly extremely aggravating, that about forty years before in Weil der Stadt the Kepler woman as a Lutheran had received communion from the Papists! As a very suspicious consideration he further cited that the Kepler woman when face to face with the witnesses did not look at them, but rather turned her glance away, just as at all speeches, especially at the oral reading of the theological articles, she had not wept. From the keepers he had further learned the incriminating fact that the prisoner (who no longer had any teeth) used a knife in eating, even though she had not been given any, and so forth. This masterpiece was followed in January, 1621, by the hearing of new witnesses, such as the deacon, Johann Bernhard Buck of Leonberg, who already had long been denying the Kepler woman the sacrament, the magistrate, Einhorn, and three municipal officers from Leonberg, whose statements were set down in a voluminous protocol. In May, the defending counsel submitted an exception and defense, in which the reasons recently brought forward against the Kepler woman were refuted with numerous citations from authoritative legal literature. Thereupon the agents for the indictment answered in August with a detailed statement of deduction or rather repudiation. Because the defense did not want the argument that the accused does not cry and turns her eyes away to be counted it was thrown in the scale afresh by the learned plaintiff referring to authorities, specially to the then famous French lawyer and publicist, Jean Bodin. (The poor woman had answered a reproach relative to this only by saying she had cried so much in her life that she could not cry any more now.) The defense ended the proceedings. On August 22, 1621, it presented a concluding statement. This strong 128 page document was mostly written in Kepler's own hand. He had composed it with legal assistance. It must be admitted that in the entire procedure the defense was permitted to speak a great deal. But Kepler did not let himself be pushed aside. His bearing at the trial is most briefly and best characterized by the city scribe remarking in the protocol of a legal session in August: "The prisoner appears, alas, with the support of her gentleman son, Johannes Kepler, the mathematician." Frau Kepler's persecutors had every reason to say "alas." For it was Johannes Kepler who snatched from them the offering which they had in their claws.

It takes a legal mind to find delight in the flinging hither and thither of the grounds and counter grounds. The layman does not grasp the business and shakes his head when he leafs through the papers. Every document talks of a drink offered many years before by the Kepler woman to this one or that, of a cow which went mad in the stall, a pig which expired twenty-five years before under peculiar circumstances, of pains which a young girl noticed in her arm, a butcher in his thigh, of a schoolmaster who was lamed, of the death-skull of her father, of the utterances of Henry, her hot-tempered son, and similar things. The lawyers battled about these; therefore the poor old woman had to languish for more than a year chained in prison, the rack before her eyes; therefore Kepler had to interrupt his scientific activity. How many cares, sorrows, tears are hidden behind these events!

The close of the trial is quickly related. It is not quite so bad as expected. The duke arranged that the documents be sent to Tübingen and the decision of the juridical faculty asked. There sat men who could examine the case dispassionately and give their opinion of it and were certainly not inaccessible to Kepler's private notions. An especially close connection existed between him and the well known jurist, Christoph Besold. The Tübingen college, in a "missive of consideration," under the date September 10, established that the asserted circumstantial evidence was, indeed, not sufficient for the use of the rack, but also does not permit complete acquittal. The faculty therefore suggested the decision: it is found according to law, that the accused should be tortured for the purpose of learning the truth. But at the same time it made known the opinion that this sentence should not be carried out but the prisoner should only be led to the place designated for the torturing, shown the hangman, and his instrument, but not handled nor tormented.

On September 28, 1621, this horrible procedure was carried out. The poor woman was brought into the torture chamber and shown the instruments of torment which would await her if she did not confess. Yet she remained firm. "Do with me what you want. Even if you were to pull one vein after another out of my body, I would have nothing to admit." Then she fell to her knees and said a Pater Noster. God would, she said, bring the truth to light and after her death disclose that wrong and violence were done her. He would not take His Holy Ghost from her and would stand by her. After this procedure she was led back to her prison. Hereupon the duke gave the order that, because she had invalidated the circumstantial evidence by the terror she had endured, the Kepler woman should be absolved by judicial verdict from the complaint and discharged as soon as her family had paid the expenses incurred by her or had given security. On October 4, 1621, after a custody of fourteen months, the woman was set free. She was not permitted to enjoy this freedom much longer. On April 13, of the following year she passed away, presumably in Heumaden. She certainly never returned to Leonberg because she was threatened with being beaten to death by the inhabitants when, after the verdict of discharge, the city, its magistrate, Einhorn, and the Reinbold woman were also sentenced each to pay a part of the trial costs.

Thus Frau Katharina Kepler, who had given the intellectual world such a wonderfully radiant luminary, her son Johannes, ended her own life in obscurity.

Immediately after the end of the trial Kepler had the reason for his long absence attested by the ducal chancellor and in November returned to Linz via Regensburg, where he called for his family. During the pause in the legal procedure, he had visited Regensburg from January to March, when his wife gave birth to a little daughter, Cordula.

The misgiving Kepler had entertained, that he would have to leave Linz, had proved unfounded. From the distance he saw his second fatherland, Austria, "with the halter of a hard tyranny about its neck, in utmost danger." The Bavarians ruled strictly there. Kepler feared the representatives, in whose service he stood, would be dissolved. He had heard that the Bavarians had forbidden the payment of so much as a heller[1] to a creditor or a salary recipient. And yet the maternal inheritance of his two older children and his remaining wealth were invested there. Therefore he had written to his friend Bernegger for advice and help. Certainly, he hardly desired to go to England, whither he had been invited. "I should go over the sea, whither Wotton invites me? I a German? I who love the continent and shrink back from the narrow confines of an island, whose dangers I perceive beforehand?"

For all that, some Linz people were surprised when their mathematician reappeared. During his long absence all sorts of rumors about him had circulated. For understandable reasons he had kept the cause of his departure strictly secret. Not even his assistant, Gringalletus, who accompanied him in order to travel on to Strasburg, was informed about it. Since Kepler suddenly departed immediately after the entrance of the Bavarians and took his whole family along, as

[1] ED. NOTE. Small copper coin worth about a quarter of a cent.

well as his "mathematical furnishings," his trip was interpreted as flight and reasons for it were sought. Some said he had gone to England, since something had evidently been learned of the invitation issued to him and Henry Wotton's visit. Others were of the opinion he had permanently moved to Württemberg. It was whispered that he had incurred the emperor's wrath, indeed that the emperor had even set a large sum on his head. After his return home, he wrote to a friend in Dresden that "it is unbelievable with what tortures my poor reputation had been tormented during my one year's absence in Württemberg." He added the request that his friend tell him what he had heard, to enable him to uncover blemishes that might have tarnished his reputation. The rumor of the imperial displeasure at any rate has its basis in that apocalyptic threat which Kepler had pronounced in his calendar for the year 1619 against an "animal resplendent with roses," and which might have been erroneously applied to the emperor. We shall have more to say about it later. At any rate, at the critical time the prophet found it necessary to explain to Paul Guldin, the Jesuit Father, that he had not intended any ruling house.

Yet certainly other causes are concealed behind these rumors. Indeed, in the year which Kepler passed in Württemberg, significant things happened in the domain of high politics. On November 8, 1620, in the battle on White Mountain, Frederick of Bohemia, the opposition king, had been decisively beaten. He had to take flight and was outlawed. The Bohemian venture, which had started so hopefully, collapsed wretchedly. Emperor Ferdinand utilized the victory for ruthlessly rooting out protestantism in Bohemia and took relentless revenge on Frederick's partisans. In the bloody "Prague execution" on June 21, 1621, twenty-seven ringleaders, prominent noblemen and burghers, had been publicly executed in the market place. These victims included men who came in close contact with Kepler during his time in Prague, such as his previously mentioned friends Jessenius and Budowetz von Budow. A great many other men of rank and name were punished by many years of imprisonment and confiscation of their property. The persecution, although not in this bloody form, naturally, reached as far as the circles in Upper Austria which had conspired with the rebels in Bohemia. There, because he had been accused of taking part in the venture against Ferdinand, Pastor Hitzler, too, was kept in custody for several months by Count Herbersdorf, the Bavarian governor. Since the identity of Kepler's friends in Prague and his outspoken sympathy for the exiled rival king's father-in-law were known, it is understandable how, in a time of restlessness, rumors could form about Kepler's person. At any rate, it was good that he was away during the months of the greatest excitement. Yet, after he had returned, not only did nothing bad happen to him, but, on the contrary, a few weeks later, on December 30, 1621, Emperor Ferdinand confirmed him in his position as court mathematician. Also, when in the following year the Counter Reformation measures led to banishment of the "preachers and non-Catholic schoolmasters," an exception was made for him and he was able to dedicate himself without further disturbance to his scholarly works.

7. *Last unavailing attempt in Tübingen to obtain admission to communion*

Naturally, the description of the criminal trial does not present all the events in Kepler's life during those years. Another experience also weighed oppressively on his mind and caused him much trouble in this period. At the same time as the legal tragedy, the closing act in his settlement with the leaders of the church to which he belonged was enacted on a nearby stage.

In his last letter to the Württemberg consistory, Kepler reserved for himself the right to come forward again at a favorable opportunity with his request for admission to communion. He says he was of the opinion that the answer of the consistory was founded on erroneous information. In the fall of 1617, when he tarried in Württemberg because of the trial and also visited Tübingen acquaintances and friends, he considered the correct moment had come to make a further attempt. Since he had been unable to achieve his goal by the official path dealing directly with the consistory, he now hoped, by his request, to prevail on the theological faculty. Its intercession would greatly influence the consistory; indeed, its recommendation would suffice in Linz to have him readmitted to the community there. But, in the faculty's decision, that good will, which they entertained toward him, would also weigh heavily. First of all, there was his former teacher, Matthias Hafenreffer, now chancellor of the university, who at all times had shown a special liking for him. "This man was filled with such a great love for me, that the pain over that, which he must at some time have heard about me, had not yet withdrawn from his heart," relates Kepler in a report of his visit to Tübingen. Here consequently his inquiry would be discussed in a more favorable, warmer, more personal atmosphere. Oh, he should only experience a so much the more bitter disappointment.

He proceeded carefully. Evidently in order to obtain something in writing, instead of first speaking with Hafenreffer, he wrote him a few lines from Maestlin's house, inquiring whether Hafenreffer was in favor of his joining the communion table before his departure. Had not Helvicus Garthius, the pastor, offered him communion long before in Prague? He did not achieve what he wanted. Hafenreffer asked to postpone his answer. Kepler shifted his departure. There was an oral discussion. The chancellor declared he could not by himself decide the question of admitting Kepler; he would have to deliberate it in the faculty and with the consistory. He would send him a written decision to Linz. From this Kepler drew the hope that Hafenreffer would write to Pastor Hitzler; after his return he repeatedly asked the latter whether he had received any letter from Tübingen dealing with him. For a whole year he waited in vain for an answer. Then he picked up his pen. There were two dramatic duels with letters as the weapons. The two men opposed each other with painful emotion and passionate irritation.

Kepler's first letter of November 28, 1618, begins with the declaration that from Hafenreffer's long silence it is easy to draw a conclusion about the contents of the writing which once was under surveillance. If Hafenreffer had not yet answered, it was the fault of his friendly disposition which did not permit him to hurt a friend. But then Kepler's growing agitation broke out in all violence. "Now in God's name, you are excluding me from communication with you because of the frankness with which I recognize that in regard to the one article of the omnipresence of the body of Christ and the other views ascribed to me, the utterances, proofs and exegetical elucidations of the ancient Fathers are more conclusive than yours in the Formula of Concord! Out of the Holy Communion make me a criterion of your creed, which I insist on in this very special point concerning the order given by Christ to all laymen: 'Do this in my memory, make known my death.' You might object by saying that the servants and administrators of the secrets of God had been commanded: 'Do not abandon the sacrament to the dogs!' Granted, I may be for you, O sorrow, a spiritual dog, not because of any blemish in my conduct of life, but solely because of my frankness, in declining to speak in this one point in conformity with your Formula of Concord contrary to my conviction and to express orally different opinions than I myself believe." Still he is not to be embittered by the treatment vouchsafed him. He wants to persevere in the love with which he embraces all parts of Christ and above all his spiritual Fathers. But also he does not think of painting up a creed to set aside his exclusion. He expresses the complete frankness and pureness of the conviction which fills and guides him in his behavior when he continues: "But I do not hold this exclusion in contempt nor do I rejoice over it. But I ask God to support me with his Holy Ghost, so that I never make myself guilty of anything by which I would earn the exclusion." Without expressing a request for the lifting of the order excluding him, Kepler ends his letter by asking that the accustomed candor please be resumed.

Hafenreffer was struck with sadness by the vehement tone of this letter. He does not speak of the reason for his long silence. Nor does he explicitly confirm the exclusion, but he tacitly approves Kepler's assumption that he was in accord with it. Again must Kepler hear the warning to keep his hands off theological questions; here anyone however wise would have to become a simpleton. "If you love me, if you ever loved me, then get rid of the passionate excitement." Kepler should ponder the three words: *Verbum caro factum* (the word has become flesh). "Either I am a simpleton and out of my mind or you understand how foolish, irrational and insane we are in wanting to measure the divine mysteries by the scale of our stupidity."

In the next letter Kepler's tone changes once more. He had offended his friend by his violent harangue in the first part of his previous letter. That he had not wanted to do; he, with his gentle disposition, could not bear that. He had caused pain to the man whom he loves. He wants to erase it; he must do so, because otherwise he is not at ease. He must have the feeling that the man whom he reveres is satisfied with him. Indeed, in the friendship between him and Hafenreffer, no matter how much older he has become, he can no more dislodge the sensation of deferential dependency which had inspired him as a student than can Hafenreffer set aside the sensation of superior virtue, which his office confers on him. Kepler had not begged for an exchange of letters for entertainment. That would be punishable arrogance toward a man in such a high position. No. "I had waited for consolation for a whole year, which was not pleasant entertainment. That I had to be deprived of consolation, consolation ever since, that was what hurt me." And so in his deep need he implores the other: "I adjure you, let a friendly, bright face shine on this letter, so that my wounded heart, warmed and strengthened thereby, is reassured to some extent." Cannot the theologians understand why he cannot subscribe to the Formula of Concord? Why is he not allotted the same benefit as the simple laymen, who are admitted to communion without ratifying it? "To what extent am I a worse sinner than these? It is necessary to become a simpleton, you say. I act accordingly whenever I prepare myself for divine service and cut out all captiousness." "But it is no captiousness, rather brotherly love, when I do not want to pass sentence on those who pay heed to the ancient doctrines but prefer to follow them rather than the Formula of Concord in the one article about the person of Christ." Then Kepler gets to his friend's request and by lengthy arguments propounds his ideas about the scriptural passage, "The word has become flesh," referring to the old teachers of the church whom he had studied in that respect. Finally, in moving words, he asks his high and mighty friend to recommend him to the theological faculty and the consistory: "In this hope I stand always still submissive before you and in this sacred matter for the sake of Christ, a part of whom I hope to be and also wish others to consider me as part of him; and I ask you to be my sponsor and to bring about a decision, which does not yield and heap other reproaches on me, but relies on what I have made known as my view."

Did not such a humbly expressed request have to be heard? Hafenreffer laid the letters before the theological faculty and the consistory. Because he had the answer signed by all members of the faculty, this is no longer a personal letter, even if in it he addresses Kepler as his dearest friend; it becomes an official document. Accordingly, it was also composed more in the style of an official verdict. It comes to a climax in the joint solemn admonition: "Either you will cease your erroneous and entirely wrong chimera and embrace the sacred truth of humble belief or else steer clear of the congregation of our church and our creed." Christ, as the letter further states, does not permit anyone to mock Him, nor does He, the purest bridegroom of His church, share His love with vain and blasphemous opinions. Kepler should finally give notice to his reason, which in divine matters is blind and wrong, and surrender himself to the simplicity of the fisherman of humble heart. If he did that, his conscience would find peace, his soul salvation. If he did not, then it was to be feared that God would deliver him to a depraved mind, as an impudent scorner of his word, and eternal damnation would be his lot.

That was the last word which Kepler heard from the heads of his church. The gate was shut. The exclusion was sealed. He who wrestled with God as Jacob had once done with the angel was accused of blasphemy. He who with fervent heart sought the truth was expelled by those who fancied themselves in possession of the truth. They wanted no intercourse with the man who offered brotherly love to all Christians. And he was forced to stand branded with the mark of the heretic as living scandal among his brothers. The main motive for his action with the theological faculty had been to avoid this.

To read the correspondence between Kepler and Hafenreffer evokes great sympathy. The tragedy in the conflict of Kepler with his church is here touchingly expressed, because the contrast in creed comes between two men who in sincere admiration have a great liking for each other. There is no doubt that it was very difficult for Hafenreffer to have to inform his younger friend of the decision refusing him. He was in bitter earnest about his concern for the other's salvation. But it caused Kepler deep pain to bring grief to his revered teacher by something which his conscience demanded of him. Certainly it was no idle wrangling about empty formulae which separated the two, no scholarly squabble between two incompatible heads, each of which obstinately wanted to be right. For both the most sublime and ultimate things were at stake. Both were permeated and filled with the strength and truth of the scripture: The word has become flesh. Both revere this sacred mystery. Both also know that this is a question of a mystery which the human intellect can never fathom but must include in its creed. Both are convinced that an authority decides and is binding in this question of creed. But they diverge on the question of to whom this authority belongs. For Hafenreffer the authoritative pattern is the Formula of Concord, for Kepler the tradition. Thus the fathers of the Formula of Concord and the ancient teachers of the church oppose each other. The contrast between these regarding the question of the person of Christ was of necessity transferred to the relationship of the two men. That is the final cause of their estrangement. The personal side of the tragedy draws to a close by this last-mentioned letter really containing the last word directed to his friend by Hafenreffer. For, barely three months later, he died. So his last word to Kepler was a very bitter one.

There is still to be mentioned a vicious word that fell in between. Even if Hafenreffer advocated his cause with dignity from the point of view of the churchman and did not deny esteem and approbation to Kepler's person and accomplishment, still the same cannot be said of the consistory, to which the case had been submitted for an expression of opinion before the decision. A member of this ecclesiastical authority, Erasmus Grüninger, had couched his description of its decision in a letter to Lukas Osiander, professor and superintendent at the seminary, in the words: "As regards Kepler, by this time one has long dealt with that same swindle-brain, but in vain, and he does not listen to what you say. We did not want to neglect to communicate to their honors, the Tübingen theologians, what had been addressed to him by the consistory some years ago about this same subject (*de hac ipsa materia*) as it might please the gentlemen to deal with him in the same vein, since one cannot change one's opinion for the sake of his sick head." With these words those churchmen passed sentence not on Kepler but on themselves. It is possible to understand their decision in the question of dogma. What, however, one cannot comprehend and for which one must reproach them is that they themselves show not a trace of understanding and esteem for Kepler's towering intellectual and moral greatness. They had received his letters to read. From these the sincerity and frankness of his way of thinking, the honorableness of his struggle for his faith shone forth in all clearness. They had perceived the cry of distress of his wounded heart, heard his humble request. Yet all that makes so little impression on them that without hesitation they push out of the door the man who had revealed his very soul to them. They could not show their narrow spiritual arrogance more plainly. In truth, Kepler was badly received by the highest authorities of his native land. The chief council had involved his mother in diabolical proceedings and the church counsel had sent him about his business and excluded him with indignity.

What especially angered and offended Kepler's theological adversaries in Stuttgart and Tübingen was that he repeatedly predicted evil for the Württemberg church because of its rigid uncompromising attitude. He saw clearly that such an attitude could lead to nothing good and that the outcome would be acts of violence, if the separated confessions could not come to an understanding. He expressed the apprehension that the putrified flesh around the wounds of the three-way split church would have to be cut out with a knife of a universal German religious war and be burned out with the fire of many hardships and sufferings. "Then would the hardship," he wrote to Hafenreffer, "give you understanding for many things which are now concealed from the eyes of your young people growing up for the service of the church." How quickly did the misfortune appear, which he had predicted with prophetic eye! Just as he gladly made use of his calendar to exhort both the high and the low, so in an apocalyptic form also he expressed his special warning to the Württemberg church, without explicitly naming the latter, in his prognostication for the year 1619. There in a long passage, it says: "I know an animal of the neuter gender that sits and is resplendent in the roses, looks only at another animal, its enemy, when it will perish, and is not afraid of anything in the least." It will not recognize its errors and shortcomings: they are hidden from its eyes; it is indeed satisfied, relies on deep-rooted custom and scorns all sincere warnings. Will it not be spoiled by such things by which it now sins? Is it not to be feared that a butcher will come along and put a finishing stroke to the animal? "For that reason, be warned, accustom yourself to the stroke, stop pushing, bethink that you are not there for that nor for your own benefit, but for the sake of the milk." The prophecy aroused much attention and wild guessing among all his readers. Kepler was approached from all sides to express himself more distinctly. It was spoken of for several years in letters and printed matter. Some guessed at the Rosicrucians, whose order appeared just about then and was much talked about, others the Jesuits, the Pope, or, as we have already heard, the House of Hapsburg. Kepler, however, was never prevailed upon to reveal the secret, to disclose the meaning of his riddle. All he said was that those to whom he referred knew it and others, whom it did not concern, did not need to know. Yet, if all relevant remarks are examined, it follows indisputably that he had wanted to direct his warnings precisely to the Württemberg theologians. It can be imagined that they were not exactly put in a milder and more conciliatory mood by being so addressed. But, after the treatment which they had given the uncomfortable admonisher, they were in no position to complain about the unusual form of his answer.

Since it was hard for Kepler to stand before friends and acquaintances as a heretic, in the year 1623 he had a justification printed in which he concisely presented his religious convictions in frank and straightforward words. Many of the places cited above to characterize his attitude in the religious dispute are taken from this work. It is called: *Profession of belief and excuse of various unkind rumors which occurred for that reason.*[1] This pamphlet appeared anonymously and was secretly printed in Strasburg with his friend Bernegger as intermediary. The few copies which were issued were distributed in the narrow circle of the author's friends. When two years later, at the time of another conflict which the warlike gentlemen had to fight, the Tübingen theologians committed the indiscretion of publishing Hafenreffer's last letter to prove their unanimity, Kepler felt obliged to take up pen once more and justify himself publicly in a small pamphlet. In separate footnotes to that letter he made his dogmatic convictions precise; in the introduction he gave a short survey of his religious growth, giving many explanations. Thereby the deeds in the fight with his coreligionists are conclusively closed. But Kepler was to feel required to guard his standpoint against the other, the Catholic side.

8. Harmonice Mundi; *and the third planet law*

But before following him further in this path, it is necessary to turn to the scholarly results of the years whose events are described above.

---

[1] *Glaubensbekandtnus vnnd Ableinung allerhand desthalben entstandener vngütlichen Nachreden.* (Title from text.)

Was it not to be believed that these events, the witch trial, the religious dispute, and war, had crippled Kepler's strength and drawn him away from scholarly work? But now once again the wonder of his personality appears. In these very years this unusual man, by the number of his published works and the significance inherent in them, mounted a high peak of his intellectual activity, exactly similar to that which he had reached in Prague with the *Astronomia Nova* and the *Optics*. The work which crowns this peak is the *Harmonice Mundi*. There is a curious plant, *Cereus nyticalus*, which, growing for a long time, one night suddenly opens a wonderful blossom. The trial which threatened his mother with death, his name with shame, the religious controversy which shook and clouded his innermost being, the war which stood like a threatening storm in the sky—did not these happenings encircle Kepler with darkness? In this darkness, further deepened by the death of a child, there suddenly broke forth after a long period of growth, yes suddenly, the bloom of the *World Harmony*. Happily that sad chapter lies behind and it is permissible to turn to the contemplation of this intellectual flowering.

We have seen that Kepler was very busy with his *Epitome* and with the calculations of the *Ephemerides* when he was forced to journey to Württemberg in the fall of 1617 because of the trial of his mother. After his return in the first weeks of the following year he resumed these works with unswerving zeal. One event occurred, as he relates to us, which induced him to alter his plan of work. On February 9, his little daughter Katharina died. The father was oppressed by sorrow over this loss. "I set the *Tables* aside since they required peace, and turned my mind to the contemplation of the *Harmony*." We recall that in the year 1599, while still in Graz, he had drafted the plan of a work on this subject; then as now he was saddened by the death of a child and he was much troubled by concern for his future. At that time of need he had also found solace and strength in his harmonic speculations. Many years had passed since then, without the old plan being carried into effect. New works had pushed it into the background, brilliant results had deepened the insight, widened the views. But at every opportunity thoughts from the extensive complex questions of his *Harmony* emerged in Kepler's mind. They ripened in him as his knowledge grew in extent and content. The fruits of his works in astronomical, mathematical, philosophical domains filled and nourished his supply of harmonic ideas, served them for cleansing, correction, widening and deepening, offered supports and presented new combinations of thoughts. In the conviction "that the geometrical things have provided the Creator with the model for decorating the whole world," he studied the tenth book of Euclid, which deals with the various kinds of irrational quantities which he needed for the erection of his thought structure.

To give the musical section of his reflections a sound basis he drew on what was offered by the ancient Greeks as well as by more recent authors. After Kepler had long tried to get hold of a Greek manuscript of Claudius Ptolemy's *Harmony*, Herwart von Hohenburg gave him one to read. This was a notable experience for Kepler because the perusal of this work afforded him an "especial increase of his passionate desire for knowledge and encouragement of his purpose." He was much amazed when he ascertained that this ancient astronomer had been occupied with a similar contemplation of the heavenly harmony as he himself, even if this agreement was, to be sure, evident only in the formulating of questions, not in their solution. Yet Ptolemy impressed him "as though he would sooner report on a pleasing dream with Scipio in Cicero ['s work] than promote philosophical knowledge." None the less he felt strongly stimulated and encouraged in his harmonic researches by the similarity of the formulating of the problem; for him it was frankly God's indication "that the same thought about the harmonic formation had turned up in the minds of two men (though lying so far apart in time), who had devoted themselves entirely to contemplating nature," although neither was the guide of the other in walking the path. Kepler was further very strongly stimulated and corroborated by Proclus, the Neo-Platonist, who, in his famous commentary on the first book of Euclid, expressed the same opinions on the existence and reality of mathematical things which he himself had formed. Along with these historical studies, he also promoted his research into that side of his formulating of questions which concerned psychology and the theory of knowledge, since it was clear to him that what he had to say ultimately had its root in the mind. Many arguments in earlier writings, such as in his book about the new star and especially in his *Tertius Interveniens* give evidence of the intensity and growth of his struggle to develop his picture of the world in accordance with the laws of harmony. In letters to David Fabricius, Johann Georg Brengger, Joachim Tanckius, professor of medicine at Leipzig, and Christopher Heydon, the Englishman, he expressed himself with warmth and enthusiasm about that which was close to his heart. Before the end of 1616 he was seeking in his astronomical calculations, too, the law connecting the periods of the planets with their distances from the sun, that law which crowned his new book.

Now the moment had come when the waters, dammed up in him, broke through, when he recalled everything which he had ever thought about the great burning desire of his heart, when he was struck by the pictures and faces which had been dormant. He collected the accumulated materials. He had already compiled much for himself at various times while awaiting the correct hour, which must some day come. He arranged, compared, edited, augmented, and expanded what he found. Kepler in those weeks and months must be pictured in a condition of greatly increased excitement, such as seizes an artist who has an idea which strives to be shaped. He forgot the world round about him and its many tribulations. The shrill dissonances, which resounded in this sphere, did not reach his ear. From higher provinces his mental ear perceived a music, which lifted him above earthly suffering and earthly difficulties. The magic of the word harmony transported him to another, a pure, paradisiacal world. Just as the Greeks, who created the word, set the idea of harmony in the center of their cosmology and from there sought to advance to the root of existence, so was all his thinking filled and ruled by this idea. The wall of sense matters became transparent like glass. Here in the idea of harmony he found unity in the many-sidedness of phenomena, the essential kernel of nature, the supporting principle of regulation, which really and actually and in fact makes the world into the loveliest possible, the key to the understanding of relationship, connecting the created spirit with the original spirit, with God. His visions inspired him. "I feel carried away and possessed by an unutterable rapture over the divine spectacle of the heavenly harmony." Having, moreover, on May 15, 1618, discovered the third of his planet laws, which had still been lacking in support of his views, he flung out a few days later the dithryambic sentences: "Now, because eighteen months ago the first dawn, three months ago the broad daylight, but a very few days ago the full sun of a most highly remarkable spectacle has risen, nothing holds me back. Indeed, I give myself up to sacred frenzy. I sneeringly defy mortals by the following public avowal: I have plundered the golden vessels of the Egyptians, in order to furnish a sacred tabernacle for my God out of them far from the borders of Egypt. If you pardon me, I am happy. If you are angry with me, I bear it. Well, then, I cast the die and write a book for the present time, or for posterity. It is all the same to me. It may wait a hundred years for its readers, as God also has waited six thousand years for an onlooker."

However, nothing would be more wrong than to assume that by his harmony speculations Kepler lost himself in a nebulous reverie. He did not enjoy the resounding tinkling of words or the torrential flow of vague speech. What he presents to us is an edifice of thoughts, well proportioned in all parts, built according to a clearly recognizable plan, whose uniform organization and unbroken lines bear witness to the mathematical stamp of the architect's mind. His rapture is not the exaltation of a dreamer; it is the enthusiasm of the discoverer, of the mathematician, of a mathematician, of course, who does not want to play with self-created notions in accordance with self-created rules, but recognizes the foundation of the mathematical notions in the highest and ultimate existence and knows of what he talks. His enthusiasm borders on sublime rapture, even on the ecstasy of mysticism; mysticism, indeed, in the lucid brightness of its contemplation, comes closer to the transparency of mathematical observations than is known to or suspected by the many representatives of a meager and obscure interpretation of that intellectual domain. What comes first is not emotion seeking expression, but the clear thoughts, which rouse and fire emotion. Kepler's enthusiasm is only now and then expressed in words. For the most part his manner of speech is dispassionate and objective. It is only in the background that one senses the warmth, the fullness of feeling, the exaltation and happiness in the knowledge of being allowed to pull aside the veil before the divine plan of creation. All parts of the work deal with numbers. Where numbers appear, there will be calculating, a dispassionate activity. Yet Kepler knows that everything in nature is arranged according to measure and number. Many already have heard the word and repeated it. At this time, toward the end of the Renaissance, moderate arrangement and harmony were still chiefly in operation as determinant characteristics of the beautiful. Now he did not want to stop at general words and views; he wanted to be serious about the complete carrying out of this principle, pointing out its validity and objective realization in all details. That was his exalted aim, a task which could be taken up and solved only with enthusiasm.

What about this harmony, which Kepler makes the central idea of his thought structure? Earlier (p. 91 ff.) when he expressed himself for the first time about this subject, an introductory answer to this question was given and the whole complex of questions, which concern him in his *Harmony*, was sketched. It is known that he was dealing not primarily with a musical but rather with a mathematical idea, with certain distinguished relationships of numbers, which in accordance with exact, well-defined postulates furnish him with the divisions of a circle which can be worked out with ruler and compass, the "knowable" regular polygons. But now he does not pursue this, to some extent objective, side of the harmony idea only with all the paraphernalia of the keen logician and mathematician which he is, but asks himself, wherein consists the existence of the harmony from the point of view of the discerning subject. Answering this offers him the occasion for developing his idealistic theory of knowledge. This is the path to be followed next.

Kepler distinguishes between harmonies of the senses and pure harmonies. The first are especially evident in the musical sphere, which supplies the best instance of it, but in addition in all matters of the senses, which fall in the category of size, so that they can be compared with one another in respect to their quantities, as for example the radiancy of the planets, their rates, works of architecture, metric rhythm and so forth. Now to realize a harmony of the senses two such comparable sense matters must first be at hand, perhaps two tones. But it would be erroneous to think of harmony as sticking to these sense matters. Kepler very distinctly works out the thesis, that it is the soul which creates harmony. For, he says, harmony falls under the category of relation. "But what is referred to is not what it is called, if no spirit is accepted connecting the one with the other." If the comparing soul is taken away, the sense things, to be sure, continue to exist as such, but they do not form a harmony in so far as this is a thing of reason. Thus it is the soul which lends reality to the harmony. To discover a harmonic proportion in matters of the senses is for Kepler, accordingly, "to uncover, to comprehend and bring to light the similarity of the proportion in sense matters with a particular prototype of a real and true harmony, a prototype existing inside in the mind." If these prototypes, that is the pure harmonies, do not exist, the proportions would not be called harmonic, since they possessed no power whatsoever to arouse the feelings.

In regard to perception, that is the reception of sensory matter in the soul, Kepler followed the scholastic doctrine of species. From sensory matter issues an immaterial species, a form (in the Aristotelian sense), which affects and informs the sense organs. Taken from the forecourts or mouths of the senses, the species is accepted within by the imagination and is recognized by the common sense (*sensus communis*), preserved by memory, brought forth by recollection, distinguished by the higher faculty of the soul. The highest faculty of the soul that understands number and comparison can then form a single species of relation, arrangement, and comparison out of several intellectual species of things. Discursive thinking is in no way necessary for the soul to form a harmony. Perceiving the melodious sound of two tones an octave apart, the number of their vibrations being as $1:2$, it is not necessary first to establish this relationship by discursive comparison. On the contrary, inside souls there dwells a lower power, an instinct, with which they can receive the proportions in sensory matter and comprehend it as harmonious. So it happens, Kepler supposes, that children, the uneducated, peasants and barbarians, indeed even animals, feel the harmony of the tone, even if they know nothing at all about harmonic science. The perception of harmonies by the lower power of the soul is, to be sure, only dull and dark; it lies, so to speak, "under the cloud of ignorance," without the soul, however, maintaining herewith a purely passive attitude. Rather it is a question here too, of an activity of the soul, in so far as the latter works on itself by natural excitement and rouses itself.

How does it happen, then, as experience shows, that the soul reacts only to definite proportions, so that only when these occur it says: here is harmony? Kepler finds that answer in the axiom that by its own nature the soul contains the pure harmonies as prototypes or paradigms of the harmonies of the senses. Since these pure harmonies are also a matter of proportion, there must be present corresponding members which can be compared with one another. These corresponding members are the circle and those parts of the circle which result when arcs which can be constructed are cut off from it. The circle then is, as all mathematical ideas, something which, in accordance with its ideal existence, occurs only in the mind. The circle which we draw with the compass is only an inexact copy of an idea which the mind carries as really existing in itself; the reverse expression that the idea of the circle is gained from the physical figure is to be rejected. Kepler puts the greatest emphasis on establishing this thesis. He very determinedly turns against Aristotle's view that mathematical things never at all exist apart from sense things and that the ideas of these things are gained from sense things by abstraction. Yet, when Aristotle says that the mind is a blank tablet on which nothing is written, even nothing mathematical, but on which everything can be written, then Kepler believes that in the name of the Christian religion he has to raise a protest against this. He is a follower of Plato, according to whose theory the human mind learns all mathematical ideas and figures, all axioms, all solutions about these things out of itself; by the physical signs it only remembers that which it knows out of itself. As chief witness for his view Kepler, in this connection, draws on Proclus, from whose commentary on Euclid he cites long passages. According to it the mathematical ideas are the essence of the soul and inversely the soul is the essence for them. Thus the corresponding members of pure harmony are not supplied by the circle vouchsafed by the senses, but rather by the purely intellectually thought circle of pure harmony. Of itself the mind conceives equidistance from a point and out of that makes a picture for itself of a circle without any sense perception whatsoever. The senses turn toward the mind, not vice versa. Out of this insight Kepler feels driven to the profession which was very significant for his theory of knowledge: "If the mind had never shared an eye, then it would, for the conceiving of the things situated outside of itself, demand the eye and prescribe its own laws for forming it (provided it is clean and healthy and without encumbrance, that is if it only is that which it is). For the recognition of quantities which is innate in the mind determines how the eye must be, and therefore the eye is so, because the mind is so, not vice versa. Why use many words? Geometry, being part of the divine mind from time immemorial, from before the origin of things, being God Himself (for what is in God that was not God Himself), has supplied God with the models for the creation of the world and has been transferred to man together with the image of God. Geometry was not received inside through the eyes."[1]

Now as the idea of the circle is rooted in the mind, so it also issues from the nature of the mind that one kind of polygon can be constructed by compass and rule, but not others. Thus from time immemorial the truth is founded in the mind of God, that the square of the side of a square is half the square of the diagonal. Now it is the polygons which can even be constructed with the help of straight lines and circles, that supply Kepler with that relationship which he designates as harmonious. They can be simply representable or knowable, whereas the others, like the seven-, eleven-, thirteen-angled figures are designated as non-existent (*non-entia*). For all purposes these can be drawn as exactly as the others. But that is not the question. Imagine that watching from a point on a circle you see a second point on that circle move, then this second point will sometime reach a point which marks off an arc of one-seventh of the circle. Yet, as Kepler says, one cannot know this point. Similarly, as bearer of pure harmony, a circle is to be imagined on which the corner points of a polygon that can be constructed are marked, as it were, moving out from any point.

Kepler's thought process sets a marked significance on the circle above all other curved lines, such as is unknown to today's mathematics, where the circle figures as a second-degree curve along with others of the same degree and curves of higher degree. This special significance consists not only in the fact that, in addition to the rule, the compass, in its primitive form as a piece of string, is handed to man, as it were by

---

[1] ED. NOTE. *Johannes Kepler Gesammelte Werke*, VI, 223 (*Harmonice Mundi*, Book IV, chapter I). This passage, in Latin and in English translation, can be found in W. Pauli, *The Influence of Archetypal Ideas on the Scientific Theories of Kepler*, translated from the German by Priscilla Silz and included in *The Interpretation of Nature and the Psyche* (Bollingen Series, LI, New York: Pantheon Books, 1955), pp. 165–6.

nature, as a most simple mathematical instrument, and was used by him at all times before any theoretical activity with geometrical things. But Kepler also assumes distinctions of rank in the individual geometrical forms according to their higher or lesser degree of perfection. In the plane the circle is the most perfect form, in space the sphere. Correspondingly and further, the mathematical forms become symbols for him. He finds that the quantities form a wonderful and positively divine state and express the divine and the human symbolically in the same manner. Here he now utters a favorite thought, which Nicholas of Cusa's view of the world had urged upon him, and to which he returns in the most varied periods of his activity in innumerable connections: for him the sphere is the symbol of the Holy Trinity. The center is God the Father, the surface, which expands from the middle point, God the Son, and the equality of the distances of the points on the surface from the center symbolizes the Holy Ghost. Just as the idea of the sphere is abolished if one of the three pieces is taken away, so also is the one nature of God denied if one of the three personages is denied. But the straight line is the element of bodily form: if a straight line is rotated it describes a bodily form, generating a plane. If the sphere is cut by the plane, the cut results in a circle, the true image of the created mind, which is assigned to govern the body. Just as the circle lies at the same time on the sphere and in the plane, so is the mind at the same time in the body, which it instructs, and in God as a radiation which, so to speak, flows from God's countenance into the body from which it obtains its noble nature. Since now Kepler conceives the circle as the bearer of pure harmonies, but on the other hand designates these harmonies as based in the nature of the soul, the symbol finally becomes for him the object itself, the soul becomes a circle supplied with the marks of the possible divisions, an infinitely small circle, a point equipped with directions, that is a qualitative point. By establishing pure harmonies the soul institutes a comparison between its own parts. The soul itself becomes harmony and harmony becomes soul.

There are three fields in which Kepler moves in particular in developing his harmony theory: geometry, music, astronomy. The mind, as God's image, carries in itself the forms and laws of geometry and geometry supplies the relationships of numbers for the harmonies. In music, man plays with these harmonies for his pleasure, and it is no longer to be wondered at that the relationships of numbers which lie at the basis of the musical melodious sounds agree with those supplied by geometry. The motions of the stars, which take place in such sublime order, on the other hand have been harmonized by God Himself according to the relationships of numbers supplied by geometry, so that the world would be the best and loveliest, the most like the Creator.

In the domain of geometry, Euclid is his master and guide. To him he pays the highest praise. His *Elements* are for him an unequaled master work. He greatly admires the way the whole structure consistently rises proposition after proposition on the foundation of definitions and axioms, until finally in the thirteenth book the five regular solids appear as the towering cupola. When the mind, as he is convinced, has high-mindedly struggled through to an understanding of the deeper plan of this work, "then he becomes aware that he walks in the light of truth; he is seized by an unbelievable rapture and, exulting, he here surveys most minutely, as though from a high watchtower, the whole world and all the differences of its parts." The meaning of these poetic words becomes clear on considering the role Kepler still assigns, as earlier, to the five regular solids in his world picture. Indeed, he sees (with Proclus) in the introducing of these bodies the keystone of Euclid's thought structure, the last goal, to which absolutely all propositions of all books of the *Elements* (with the exception of those on the theory of numbers) refer, so that for him that, which would remain over after striking out the last book, would appear only as a shapeless heap of propositions. Since not long before his time, Peter Ramus, the French mathematician and philosopher, had submitted Euclid's work to sharp criticism and thereby founded a school, Kepler at the very beginning of his researches directed a violent attack against this man and determinedly maintained his idealistic standpoint in opposition to the other's nominalistic point of view. The controversy of universals, which had divided the medieval philosophers in two camps, here comes to light in the domain of mathematics revived in the contrast between the two men.

With the great thoroughness which was consonant with his manner, Kepler investigated the quality, of being such as can be constructed, of the individual regular polygons. With his mathematical aids, he certainly was not able to furnish a proof of impossibility, even though he once started out to do so for the heptagon. He also was not satisfied simply to count those polygons which can be constructed. That would have occurred soon according to the contemporary state of mathematical knowledge. He needs more; for his theory of harmony he needs an arrangement by rank of these polygons that can be constructed. He is furnished with these, primarily by the character of the irrational ratio of the side of a polygon to the radius of the circumscribed circle. The aid for this classifying is now again supplied him by Euclid who, in the tenth book of the *Elements*, as the most brilliant performance of Greek mathematics, introduced and distinguished thirteen classes of irrationalities. In this manner Kepler investigates the three-, four-, five-sided figures and those polygons which result from these by continued doubling of the number of sides. The fifteen-sided figure can, it is true, also be constructed. Since he cannot, however, use it for his harmonies, it is put aside with the remark that its construction is an "improper" one, insomuch as it is produced in known manner with the help of the six- and ten-sided ones. That in addition to the polygons with nothing but salient corners, Kepler also examines as independent figures those regular star polygons which can be constructed and which are obtained by a continuous line, that is not made up of sections, deserves special mention as a mathematical accomplishment. The calculation of any polygon sides with the help of "Coss," as algebra was at that time designated, displeased him. He enters into it in detail and criticizes this procedure, whilst putting his finger on the difficulties which arise when an attempt is made to grasp continuous dimensions by discrete numbers.

Besides the quality of being such as can be constructed and the greater or lesser degree of the knowableness, that is the character of irrationality, there is still another characteristic which makes it possible to distinguish the regular polygons. Kepler calls it congruence. It deals with the question whether and how far a regular polygon is suitable, with the same or other regular polygons, to fill out the plane around a point without gaps and, with the same or other regular polygons, to form enclosed figures in space. It is a joy to watch him playing with visible pleasure with the building blocks of the regular polygons and unfolding a profusion of charming figures. The task which he has here set himself has much occupied mathematicians to this day. Since he pays special attention to whether a plane congruence can be continuous, that is if the whole plane can be filled up without gaps, and also draws the star polygon into the domain of his inquiries, he reaches results which deserve the special interest of mathematicians. In erecting solids out of regular polygons, which he follows systematically and by exhausting all possibilities, he reached a significant discovery. He was the first to continue the series of the five regular polygons by introducing two regular star polyhedra. However, the noteworthy achievement, which he thus accomplished, remained unnoticed for two and a half centuries. In 1810 Louis Poinsot, the French mathematician, independent of Kepler, rediscovered these two polyhedra along with two additional ones, and in the following year his countryman, Augustin Louis Cauchy, furnished the proof that these four-star polyhedra are the only ones there are. Kepler's prior claim was, notwithstanding, only noted a half century later and recognized as he deserved.

From the greater or lesser qualification of regular polygons for the forming of congruences, Kepler traced a classified list for them, which, however, does not exactly coincide with that supplied him by the knowableness of the sides of the polygon. A noteworthy difference also exists especially in the fact that the number of the polygons which can be constructed is infinitely great and the number shared by congruences, on the contrary, finite. All polygons of the latter group are, however, included in the former.

Since Kepler wants to base the harmonic ratios on the regular polygons, but there are infinitely many of them, whereas experience bears witness to only a finite quantity of harmonies, the question arises: how can the finite ones corresponding to experience be separated out of that infinite multitude? This he succeeded in doing in the following manner. Imagine a circle divided by the angles of all possible polygons that can be constructed (omitting the fifteen-sided) and pick out all those divisions for which the ratio of the two pieces formed by a division also corresponds to a figure that can be constructed. The finite group of divisions thus gained is the one sought. If, for example, the circle is divided in two by the side of the five-sided figure, then the two parts are to each other as $1:4$. Since the quadrilateral can be constructed, that division is, according to Kepler, harmonic. But if the circle is divided by the side of the octagon, then this division is not considered harmonic, since the pieces of the circle are to each other as $1:7$, and the seven-sided figure cannot be constructed. It is a different matter if the circle is divided by the two corners of the diagonal crossing the octagon. The division is harmonic, because the parts are to each other as $3:5$ and the five-sided figure can be constructed. If this principle is followed through, the result is the group of seven ratios: $1:2$, $1:3$, $1:4$, $1:5$, $1:6$, $2:5$, $3:8$ or their compliments: $1:2$, $2:3$, $3:4$, $4:5$, $5:6$, $3:5$, $5:8$. There are no further ratios which fulfill the stated requirement. Next imagine the circle drawn out in a line; then to the divisions according to these ratios there correspond precisely those seven concords of two tones, which we perceive as melodious sounds or consonances, namely the octave $1:2$, the fifth $2:3$, the fourth $3:4$, the big third $4:5$, the little third $5:6$, the big sixth $3:5$, the little sixth $5:8$. Kepler believed that here, in the deepest foundations of geometry, he had discovered the causes of these consonances. In accordance with the following scheme he set up a genealogical tree for these original harmonies. Starting with $1:1$, each following fraction results from the previous since, with the numerator unchanged, the sum of numerator and denominator of the previous fractions is placed in the denominator and then the complement to 1:1 is formed. The procedure is carried on until a number appears which is characteristic of a figure which cannot be constructed.

$$
\frac{1}{1}\left\{\frac{1}{2}\left\{
\begin{array}{l}
\frac{1}{3}\left\{
\begin{array}{l}
\frac{1}{4}\left\{
\begin{array}{l}
\frac{1}{5}\left\{
\begin{array}{l}
\frac{1}{6}\ldots\ 7\\
\frac{5}{6}\ldots 11
\end{array}\right.\\
\frac{4}{5}\quad\ldots\ 9
\end{array}\right.\\
\frac{3}{4}\quad\ldots\ 7
\end{array}\right.\\[2em]
\frac{2}{3}\left\{
\begin{array}{l}
\frac{2}{5}\quad\ldots\ 7\\
\frac{3}{5}\left\{
\begin{array}{l}
\frac{3}{8}\ \ldots 11\\
\frac{5}{8}\ \ldots 13
\end{array}\right.
\end{array}\right.
\end{array}\right.\right.
$$

In judging Kepler's presentation it is necessary, in the first place, to keep clearly in mind the mystic connection which exists between the psychic original experience of the melodious sound and certain numerical relationships. How does it happen that the tones of two identically constituted and identically stretched strings are perceived as a melodious sound when their lengths are somewhat like 2:3, but on the contrary as a discord when this relationship is like 4:7? Yet the statements, to discover a concord and to have this or that length, belong to two basically different domains of experience. This disagreement caused Kepler's desire for knowledge to catch fire and his joy in discovering is understandable in view of his interpretation of the subject of mathematical things. Reference will be made to the fact that since the time of Gauss it is known that there are additional regular polygons which can be constructed, such as the 17- and the 257-sided ones. Had Kepler been told this, he would certainly have been most surprised. He would have welcomed these mathematical discoveries, just as he was receptive to all new truths; he would, however, in the next instant have reflected on how he could build this new insight into his theory. He would not have left off. At any rate, he would have said that the "knowableness" of that new polygon was one too far removed to come into consideration for the formation of consonances.

Kepler's arguments over this part of his *Harmony* found ready ears in his time, since the thoughts of the Greeks in this domain were familiar to many. Kepler also analyzed this speculation of the Greeks. Nevertheless, his is another point of view. He fights with determination against the theories of harmony of the Greeks, when Plato traces the basis for the consonances from pure speculation about the three first numbers and their second and third powers *a priori*, when he rejects the thirds and sixths as consonances, and when other Greek theoreticians seek a smallest tone interval from which all consonances can be compounded. He rejects the "tyranny" of the ancients because they "violate the natural instinct of hearing." In the establishment of that, which is consonant, he starts with experience and seeks the cause for the realities of the experience in the forms of geometry rooted in the mind. A special part in this is played by the remarkable fact that, in the geometric as well as in the musical domain, a special significance is attached to the ratio $1:2$. Because the continuous doubling of the number of sides of a polygon that can be constructed repeatedly results in polygons that can be constructed, and the halving of a part of a circle is absolutely feasible in contrast to other divisions, so the ratios $1:2^n$ also found consonances of a special kind, octaves, which Kepler terms identical consonances. By this insight he felt himself strengthened in his axiom: what is called constructing in geometry is called concording in music.

The above exposition makes the transition from the geometrical domain to the musical, which Kepler now ploughed up in its entire breadth and in all directions. First, after introducing his consonances, he made straight for the goal, "to erect the magnificent edifice of the harmonic system or of the musical scale, an edifice whose furnishing is not arbitrary, as one might think, not a human discovery, which one can alter, but presents itself through and through in conformity with reason and nature, so that God the Creator, himself, has expressed it in harmonizing the heavenly motions." Not only the theoretician of music, but also the layman, finds it very fascinating to watch Kepler erecting this edifice in detail and, in accordance with a plan well thought out and weighed out to the last, placing stone upon stone, until it stands there completed. Thus from the consonances he traces the tone steps, according to which a melody can proceed in a natural manner, the big and little whole tones and the half tone. These, too, are approved by the hearing, not as a result of habit, but by natural instinct. The natural doubling of thirds and sixths gives him the opportunity of distinguishing kinds of tones, major and minor. After the diatonic scale, he demonstrates the structure of the chromatic and the "perfect" system of a double octave in all details. He had made himself familiar with the rules of tracing a melody within a melodius composition conforming with the rules of art. He expresses his thoughts about the composition of the various styles of tones, their effectiveness, the reality of performed music, the musical character of individual vocal parts, the use of dissonances, about syncopates and cadences, about fifth and octave steps, thoughts which give evidence on the one hand of his mathematical sense of order, on the other of his sensitive musical understanding. Yes, mathematics and music are more closely connected than many believe.

There is, however, in addition to music, something else, about which the circular soul or the soul circle makes a pronouncement—stimulations which nature, seen according to geometrical laws, exerts on the soul responding to those same laws. Sun, moon, planets wander among the fixed stars, their steps on the same street, the zodiac, in eternal symmetry. One body runs faster, another slower. Now two stand in opposition, now one overtakes the other. They look at each other at various angles and from various signs. An eternal game of change. Since the beginning, the astrologer's endeavor has been to seek, behind that, one idea and one bearing on human life. Tycho Brahe clothed the thought at the base of this endeavor in the simple words: sun, moon, and fixed stars would have sufficed the world for measuring time and for adornment; now if, besides, the planets with their motions, wonderful and difficult to explain, are added, this must mean something special, a relationship between these heavenly bodies and the world and man. We have long known that Kepler, too, shared this notion, and we have also already learned the fundaments of his theory of aspects (pp. 94–5). He now again rejects, as enormously mistaken beliefs, the conventional opinions which speak of the powers of the zodiac signs, of good and bad planets, of an indirect or not indirect influence of the light rays as such. His theory of aspects rests on his theory of the soul. Just as our eye, whose functional activity is determined by the soul, sees the necessary external objects standing in spherical arrangement around the point-shaped soul, so the zodiac is not a real circle, but an image of the soul. The soul creates the zodiac as a projection, as it were, of itself. Thus the sympathy between the zodiac and the soul consists in the soul being reminded of itself when perceiving the aspects. "The effect of the configuration rests not on its own strength but on the strength of the soul. Instead of suffering something, as it is said, the soul, in reality is active, working upon itself." Thus it is understood how according to Kepler the geometrically formed soul becomes excited, when the heavenly bodies traveling on the zodiac form with each other the kind of angles which appear as angles at the center of regular polygons which can be constructed. The excitement expresses itself in the human souls at the time of the heavenly aspects instinctively experiencing a special impulse for completing the business and tasks at hand. "What the stick is to oxen, spurs or training to the horse, drum and trumpet to the soldier, a fiery speech to listeners, the rhythm of the flute, bagpipes or fiddle to crowds of peasants, that all the heavenly configuration of suitable planets is, especially if they are together. The individual is driven on in his doing and thinking, the whole becomes more willing to go together and lend each other a helping hand."

From the above it is also understood how Kepler pictured the influence of the birth constellation on the human soul. Since its own vital soul wealth then begins to act and its actuality is aroused when it is ignited inside in the lamp of the heart by being born, the soul absorbs the form of the zodiac which is characterized by the momentary position of the wandering stars in relation to one another, and also by definite points, especially such as the ascendant, that is the just rising point of the zodiac. This form of the zodiac remains impressed on the soul for the whole life. Accordingly, conclusions about a man's fate can be drawn from the birth horoscope only in so far as the soul, as a result of this formation, is excited to special activity at those times when related constellations make their appearance with the birth horoscope. In this connection he makes application to his own horoscope and rejects the assumption that his achievements and fates had been determined by the heavenly bodies. "My heavenly bodies were not the rising Mercury in the angle of the seventh house in quadrature to Mars, but Copernicus and Tycho Brahe, without whose journal of observations everything which I have up to now pulled into bright light would be buried in darkness. My ruler was not Saturn, the ruler of Mercury; my lords were the sublime Emperors Rudolph and Matthias. The house of the planets was not provided by Capricorn with Saturn; Upper Austria formed the house of the emperor. To this must be added the liberality, in unusual measure, shown to me at my request by its representatives. Here is the corner, not the angle of setting of the nativity, but the earthly corner into which I have, with the approval of my imperial lord, withdrawn from all the over-restless court and in which, through the passage of years, already approaching the end of my life, I compose my work on harmony and whatever else I have in progress. The only thing which the birth constellation has effected is that it blew the little flame of aptitude and discernment, spurred the mind on to tireless work and increased the thirst for knowledge."

In this connection, the explanation of the phenomenon of the earth soul has a special significance. Kepler cannot find enough words to say everything which crowds in on him. Yes, the earth has a soul in which the gift of differentiating the harmonious proportions was implanted at the creation of the world and which estimates the angle of two shining heavenly bodies for itself, compares it with four right angles, distinguishes the harmonious from the not harmonious and so provides harmony with its intelligible being. According to the material, the soul is for him a kind of flame, as shown by the subterranean, continual, noticeable warmth. But impressed on it as form is the picture of the divine countenance with the idea of the circle and all its connections, the idea of the sensual body, in charge of whose guidance it is placed like the idea of the whole world in which the body is supposed to live. "So therefore in return there also shines in the earth soul the image of the sense zodiac, as well as of the whole firmament as a band of sympathy between the things in heaven and on earth." Kepler pictures this earth soul or, as was said at that time, sublunary nature, poured out over the whole body of our foster mother earth and rooted in a precise part of this body as is the human soul in the heart, whence as from a hearth, a source or a central point, it goes out through its species to the ocean flowing around the earth and to the sea of air. That would be attested by a distinctly recognizable, obvious excitement of the bowels of the earth on those very days when the planets form a harmonious constellation with their rays to the earth. This being so, the earth behaves as a potential zodiac like one who, listening to the pleasing music of a singer, testifies that he understands and recognizes that which is harmonious in the melody by a serene air, by voice, by applauding and stamping with hands and feet in time with the melody. The body of the earth and the subterranean workshops in the mountains become warmed and excited at such moments by the affection of the vital faculty, so that these workshops exhale a great amount of vapor and cloud out of which, as a result of colliding with the cold prevailing round about in the upper regions, weather phenomena of every type take form. "The configurations strike up; sublunary nature dances after the fashion of the music." For the nature of the earth soul consists, like that of all souls, in action, as that of the flame in blazing. The earth soul is continually irradiated by God, in so far as it is an image of God, the actual substance.

Carried away by his flowing fantasy, Kepler compares phenomena and proceedings on and in the earth body with expressions of life of the animal body and finds analogies everywhere. Thus he here interprets the increase and decrease of the seas with the tides as the breathing of the earth body and compares this phenomenon with the activity of the fish, which draws in the water with its snout and pushes it out again through its gills. In the introduction to his work on Mars he had very clearly explained the tides as due to the moon's force of attraction. Galileo later reproved him for this and others also took offense at this explanation. Kepler had nevertheless rejected the criticism and energetically advocated that phenomenon's physical basis. Now, how does he come to interpret it as the breathing of the animated earth body? Two Keplers, so to speak, face each other. Both pursue to its final consequence one thought that they have caught hold of or which, rather, has caught hold of them. With the one Kepler it was the thought of gravitation, with the other that of the earth soul. Now they collide. What can they tell each other? The contradiction is unresolved. The mechanically thinking Kepler is silent. The animistic Kepler somewhat timidly supposes that philosophical circles would lend a friendly ear if he maintained that the earth suited her breathing to the motion of the sun and moon, as animals in sleeping and waking follow the change of night and day.

For the connection alleged by him between the position of the heavenly bodies and the weather, Kepler cites experience, which should be noted in this section in judging his enterprise. To this end he had observed the weather phenomena for twenty years. He had not taken his task lightly. On the contrary. It was precisely experience which had provided him with a phantom. What would have been nicer than if precisely those divisions of the circle, that the consonances furnish in music, had also proved effective in the zodiac; if complete agreement between the two domains had thus existed! Yet he was forced to admit that this was in no way universally the case. So he believed he would have to conclude from his long observations that the division marked out in the domain of music in the ratio 3 : 5 is almost useless with the aspects, whereas the configuration, in which two heavenly bodies are one-twelfth of a circle apart produces the desired effect, even though this division is rejected in music. He pictured himself confronted with the task of seeking new reasons for these facts of experience, naturally in geometry. "Meteorology and music are so to speak different peoples, both stemming from the common fatherland, geometry." The two peoples live by different laws. Whereas with music Kepler let the knowableness of the figures be decisive for the number and arrangement of the consonances, with aspects he put the congruences into the foreground. He was never at a loss for reasons. So he arranged an extremely complicated system of axioms and propositions, in which he established what he needed: a restriction of the number of the aspects corresponding to his observations and an arrangement by rank of these configurations according to the degree of their effectiveness.

But now Kepler comes to the most sublime and final statement for him to make in his *Harmony*. He had shown how man unites the innate harmonies of his mind with that wonderwork, the scale, and brings forth in music the profoundest means of expressing the life of his soul. He had shown how everything which has a soul receives shape and impulse from the motions of the heavens and thereby is projected into the sequence of history. Now he no longer talks of man's work and man's lot. His glance rises above terrestrial events and turns to the contemplation of the cosmos. God, Himself, in His wisdom and goodness created the cosmos when building the heavenly world and tuned it in accord with harmonic relationships, a wonderwork from which the highest, that human mind and human hand creates, borrows measure and rule in accordance with hidden connections. His intention is to demonstrate to the least detail the plan which God has realized by His work of creation. The way to this goal is pointed out to him by his sure and fast conviction that "the Creator, the source of all wisdom, the permanent preserver of order, the eternal, supernatural source of geometry and harmony, that this heavenly craftsman all by Himself has connected the harmonic proportions resulting from the plane figures, with the five spatial regular figures, in order to form out of the two classes of figures a single most perfect model of the heaven. In this model on the one hand the ideas of the spheres carrying around the six heavenly bodies were expressed by means of the five spatial figures, and on the other hand the sizes of the eccentricities of the individual orbits were contained by means of the descendants of the plane figures, the harmonies, for the purpose of a corresponding regulation of the planet motions. Out of these two parts, a unified, balanced system was to be made." The mind of the enthusiastic scholar, thus set in motion, expressed itself of its own accord in his pious heart in an introductory prayer to the Father of the spirits, the Donor of mortal senses, the Founder of the heavens.

According to Kepler, there are two pillars on which the divine plan of the world rests: the five regular bodies, by which the number of the planets and their distances from the sun are determined, and the original harmonies or consonances, by which the eccentricities and the periods are causally explained. The first principle is known to us; it forms the basis of his work as a young man. He still sticks to this idea. It had, of course, always bothered him that the distances of the planets from the sun are not presented exactly by inserting the regular bodies between the spheres of the planets. But now he will find an explanation for this disagreement.

Twenty years before, he had tried to point out the musical harmonies in the motions of the planets. But now he uses a different approach. He wants to point out these harmonies, not in the semi-diameter of the orbit, not in the period, not in the orbital velocities but in that which an observing eye or impressionable soul perceives, namely in the angular velocities as they appear looking out from a definite designated place in the universe, that is from the body of the sun, which is the central point of the universe and the source of the motions for all planets. For each planet these angular velocities are least at aphelion, greatest at perihelion. Now it is the relationship of these extreme values to a harmonic system on which Kepler fixes his eye. He made himself a little table of these values, accurately furnished for him by Tycho Brahe's observations. And behold, it is evident to him not only that the ratios of the two extreme angular velocities for the individual planets come very close to harmonies, but that furthermore, by comparing the extreme velocities of two different planets at a time "immediately, at the first glance, the sun of the harmonies" breaks forth "in all clarity." That the numerical values do not agree exactly, made no difference to Kepler. On the contrary, he is going to show that these little divergencies are physically necessary.

There is a great difference between the harmonies for individual planets and those for pairs of planets. The former do not materialize in a particular point of time, because the maximum and minimum velocities are separated by about half a period of revolution. On the other hand, the latter indeed appear in a particular moment, when two planets are in one of their apsides at the same time. "Just as the simple or monodic song, the only one known to the ancients, is related to the polyphonic, elaborate music, a discovery of the last centuries, so also are the harmonies which the individual planets form related to the harmonies of the pairs of planets."

At closer investigation Kepler now discovers the welcome fact that the steps of the scale are expressed in the ratios of the apparent planet motions if he reduces the extreme velocities from two, as it were, to one single octave by dividing by a suitable power and establishes from one basic note the notes corresponding to the numerical proportions so obtained. If he co-ordinates the slowest motion of Saturn with the subcontra-*g* as basic note, then the most rapid motion of Mercury corresponds to the five-staffed *e*; thus, after the reduction nearly all notes of the major scale will be indicated by the extreme values of velocities. But still further. If the note *g* is assigned to the motion of Saturn at perihelion and proceeds in the same manner, then all notes of the minor scale will be expressed by means of the limits of the velocities, so that then also the kinds of notes are traced in the heaven. On contemplating the harmonies existing between the limiting velocities of different planets, it is apparent that even at definite points of time collective harmonies of all six planets can appear, and indeed in both kinds of notes. To be sure, the instances in which collective harmonies appear are separated from each other by very long intervals of time. Would it not perhaps be possible to determine the age of the universe from this, since it must still be admitted that at its creation a distinguished collective harmony would be started? More frequently harmonies of five or four planets appear. "Thus the heavenly motions are nothing but a continual music of several voices (which can be comprehended by the intellect, not the ear), a music which, by discordant stretching, as it were by syncopes and cadences throughout, makes straight for definite, designated, little clauses, even of six parts, and thereby puts differentiating attributes in the immeasurable passage of time."

Thus the mighty heavenly organ plays its eternal melody throughout the whole time of the world. In earthly music, heavenly music is only mirroring itself. When humans make music, they do so by virtue of the harmonies rooted in the soul only in imitation of this heavenly music. Soprano, alto, tenor, bass become expressed by individual planets. "It is, therefore, no longer to be wondered at that people arrange this very distinguished system of notes or note steps in scales, when it is seen that in doing so they really play only the part of imitator of the divine Creator and so to speak dramatically build the organization of the motions." The same is valid for music for more than one voice. "It is no longer remarkable that man, the imitator of his Creator, has finally discovered the art of polyphonic music that was unknown to the ancients. He wanted to perform the continuous duration of the time of the world in a fraction of an hour in terms of an artful symphony and thus taste the pleasure which the divine master craftsman takes in his works so far as possible in the very agreeable feeling of bliss, afforded him by this music in the imitation of God."

Yet Kepler is still not finished. His main idea only comes now. Although up to here he had proceeded inductively by basing his deliberations on the numerical values put forth by nature, now the deductive thinker appeared in the arena. Why was everything made just so and not differently? That was the question he had already asked himself in the *Mysterium Cosmographicum*. Now he approached it again. As he at that time arrived at the assertion, this or that regular body must have been inserted between this or that pair of planets, so he says now, this or that harmony must have been put between the velocities for this or that planet. What reasons make such assertions possible for him? They are based on the hierarchy in which he arranges the planets, the solids inserted between them and the harmonies, on the greater or lesser relationship which he establishes between the individual regular solids and the harmonies, as well as on the principle of symmetry or that of contrasts. Since now, however, as he shows with great acumen, the total number of harmonies cannot be accommodated with complete exactness, if by harmonizing the limiting motions of the six planets total harmonies, in both types of note, are to be made possible, then small, imperceptible corrections must be fixed up for individual ratios. He even draws on those reasons again, when a little must be added to or taken away from the numerical ratios, and explains why the change is precisely in this or that place. It is hard to imagine the effort he spent in carrying these demonstrations out to the last details. The situation was like that with his calculation of the orbit of Mars where, also, nothing was too much for him. Before him lay an almost impenetrable thicket. But he believed in the goal behind it and in the path which led there. At the end of his wearisome way, in sight of his goal, he exultantly calls out: "In this manner did He who is before all time and in eternity adorn the wonder works of His wisdom. Nowhere is anything too much, nowhere too little. Everything is doubled, one thing opposite the other. None lacks its counterpart. To each one He meted out its qualities. Who can ever tire of contemplating their splendor!"

There is still another important conclusion. From the ratio of the extremes of the velocities of a planet, there results (according to the area proposition) the ratio of the extremes of their distances. This gives their numerical eccentricity. Just as he once, in the *Mysterium Cosmographicum*, thought he had shown the number and distances of the planets from the regular solids *a priori*, so now he was convinced that he had succeeded in the same way for the eccentricities, too, with the help of the harmonies. In the work of his youth he had expressed the hope that the day might sometime come when this secret also would disclose itself, since indeed God had not meted out the eccentricities in these sizes to the individual planets at random and without reason. The day had come, the goal was reached.

The reader might be amazed that nothing has been said up to now about the so-called third planet law, which we have designated as the crowning of his contemplation of the cosmological harmonies. How did Kepler arrive at this law? What part does it play in the continuity of his researches? Why did he see in this discovery a triumph of his research? The third chapter of the fifth book of the *World Harmony* is entitled "The main propositions of astronomy necessary for the contemplation of the heavenly harmonies"[1] and, after describing and comparing the hypotheses of Copernicus and Tycho Brahe, has thirteen divisions. In the eighth is the following information: "After I had discovered true intervals of the orbits by ceaseless labor over a very long time and with the help of Brahe's observations, finally the true proportion of the periodic times to the proportion of the orbits showed itself to me. On the 8th of March of this year 1618, if exact information about the time is desired, it appeared in my head. But I was unlucky when I inserted it into the calculation, and rejected it as false. Finally, on May 15, it came again and with a new onset conquered the darkness of my mind, whereat there followed such an excellent agreement between my seventeen years of work at the Tychonic observations and my present deliberation that I at first believed that I had dreamed and assumed the sought for in the supporting proofs. But it is entirely certain and exact that the proportion between the periodic times of any two planets is precisely one and a half times the proportion of the mean distances."[2] In the technical language of today the wonderful law, which Kepler announced to the world hidden away in his book, is expressed: *The squares of the periodic times are to each other as the cubes of the mean distances.* With the ellipse proposition, which defines the shape of the orbit of a planet, and the area proposition, which defines its form of motion, this law has created an entirely new foundation for astronomical calculation. It is a great stride beyond Copernicus, who had been the first to show how to obtain the relative distances of the planets from the observations. For, while with him the ascertained values stand unrelated side by side, they now appear connected with each other by the periodic times, so that the entire number of the planets in truth presents itself as a system arranged by law. For Newton the law led to the statement of the law of gravitation. It was a momentous month, this May, 1618. Kepler, the man of peaceful work, ignited the torch of a new truth. Eight days later the men who make history gave the signal in Prague for the outbreak of the Thirty Years' War. And not forgetting the cares which crept about the discoverer: in the same May there finally ensued the first hearing of witnesses in his mother's trial.

Kepler had already shown his genius in the work of his youth by asking the question about the connection between the distances and periodic times (cf. p. 67). In this he had been guided by the thought of a force going out from the sun, causing the motions of the planets and

[1] ED. NOTE. *Svmma Doctrinae Astronomicae, Necessaria Ad Contemplationem Harmoniarvm Coelestivm.*

[2] ED. NOTE. *Johannes Kepler Gesammelte Werke,* VI, 302.

becoming weaker as its distance from the sun increased. The solution by approximation, which he had found at that time, did not satisfy him. As is apparent from the preceding report, he had now found the true law simply by trial, by comparing the powers of the known values of the distances with the periodic times. In the *Epitome* he later expresses himself more fully about his thought processes. We know that he explained the motion of the planets around the sun by the rotating sun pulling the planets around by the rays of force extending from it. According to his interpretation, these bodies offer a resistance to this motion, and this resistance is so much greater, the greater the mass of a planet. From analagous reasons, for which he is never at a loss, Kepler now assumes that the masses of two planets are to each other as the square roots of their distances. (In his time experience could teach him nothing better.) So, in the same relationship, the motions of the planets become slower, the further away they are from the sun. But since, in addition, the paths increase in simple proportion to the distances, by putting both proportions together, he obtains exactly that which his law asserts. There is no doubt that Kepler only explained this deduction afterwards.

Moreover, what part his new law plays in his *Harmony* is quickly told. From the angular velocities designated by the harmonies he obtains the mean velocities of the individual planets. The ratios of the mean angular velocities of various planets furnish him with the ratios of the periodic times. From these ratios he calculates, by his new law, the ratios of the distances. Now it has been said above that his world structure rests on these two pillars: on the insertion of the regular solids which supply the relative distances, and on the introduction of the harmonies furnished by the limiting angular velocities. Since from the angular velocities he can now calculate the distances, the new law, which he found, makes a bridge from one pillar to the other. That was his triumph. We understand the joy of discovery which filled him as he viewed his work from the arch which curved over the two pillars. Since everything united so nicely, he felt strengthened and fortified in his belief in the truth of his *a priori* premises.

Now the last remaining bit of disagreement is also explained. We know that he is always worried by the fact that the insertion of the regular solids does not present the distances of the planets exactly. Now it is clear to him: the existing discrepancies must be there; they are the necessary consequences of the dominant principle of harmony, which was to have been presented in the world structure and which hindered the perfection of the insertion. In order that the harmonies could be expressed in the motions, the values of the distances supplied by the regular solids had to undergo little changes. "The geometric cosmos of a perfect insertion had no place next to the other cosmos which was the most harmonic possible." The regular solids are material, harmony is form. The former describe the raw masses, the latter prescribe the fine structure, by which the whole becomes that which it is, a perfected work of art.

As Kepler had begun with a prayer, so at the close, in a singularly lovely prayer he turns to the Father of Light and thanks him for having given him pleasure in the work of his hands. May what he has written contribute to the glory of God and to the welfare of souls.

Harmonices Mundi Libri V is the title of the work in which Kepler communicated his researches to the public.[1] The printing, attended to by Johannes Plank, was completed in the summer of 1619. Twenty years before, Kepler had intended to dedicate the book to King James I of England. When the sheet with the dedication, which was not delivered until the beginning of 1620, appeared, James' son-in-law, Elector Frederick of the Palatinate, had entered Bohemia as counterking. Even if no essential facts have come to hand for the assertion advanced on several sides that the dedication had been forbidden by the censor, still it is understandable, after what has been said about the political situation of the time (p. 251), that it was such as would put Kepler in an awkward position. Indeed, in some of the extant copies, the dedication is lacking. In the production of the lovely tables of figures, Kepler made use of the skillful hand of William Schickard, later to become professor at Tübingen, with whom he had become acquainted on his trip to Württemberg in 1617 and with whom he later entered into correspondence. Today, copies of the *Harmony* are among the most valuable works by Kepler.

We have intentionally rendered the contents of the *World Harmony* with some completeness. Certainly for Kepler this book was his mind's favorite child. Those were the thoughts to which he clung during the trials of his life and which distributed light to him in the darkness which surrounded him. They formed the place of refuge, where he felt secure, which he recognized as his true home. However much we have said, of course our arguments still give only a weak conception of the wealth of the contents, especially of the penetration with which the author pursues his successions of thoughts to the furthest roots and ramifications. He does not paddle around in his

---

[1] AU. NOTE. Many authors who mention Kepler's book call it *Harmonices*, as though this word were a plural. How, in their opinion, the singular is expressed, they do not disclose to us. Naturally, "harmonices" is the genitive of ἁρμονική which Kepler has taken over in his Latin and which signifies theory of harmony. Let us hope that this very disturbing error will finally disappear!

ideas; he drains them off. To become acquainted with the entire splendor of this unusual flower, it is necessary to take hold of the work itself. The contents are only very inadequately rendered by most of the biographies, because their authors do not know them or cannot evaluate them. The third law is mentioned without an explanation of the context in which it appears in the book. That is to say, the pearl is taken out of its mounting, where, however, its whole charm first becomes important. But the style of this mounting does not correspond to the materialism of our time; it is full of ornamentation which is rich in references and with whose symbolic loveliness many do not know how to begin anything. It is trivial to object to Kepler's conception on the grounds that there are not six planets only, that in later times two or rather three[1] additional planets outside Saturn's orbit and many hundreds of little ones between Mars and Jupiter have been discovered. As if every scientific system, in which we frame the phenomena of nature, did not correspond only to the position of research of the time and could not be overturned the very next day by the discovery of new experiential facts!

Nevertheless, this is not the only question involved in the critique of the *Harmonice*. Who asks what is true and what is false in Plato's *Timaeus*? So also the measure of the positivists should not be used to value Kepler's book, nor should an attempt be made to weigh its contents by the scale of the modern physicists, even if its weight is significant by such test, also. In truth, if a work presents science with such a valuable contribution as the third planet law (not to mention the mathematical and musical fruits), then a critic must seek the lack in himself if he does not achieve an understanding of the manner of contemplating nature out of which the work has arisen. This lack consists in the thinking being stuck fast in a rut of a one-sided contemplation of nature. It has been forgotten that that which is visible is a symbol of that which is invisible. Therefore, the poet, the artist, brings us closer to nature and can convey more and profounder and better things about it. Someone who once has been plunged in the cosmos of Platonic philosophy lives from this truth. That was the case with Kepler. Therein rests his conviction that "all nature and all heavenly gracefulness is symbolized in geometry." Besides, Kepler understands the debt owed to the accuracy of scientific research and knows what he owes to experience. "These speculations may not *a priori* offend well-known experience, but must be brought into agreement with it." But to establish matters of fact is never his final goal. He lives and

---

[1] ED. NOTE. Uranus, Neptune and Pluto. This is not the place to discuss whether Pluto is really a planet.

works on another plane. With the accuracy of the researcher, who arranges and calculates observations, is united the power of shaping of an artist, who knows about the image, and the ardor of the seeker for God, who struggles with the angel. So his *Harmonice* appears as a great cosmic vision, woven out of science, poetry, philosophy, theology, mysticism, a vision risen from the abyss of the human mind, seen as a radiation from the countenance of God, nourished from the supply of the senses, molded in the belief in ratio, inflamed by the inspiration of the prophet. It belongs to the most sublime, which has been thought and devised by the human intellect, locked in the material world, and desiring to lift itself out of it. It is a grandiose fugue on the theme "world, soul, God" with a maestoso finale. By the thoughts on which it is fed, by the shapes according to which it is molded, it is the *summa* of the Renaissance.

9. *Controversy with Robert Fludd*

With his *Harmonice*, Kepler got into a controversy,[1] which deserves explanation because it throws a spotlight on the background of the time in which it happened. In a short appendix he had taken up a position against a recent work about the macrocosm and microcosm by Robert Fludd or de Fluctibus, the Oxford physician and theosophist. Since a "Musica mundana" and similar things were also talked of in this powerful folio volume, Kepler complied with the request directed to him by a third party and analyzed, by a concise comparison, the "colossal difference" which exists between his and Fludd's theory. The latter felt irritated by Kepler's purely factual argument and offended, in the knowledge of a high mission to which he imagined himself called as announcer of secret science and deep wisdom. Therefore, to defend his confused theory, he attacked his adversary in a long, sharp, polemical pamphlet. In the arrogant pose of the esoteric and mystagogue he lectured to Kepler, reproaching him for crass ignorance and ambition. Kepler's science, in Fludd's opinion, refers only to the outside of things. A distinction must be made between vulgar and formal mathematics. Only the chosen sages, skilled in formal mathematics, perceive nature truly; to the representatives of vulgar mathematics, among whom he also counts Kepler, and whom he calls bastards and stunted people, it remains invisible and hidden. These measure only

[1] ED. NOTE. For the controversy with Fludd see W. Pauli, *The Influence of Archetypal Ideas on the Scientific Theories of Kepler*, translated by Priscilla Silz, in *The Interpretation of Nature and the Psyche*, Bollengen Series, LI, New York: Pantheon Books, 1955, pp. 190–208.

the shadows instead of the reality of things. Fludd compares Kepler's astronomy to a "mystical astronomy." While Kepler stopped short with the outer movements of nature, he himself contemplates the inner and fundamental acts, which flow forth from nature. Chemistry is the true science, by which its expert earns the title of true philosopher and of natural magician, because he penetrates not only into nature but also into the inner holiness of nature and into the most hidden depths. Only the chemist perceives how nature, as infinite power, brings forth like from like, with what power it itself lets everything grow and nourishes everything, with what help, animal, stone, tree and all visible bodies arise from her, why it should be called the bond of the elements. Fludd rejects Kepler's doctrine of harmony and attacks him for rejecting the Pythagorean-Platonic number speculation. He gives scholars the advice, not to give up the Ptolemaic foundations of astronomy lightly, so that they should not take a frog in exchange for a good fish. So it goes on, on fifty-four thickly printed folio sheets.

These samples from Fludd's pamphlet are characteristic of the intellectual temper of that epoch. One who looks about in that departed era of writing and printing is astonished at the flood of astrological, alchemical, magical, cabbalistic, theosophic, mock mystic, and pseudoprophetic writings which held the intellects in a spell. The vaguer their content and the richer the promises they ventured in predictions, in communication of secret knowledge and abilities, the more readers they found. What was always being proclaimed under the name of Hermes Trismegistos passed for revelation, whereas imitation of the ideas of Paracelsus passed as the highest wisdom. This abnormal inclination made it possible for something like a secret league, the order of the so-called Rosicrucians, to arise. A couple of years before the appearance of the *Harmonice Mundi* four anonymous works caused a prodigious sensation, urging joining in a union of that name. There is good cause to assume that behind these writings stands a younger close compatriot of Kepler's, who had completed the same course studies as he, the theologian, Valentin Andreae, a grandson of Jacob Andreae, the father of the Formula of Concord. But in the straying and confusion of that time many restless heads had not understood that those writings dealt with this author's mystification. In a satire on the vices of the world and the world improvers, he wanted to hold a mirror before his contemporaries and, with a chaos of fantastically adorned nonsense, wanted to make fun of the Paracelsists, the alchemists and the fanatics of every kind. His writings were taken seriously and on all sides secret conventicles were founded. Fludd was the man who introduced this movement into England and transfused their teaching into his so-called fire philosophy.

Only unwillingly did Kepler reply to Fludd's foolish attack. This he did in his *Pro suo Opere Harmonices Mundi Apologia* in a form which revealed the skilled dialectician and the brilliant mind. It was not difficult for him to defend his own opinions and to disclose the weaknesses in his opponent's absurb assertions for which, as always at such an opportunity, Kepler's ready humor furnished him choice phrases. When Fludd, in the delusion of possessing deeper perception, held forth that he himself had the head in his hands, Kepler only the tail, then the latter replied humorously: "I hold the tail but with the hand; you clasp the head, if only it does not happen just in a dream." The widely disseminated writings, aiming to found and extend the order of the Rosicrucians, were naturally also known to Kepler. Yet he wanted to have nothing to do with a secret organization which feared the light. He urged the Brothers of the new order not to turn only to the "children of the truth," but also to go and to talk in the meetings of people, on the mountains and in public places, so that people would get to know their true doctrine. In another little writing from those years he turns once more against Rosicrucianism with the words: "What kind of unbalanced swarm of mental freaks now came flying with the *fama fraternitatis* [title of one of those writings], because in that same work this sentence was to be found: Germany is pregnant with a new birth, for which it needs a lot of godfather's money. When, by God's decree, the devil tries to play blind man's buff with humans, then he needs such a cloak of fanatic opinion with which to cover the eyes of reason."

Thus Kepler had drawn a sharp line of distinction between his world of thought and the occult literature of his time. "I hate all cabbalists," says he in a letter written in those years. And just so determinedly he rejected the "intellectual number prophets" who, like Johannes Faulhaber, the Ulm mathematician, and Paul Nagel, the Leipzig theologian, offered their secret goods in empty play with numbers and imagined themselves able to predict the future and the near end of the world from the Apocalypse and the Book of Daniel. In the face of all such pseudoscientific efforts, Kepler most strikingly characterized his manner of thought and the goal, which he also pursued in the *Harmonice*, when he says about his connection with Fludd: "One sees that Fludd takes his chief pleasure in incomprehensible picture puzzles of the reality, whereas I go forth from there, precisely to move into the bright light of knowledge the facts of nature which are veiled in darkness. The former is the subject of the chemist, followers of

Hermes and Paracelsus, the latter, on the contrary, the task of the mathematician."

Fludd answered Kepler's apology once more. The latter, however, did not want, as he says, to press this issue any longer and was silent. "I have moved mountains; it is astonishing how much smoke they expel."

10. Epitome Astronomiae Copernicanae

The mathematician, which Kepler considered himself, appeared in bright resplendence in a further work, born in those years and which must now be reviewed, his *Epitome Astronomiae Copernicanae*. The work is not unknown to us. We have seen its author occupied with it before his first journey to Württemberg and also learned that the first portion, comprising books one to three, was printed in 1617 by Johannes Plank in Linz. After completing the *World Harmony*, Kepler once more took up the work; at that time he was full of energy. In the summer of 1620 the fourth book went to press. Daily the author went back and forth between his work table and the printing establishment. Besides, the entrance of the Bavarians in July and the unrest accompanying the military occupation of the city made the work more difficult, but did not substantially impede its progress. It was worse when the threatening reports about the state of his mother's trial necessitated his presence there in person and he had to travel to Württemberg in September. He packed up a number of copies of the just-completed fourth book, which lacked only title and foreword, and took them and the manuscript of the still missing books, five to seven, with him. Since the trial, which had entered the critical stage, was only sluggishly conducted, Kepler seized the forced pauses to complete his work. In spite of the heavy psychological burden which oppressed him, he was sufficiently resilient to concentrate on the scientific questions which he had in mind, at the same time as on the negotiations and written work which the defense of his mother entailed. His physical theories caused him continually to supplement and change his manuscript. To satisfy his own demand for lucidity in the methodical treatment of the extensive material, he repeatedly had to file and refurbish. In scientific discussions with Maestlin in Tübingen, whither he repeatedly repaired, some thoughts with whose presentation he was not yet satisfied clarified themselves. There he also had the first draft of the title sheet for the fourth book printed. From January to March, 1621, he made use of a pause in the trial to visit his family in Regensburg. During this time his wife gave birth to a little daughter. But even in Regensburg the work did not let go of him. In Munich, where he stopped for a few days on the way through, he occupied himself with the calculation of eclipses. In June–July he passed some weeks in Frankfurt, where he carried on negotiations with Tampach, the publisher, and had the printing of the three last books of his *Epitome* attended to. Abbot Anton von Kremsmünster, the president of the prince's private exchequer, had subsidized it. The printing was completed in the autumn of 1621, at the same time his mother's trial for evil-doing terminated. Thus the completion of this work was closely interwoven with the vicissitudes of its author's life.

In compass the *Epitome* is the biggest work for which we are indebted to Kepler. However, it does not look so big because the print is so small. It stands as a thick volume in the little octavo format next to the thin folio volumes of the *Astronomia Nova* and the *Harmonice*. The material is divided into questions and answers in the form of a catechism. Here one becomes acquainted with Kepler the didactician who, moreover, found it was difficult to restrain the torrent of his thoughts in continuous conversation and to attain the clarity of style for which he strove. But since the questions comprehend divers conclusions from or objections to previous statements, one often has the impression of being present at a dialogue. The Keplerian eloquence, which keeps letting flashes of thoughts shine forth and is never at a loss for fascinating expressions, keeps the didactic form from being taken as schoolteacher coercion.

In the three first books Kepler deals with spherical astronomy. He understands how to win a new point of view for the material here presented, which, indeed, is treated in all relevant textbooks, since he pursues the phenomena in all directions and often confronts the reader in surprising manner with questions which elsewhere are passed over. Of especial significance are the arguments in which he establishes the rotation of the earth and exquisitely refutes the objections raised against it in his time. How is it possible with a rotating earth for a stone thrown vertically up to land again on the same place on the earth? For the range of a cannon ball to be the same whether it is shot off to the east or the west? For a movable body, not to be flung away as a result of the rotation like a stone placed on a rotating wheel? These and related questions, which Galileo later treats in detail in his dialogue on the systems of the world, and the answers to which are customarily particularly credited to the Italian, are already correctly solved here by Kepler. Heaviness and lightness, as Kepler in all clearness establishes in contradiction to Aristotle, are not opposite qualities, but different degrees of one and the same quality. When he compares the earth, rotating about a uniformly directed axis, with a top, a promising view of the motion of the earth is ushered in. So there was a quantity of new thoughts which he knew how to present to his contemporaries to confirm the Copernican theory.

But still more important are the following books, in which Kepler treats theoretical astronomy. Here he moves entirely on the new foundation laid by his discovery of the planet laws. Here he drains the rich store of new knowledge gained by his labor. To evaluate this part of his work, it should be compared with what the other textbooks of his time could say about the motions of the planets and their calculations. There are two distinct worlds here and there. Maestlin also had published an *Epitome Astronomiae*, which from 1582 to 1624 went through several editions and was in many hands.[1] There, even in the latest edition, which appeared after Kepler's *Epitome*, everything is presented according to the theory and precept of Ptolemy. Up to now no one had undertaken to lay the theory of Copernicus as a foundation of the description of the heavenly motions. But the Copernican point of view is not the only new thing which distinguishes Kepler's *Epitome*. Kepler had left Copernicus, also, far behind. He had created a new science; he had undertaken to conceive and to explain the motions physically; he had offered the first textbook of celestial mechanics. The entire furnishing of epicycles, which Ptolemy, Copernicus, and Tycho Brahe had employed as absolutely necessary aids for their planetary theories, had disappeared in his presentation. It had been an ingenious aid, proposed by Apollonius more than fifteen hundred years before; however, an aid that could only be useful purely for kinematic not physical treatment of the phenomena of motion. Gone was the axiomatic demand for uniform circular motions; gone the circular motions about a purely fictitious, immeasurable central point. Gone are the oscillations by which all predecessors had undertaken to explain the observed changes of latitude of the planets. Now all motions in the domain of the planet world are referred to the sun, the corporeal sun, which stands in the center of the world and leads the planets around by the moving force which issues from it, just as its light shines through and illumines the world. By tracing the interaction of this force with those physical qualities peculiar to the planets, Kepler lets the elliptical orbits take shape. The planets travel these orbits under the influence of this same force with velocities regulated according to law. And the basic law of the state in the monarchy of the solar system, to which all members are subject and which creates

---

[1] ED. NOTE. See page 46.

the order, is the relationship between the periodic times and the distances. Since Kepler also shows how to proceed by calculations in accord with the new laws, he sets entirely new tasks for the mathematicians and then solves them himself, as best he can with the aids he has. The Keplerian problem, already mentioned, is only recalled. The problem deals with the solution of the transcendental equation arising when calculating, by his laws, the position of a planet in its orbit at a given point of time. The integration tasks overcome by Kepler in his theory of the moon, which far surpassed all his predecessors, have up to now barely been heeded by the historians of mathematics.

It is understandable that we of today find much that is strange in this new world picture. We also meet here parts of this world picture which are fundamental for Kepler, the insertion of the regular figures between the planet spheres and the principle of harmony, even if these concepts recede very much into the background. Kepler's picture of the mechanism of the motion has already been noted (p. 138) in the discussion of his *Astronomia Nova*. In conformity with those notions he has the force going out from the sun decrease only in simple ratio to the distance. The animistic principle for the explanation of the motions competes with the mechanistic. So Kepler indeed believes he can explain the motions of the planets about the sun mechanically, but considers the assumption of a soul principle necessary for the explanation of the continual regular rotation of the earth, as well as of the sun. The setting forth of his world picture is widely impregnated by theological views. Final causes play a big part. The question "why is that so?" is repeatedly posed. Why are there six wandering stars? Why are the eccentricities exactly so big? Why is the earth's axis inclined toward the ecliptic? Why is this inclination exactly so big? Why do the sun and moon have nearly the same apparent size? etc. How can Kepler answer such questions? The earth's axis is inclined to the ecliptic in consideration of the people distributed over the whole surface of the earth, so that the change of the heavenly phenomena should extend to all places on the earth and consequently all people have a share in it. In the Copernican astronomy this can be most cleverly proved by the most attractive reasons. Sun and moon have equal apparent sizes, so that the eclipses, one of the spectacles arranged by the Creator for instructing observing creatures in the orbital relations of sun and moon, can occur. The earth moves around the sun to make it possible for man to get to know the world and its dimensions. This would not be possible for him, as he is created, if he were confined to one place in the great structure of the world, and, so to speak, locked in one room. But if the earth moves around the sun, then he can, so to speak, walk about in the building and like a surveyor occupy various positions and take his measurements from these. Since the world was created as the most beautiful possible and beauty is based on well-arranged proportions, Kepler seeks to trace archetypal form-relationships everywhere. Where experience and observation give him no answer to his questions, he proceeds boldly in the use of a prioristic conclusions from analogy. He thought he could in this manner determine relationships between the sizes, volumes, densities, distances of the sun, moon, earth and planets. Never is he at a loss for an hypothesis. It agrees with nature, the adornment of the world requires—by these and similar phrases he manages to found his daring thought constructions. Thus old and new, mechanistic and animistic considerations, causal and teleological principles, Platonic speculations and scholastic abstractions are interwoven into a fascinating picture that contains and combines everything created by previous astronomers, but besides, striding far ahead of all, announces a new scientific style.

Kepler gave his work the title *Epitome Astronomiae Copernicanae*, "Compendium of the Copernican astronomy." This title does not correspond to the importance which the work deserves. It is not doing justice to it to designate it as a textbook, even though it is in the form of one; it is not to be compared with the other textbooks of astronomy then in use. For with its difficult contents it was no textbook for students, but rather a handbook for the professors, for the researchers, whom it challenges to relearn completely. No, the *Epitome* ranks next to Ptolemy's *Almagest* and Copernicus' *Revolutiones* as the first systematic complete presentation of astronomy to introduce the idea of modern celestial mechanics founded by Kepler. Not the least reason that the conventional presentations of the history of science do not concede this place to the work is the fact that it is too little known. Ulrich Junius, a professor at Leipzig and a native of Ulm, who zealously collected and examined Kepler's writings, said, eighty years after the latter's death, that these works are unknown to most mathematicians. It might be believed that this reproach is valid until this very day. Junius characterizes the *Epitome* as a "treasure of immeasurable and not yet drained erudition." But the title which Kepler had given this work also bears the blame for its having been valued too low. The title suggests that the work deals only with what Copernicus had taught in his *Revolutiones*. The title gives no inkling that Kepler had erected an entirely new structure on the foundation of the Copernican theory, that he had rescued the Copernican conception, at that time disputed and little believed, and helped it to break through by introducing his planet laws and by treating the phenomena of the motions physically.

It is part of a truly Keplerian modesty that he pushes his own accomplishment, which led far beyond Copernicus, so far in the background that at the beginning of the fifth book he justifies himself, yes frankly excuses himself, for introducing his own achievement under the banner "Copernican astronomy." How very much his behavior contrasts with the custom otherwise usual in the scholarly world! Here, too, it is evident that the great man does not care for his own fame but for the substance. Only he should not have been left to suffer for centuries for his own modesty. What was close to his heart was the dissemination of the Copernican theory. "I consider it my duty and task," he writes in the dedication to the Upper Austrian representatives, "with all the strength of my mind to advocate to the outside, among the readers, what I in my innermost being have recognized as true and whose beauty fills me with unbelievable rapture on contemplation." The sister of his modesty is his piety which he expresses in the same dedication when he wishes his work to be interpreted as a hymn which he as "priest of God at the book of nature" has composed for the glory of the Creator.

While Kepler was busy with the publication of his presentation of the Copernican theoretical structure, the fight with the Church about the new theory was already ignited in Rome by Galileo's imprudent demeanor. The result was that on March 5, 1616, Copernicus' work was banned. In the summer of 1619, in the very days when Kepler had to receive from Württemberg theologians the final corroboration of his exclusion from communion, he received from Johannes Remus, the imperial physician-in-ordinary, the news that the first part of his *Epitome*, which appeared in 1617, likewise had been placed on the Index of prohibited books by the holy officialdom. That had happened on May 10, 1619. The news alarmed Kepler understandably. He feared that should censorship be granted in Austria also, he could no longer find a printer there. Yes, he pictured the situation so black that he supposed he would be given to understand he should renounce the calling of astronomer, after he had already nearly grown old in the dissemination of the Copernican theory, without anyone up to then having raised objections. If in Austria there was no longer any place for philosophical freedom, he would ultimately also have to renounce this land. Remus nevertheless quieted him. There was no ground for fear—neither in Italy nor in Austria; only he would have to keep within his bounds and rule his feelings. His book could be read in Rome and in all Italy by scholars and people learned in astronomy, after the privilege was received. And Vincenzo Bianchi, the Venetian, with whom Kepler oftentimes corresponded in those years tried to allay his fears with the observation that in Italy books by distinguished German scholars, even if prohibited, would be secretly bought and read so much the more attentively.

Since Kepler also was apprehensive for the circulation of his *World Harmony* which appeared just then, because Copernicus' theory was upheld in it, he directed an epistle to the Italian book dealers. In this he tried to meet the precarious situation which had arisen because of the theory of the earth's motion, because "the rough procedure of a few who reported on astronomical theories not in the right place and not according to appropriate methods" had caused the prohibition. He appeared in it as Copernicus' counsel and asked the censors to open a new proceeding and to examine the overwhelming new evidence which he, unfortunately too late, produced in favor of the truth of the Copernican theory. In the meanwhile, however, the book dealers were to sell his work only to the highest clergy, the most important philosophers, the most learned mathematicians, the deepest metaphysicians. These might consider whether the immeasurable beauty of the divine works should be made known to the people or rather be withheld from them and its glory be suppressed by censorship. In another place he appended to the same thought the admonition: "Once the sharp edge of the axe has hit iron, then thereafter it is no longer useful for wood. Think of that if it commences."

But Kepler had not only undermined the censorship by establishing the Copernican theory deeper and protecting it, but he had also handed to the theologians the exegetic axiom, which should have made it possible for them to bring this theory into accord with the passages in the Holy Scriptures which appeared to them to be contradictory. It is known that the theologians one and all believed it necessary to interpret those places literally, as when it says in the Book of Joshua 10, 12: Sun stand thou still, as though otherwise the sun moved, or when it is written in the 104th Psalm: God established the earth on its foundations that it should not be moved for ever and ever, or when in the thirty-eighth chapter of the Book of Job, Jehovah holds before the eyes of the martyr, pleading with him, His power which He had shown, because He had sunk the pillars of the earth and laid its cornerstones. To this literal interpretation Kepler repeatedly opposed the thesis that it is not the purpose of the Holy Scriptures to instruct men in natural things. This he had already done especially impressively in the introduction to the *Astronomia Nova* and now also did in the *Epitome*. Rather in this domain it is better to conform to the human use of language to speak with humans, in the human way in order to be understood by the humans, and to make use of such expressions only to impart to them higher and divine things and to place before their eyes the size and power of God in His mighty and glorious creation. Kepler considered it exceedingly important to gain authority for this thesis and thus to show the compatibility of the new astronomical theory with the Holy Scriptures. Earlier in a letter to Herwart von Hohenburg he had already characterized his position in regard to this with the words: "There is nothing in all the sciences which could hold me from my opinion [concerning Copernicus], nothing, except solely and alone the authority of the Holy Scriptures, which is badly distorted by some." So the theologians had to let themselves be taught by the astronomer, even though they repeatedly reproached him for knowing nothing about theology, and warned him to keep his finger out of it.

In the years 1617–1619, simultaneously with the *Epitome*, the first volume of the *Ephemerides* for the years 1617–1620 was also completed. The publication of this volume had, as he says, given more printing work to him than to the printer himself. The supplementary printed ephemerides for 1617 contained his weather observations for each day of the year.

11. Calendars and comets

Although the previously quoted works by Kepler had grown from plans and designs long held, others reflect the events of that restless and agitated time. That holds especially for the calendars, six in all, which he had composed in Linz between 1617 and 1624. Here he not only expresses his opinions about the future weather, but also reflects on the circumstances of the time, the former candidate for theology pouring out emphatic moral sermons to his readers. "Forgive the well-meant indiscretion with which I insert morals into astrology." He struck the best tone. His calendars were much in demand; for example, the one for 1619 was sold out within eight days of its appearance. For 1618 he prophesied "that in the coming spring not only the weather but also the course of the planets will make the heart of many a fresh cock bloom and give him valor." "Then truly in May great troubles will not be avoided in those places and in those affairs, where everything is ready ahead of time and especially where the community otherwise has great freedom, unless there is a wakeful eye on them." Special conjunctions in Taurus put these prophecies into the mind of the calendar man. Truly, as is known, the great unhappy war began in May, 1618, with the Prague defenestration, so that the next year Kepler could establish in his calendar that: "Last May the tinder took fire as I had warned before and particularly with reference to May."

"But if a true comet should appear in the heaven itself or an earthquake should ensue, then it would be time for us astronomers along with the politicians to sharpen our pens." So it said in the calendar for 1618. And truly, as though the calendar writer had divined it, in the fall of this year there appeared not just one but three comets in rapid succession. The third in particular, with its long tail, must have presented an impressive spectacle. The head contained a bright nucleus so that it appeared as though an aristocratic young lady was driving along in a coach. What wonder that the masses believing in the stars, who were accustomed to see in comets the divine rod of correction, were alarmed by the threefold frightening sign in the heaven and likewise asked with so much greater fear when the calamity would befall. What wonder that, like a swarm of magpies, the astrologers alit in the fertile cornfield. Kepler, too, sharpened his pen. Calling and inclination pressed it into his hand. In 1608, on the occasion of the appearance of a comet in the previous year, in a little work in German, he had expressed himself about these strange heavenly signs. Already then he wanted to publish this work in Latin also. But, for various reasons, the printing, which was supposed to be done in Leipzig, did not take place. The theological censors had raised objections to one place, and Professor Joachim Tanckius, who was supposed to take care of the printing, died. Now Kepler wanted to stretch out further. He wrote an entire book in the Latin language, and worked that first tract into it.

Observations and calculations occupy the principal part of his comet book, which appeared in 1619–1620 in Augsburg. Kepler employed great care and effort to obtain as much observational material as he could from printed writings and private communications. To be sure, in determining the orbit, he made an erroneous assumption, to which he held fast all his life. Namely he makes the assumption that a comet moves uniformly or constantly, accelerated or retarded on a straight line. The curvature which the apparent motion presents he wanted explained as the reflex of the motion of the earth about the sun. He was kept from reaching exact results regarding the orbit by the impossibility of determining a parallax from the crude observations in his possession. For him it was an understood thing that comets, however, are not atmospheric phenomena, as Aristotle and even his followers in Kepler's time believed, but travel their orbits in much higher regions, as Brahe had first established by parallax determinations. He considers that "in the heavenly, everywhere universal and empty air" comets move about in great numbers like whales in the ocean. Just as he has these occur by spontaneous generation, so also are the comets supposed to form themselves out of the celestial air by thickening, since "such fat material is gathered together, so to speak like a boil, and illuminated according to its nature and like other stars endowed with a motion." With the explanation of the formation of the tail flashes the modern concept of the pressure of rays, when Kepler says: "The sun's beams pass through the body of the comet and instantaneously take something of its matter with them on their path away from the sun." Some decades later his observations of the comet of 1607 gained special significance when they served Edmund Halley, the English astronomer, as proof that he was dealing with a periodically returning comet with a period of about seventy-five years.

Many readers of his book were, however, more interested in the question: what do the celestial signs signify? Kepler distinguishes a threefold influence of the comet. The first is natural and is demonstrated in the defiling of the air, in the unusual case of the tail touching the earth. The second rests on sympathy, and corresponds to a notion of Kepler's already known to us, which he here expresses in the words: "When something unusual arises in the heaven, whether from strong constellations or from new hairy stars, then the whole of nature, and all living forces of all natural things feel it and are horror stricken. This sympathy with the heaven particularly belongs to that living force which resides in the earth and regulates its inner works. If it is alarmed at one place it will, in accordance with its quality, drive up and perspire forth many damp vapors. From there arise long lasting rains and floods, and therewith (because we live by air) universal epidemics, headaches, dizziness, catarrh (as in the year 1582), and even pestilence (as in the year 1596)." Just as the animated earth thus is affected, so the human also is agitated from the heaven. "Man has such vivid and sensitive forces, attuned in a secret manner to heaven, that they would become disturbed and dismayed by the appearance of new comet stars even if he were blind and had never seen heaven. Thus disturbance causes not only unnatural movements of the blood and other humors and consequently illnesses but also strong emotions."

In the third place Kepler treats the ominous significance of comets, which was the most important for all astrologers. Kepler knows that astrological usage is erected on superstition, on a "universal idea of all people," when it draws its prophecies from the individual phenomena in the comet's course. And still he is convinced that these announce evil and bring misfortune in their train. He can append a series of evidences for this from history. "It is in accordance with history that with the appearance of comets there commonly arise long lasting troubles, which together with the death of a great many people also bring fear and grief to the survivors, and this not only by the disappearance of a potentate and the consequent change in government, but also by many other causes." How does that tally? Here again is a view characteristic of Kepler's thinking. The astrological rules, which are applied in the interpretation of comets, may well be superstitious. Only God accommodates His measuring to the rules as men have formed them, in order to let His special warning reach them. Some spiritual beings, perhaps the "lower servile spirits suspended here and there in the mid-air of heaven" take care that the comet appears under such conditions that people using their astrological rules read therefrom what God wanted said to them. "Therefore then this, its flight or trajectory, must first be measured and circled out by a supremely intelligent and indeed mathematical principle moving it at such a time and into such a region that it appears on earth in the previously chosen places and thus signifies that which the same high principle is willing to reveal to the human race by it." Now, Kepler himself plays with those rules, not because he believes in them, but to show that he could also guess (not to mention interpret). However, he wisely abstains from prognosticating anything definite and concrete. Political acuity forbids him that. Certainly, his position at the time of the appearance of the comet of 1618 was very delicate. The situation in the first phase of the great war, the fronts which opposed each other are known: here the emperor, there the rebellious Bohemians. The people wanted to know for whom the comets were displayed in the heaven as rod of correction. Kepler's neighbors in Linz doubtless mostly felt for the latter party. And Kepler? Now, he knew that he was imperial mathematician and therefore had to restrain himself. The people expected him to express himself as to which of the two parties would be victorious and to specify, as he says, who he was, whether fox or hare. But he did not let himself be pumped. "Each one wants me to prophesy what he desires, and inversely from what I prophesy draws conclusions about my party leanings." So, indeed, he makes a couple of remarks about the "animal" for whom he prophesies evil,[1] but for the rest he indulges in general admonitions which do not lack hidden allusions to contemporary circumstances. For he was convinced that the real purpose of the appearance of comets lies in the fact "that they are witnesses that there is a God in heaven, by whom all future fortune and misfortune is foreseen, announced, decreed, regulated, measured and governed, for the betterment of the Epicurean people, who believe in no God, for the cautioning of those who prophesy disaster and who cannot restrain

[1] See page 263. (Transferred from text.)

their desires, and for the consolation and admonition that they be patient of the small, weak, and oppressed people."

When a few years later Scipio Chiaramonti, an Italian, in a book which he entitled *Antitycho*, gave optic and geometric reasons to support the Aristotelian view of the character of comets, Kepler felt compelled once more to take up the physical side of the comet problem in defense of Brahe.[1] This he did in a temperamental polemic in a larger work with the title *Hyperaspistes*, in which he so to speak held a shield over his dead master. Since at the same time Galileo in a book voiced the same erroneous concept of the nature of comets as Chiaramonti and expressed himself unfavorably about Brahe, in an appendix Kepler also reproached Galileo for his mistakes.

A general view of the activity of Kepler in the very eventful years 1617–1621 is achieved if consideration is also given to the second edition of the work of his youth which appeared in 1621, in every chapter of which he added very extensive notes to the 1596 text. In truth, the zealous reaper gathered an extremely rich harvest during this period in spite of the gloomy climate which lay over the land in which he dwelt. By then, however, it had already become fall in the life of the fifty-year-old. Certainly, there was no rest for him and no enjoyment of the fruits which he had sown and reaped. He had to proceed on the road which had been marked out for him no matter how little sun shone over the autumnal days of his life.

12. *Wartime in Linz; the Counter Reformation*

When Kepler returned to Linz in November, 1621, after his long absence, he found living conditions there changed from before. Optimists might have believed that, with the complete suppression of the Bohemian rebellion after the battle at Weisser Berg, the war would have ended in favor of the emperor. Everywhere in the kingdom there were still too many unsettled antagonisms for the hope of peace to be fulfilled and the measures taken by the emperor in turning his victory to account were not suited to quiet the excited feelings and to remove the political tensions. Georg Friedrich von Baden, Christian von Braunschweig and Ernst von Mansfeld, followers of Frederick of the Palatinate, the defeated and exiled "Winter King," took up arms to prevent their own losses which threatened them, or, in their desire for adventure, to realize ambitious plans. The bad

---

[1] ED. NOTE. Chiaramonti's *Antitycho* (1621) used Tycho's own observations in an attempt to prove comets sublunar. See Hellman, *op cit.*, pp. 315-16, 439-40.

weather, which had broken out over Bohemia, had then settled in the Palatinate in the neighborhood of the Neckar and the Rhine. Even if the Austrian lands were not themselves thus also directly involved in the military events, still no one could foresee the further course of things, and the outlook was gloomy and dark. Besides, the emperor's Counter Reformation measures, which in Bohemia led to a complete eradication of the new doctrine, also worked themselves out in Upper Austria and especially Linz, the capital. Prominent Protestant leaders were neutralized or had had to quit the country. The exercise of the divine service as prescribed by the Augsburg Confession was most severely impaired. The irritation thereby called forth vibrated through the life in the city. Meanwhile, the Bavarian garrison, which had marched into Upper Austria shortly before Kepler's departure for Württemberg, had firmly established itself and, in city and country, was felt as an oppressing burden, due to the sharp rule it exercised.

In the period that followed, Kepler spent his days in this gray atmosphere. Continual thought of present need and of future uncertainty is a bad traveling companion. But there was one piece of luck— the latent dangers, for the time being, did not lead to great explosions. The curve of the astronomer's life shows no singularities or unsteady places for the two to three years after his return. Yet his sensitive nature was deeply grieved or menaced by the dispute that blazed everywhere. He detested the warlike reports with their "barbaric neighing." Even more than the material losses he deplored the ravages in the intellectual domain which the tragic war brought in its train. "Of what great values do the unfortunate people deprive one another by their shabby and disgraceful disputes! Having become guilty, how deeply do they sink in misconstruing their destiny! What lamentable decision forces them then to run into the middle of new fires when trying to escape one fire!" He wants to help. As cure for the ills of his time, he extols occupation with mathematical and philosophical studies, since with reference to Plato he expresses the conviction, "that these studies lead the mind from ambition and other passions, out of which wars and the other evils arise, to love of peace and to moderation in all things." In the contemplation of the heavenly works he wants to offer enjoyments, assuredly not unworthy of a Christian person, and give solace in affliction. A weak medicine, certainly, for a sick person who raves and thrashes about, however well the prescription is meant by the physician! His age was no longer to be helped by advice. Kepler turned to God, praying for peace. "The more a person loves mathematics, the more devout is his devotion to God, and the more he applies himself to thankfulness, which is the crown of virtues, that much more earnestly will he with me unite his prayers to the merciful God: let him suppress the warlike disorders, eliminate ravages, extinguish hate, lead forth the golden peace again." In the confidence in God's mercifulness, the hope for better days for the country, which has become his second home, grows up for the pious man. "When this frightful storm has quieted and the clouds are dispersed, then will the divine mercifulness again let the sun shine for the penitents, it will again lead forth peace and in Austria again create space for the arts of peace, for whose exercise it indeed never ceased caring, and in that place let a number of men arise, who learn the praise of God their Creator from these arts." Kepler's hope, to be sure, was never fulfilled in his lifetime. The psychological oppression that weighed on him grew stronger and at times called forth melancholy moods in him. How strongly the universal need rested on his heart and how strongly he felt himself united with his people is evident from the lovely words which he wrote in 1623 to his friend Peter Crüger in Danzig: "I was nearly the only one to have the good fortune that the unfortunate war events shunned me, even if the pleasure in my favorable situation becomes disturbed by the pain over the disdainful treatment by which my cherished fatherland wants to have nothing to do with me alone, even though the best [people] desire to go down together with it."

At that time the action of the Counter Reformation pushed Kepler into a situation which forced him into a defense in two directions, since the followers of both the confessions fighting each other lacked understanding for the attitude which his conscience prescribed for him. It was known that his last attempt to reverse his exclusion from the evangelical church was wrecked by his rejection of the ubiquity dogma. His coreligionists kept on reviling him as a renegade, whereas the Catholics, who wanted to win him over, perceived a chance for themselves in the exclusion. Consequently, this upright man found himself in the agonizing position of having the side to which by conviction he belonged and to which at heart he wanted to belong reject him, and the other side, which he inwardly opposed, regard him as already half theirs. In the existing circumstances, the exclusion doubtless brought Kepler certain advantages. If in spite of his hitherto official position he was left unmolested at the Protestant convent school and not expelled like the other teachers, then certainly his contrast to his coreligionists certainly contributed to this. This circumstance may also have contributed to Ferdinand's ratification of Kepler as imperial mathematician. His brothers in the faith blamed him for his favored treatment by representing the matter as though he had striven for it and only because of it had persisted in his refusal to subscribe to the Formula of Concord.

On the other hand the Catholics showed him with a certain satisfaction the persecution which he would have had to suffer from his own church. It was against such imputations and temptations that his upright intellect defended itself. To the adversaries from his own ranks he said: "Which is better and more justifiable before God and the world: should I as a service to the great masses help to push through this article [of ubiquity] so that I should become the victim with them and on their side of a whole complete persecution? Or should I in discharge of my conscience prepare myself in this article to suffer what God will send me, and for the rest willingly accept that advantage which might thus accrue? Would it not be tempting God to share in this article the errors (as I see them) of others and wilfully enter into the same danger with them?" But he reminded the Catholics that they should cease their vociferousness and reflect that it is a far worse persecution, as happened in Graz, to make and carry out laws by which a whole great community is forced either to quit the land in a few days or participate against their own conscience in the performance of the divine service, which at heart they consider error and idolatry. Yet, in the face of the Catholics he defended even the church leaders who had excluded him, by explaining they could not be accused of any persecution, if in their hearts they are truly convinced that because of their profession they have no right to deal with him otherwise than occurred. Besides, he had no intention of expecting them to do something against their conscience; he only wanted to make them understand that in his case they were stretching their conscience too tightly. Also, it was this independent and magnanimous way of thinking which even earlier had induced him not to break off personal contacts with Pastor Hitzler, for later, when this preacher was expelled and had found employment in a Württemberg church position, Kepler still exchanged greetings with him. Kepler placed freedom, conscience and peace above all else, and in his mouth the words brotherly love were no empty echo.

In this connection one more incident may be mentioned. It occurred in December, 1623. Kepler's calendar for the year 1624 (the last which he composed) was publicly burned in Graz, although it was dedicated to the Styrian representatives. A friend who asked him the cause imagined the action had taken place because the calendar speaks of "coercion to a hated divine service," which however will give rise to no very great agitation because there would be such a great scarcity and famine "that the common man would gladly exchange a chalice for a piece of bread." Kepler denied this conjecture and later answered that the burning was the result of anger over the fact that on the title of the calendar the province of Austria above the Enns was named before Styria. Were the people of Graz silly enough to carry out such an execution because of such a trifle?

13. Tabulae Rudolphinae

As always, when under the stress of need and persecution, Kepler sought and found encouragement and consolation in work. He wanted to keep his hands out of the "glue of politics" and instead stroll "in the lovely pastureland of philosophy." Now the time had come to finish the *Tables*. The impatience of those who were waiting for them had been increasing. In public and private remarks Kepler would be reminded to publish the work at last. He was also cognizant that it was being asked for "in all Europe, yes even in India, for many long years." From Italy and Frisia, from south and north, complaints came. People no longer expected to live until it appeared. David Fabricius had especially "strongly tapped" him and intimated he himself would have completed it in this length of time. Yet "he altered his tune" when he noticed that the discrepancies between calculation and observation at eclipses were still too great. All these impatient reminders Kepler had previously parried with the request: "Don't sentence me completely to the treadmill of mathematical calculations and leave me time for philosophical speculations, which are my sole delight. Each one has his own particular pleasure, the one tables and nativities, I the flower of astronomy, the artistic structure of the motions." That is what he said when he was occupied with the *World Harmony*. Now it was of great consequence to him to finish. "I am as eager for the publication as Germany is for peace."

In addition to a series of separate questions still to be cleared up, the main work was valuable for an important change which Kepler wanted to introduce in his tables. As early as 1617 he had first seen the famous 1614 work by John Napier, the Englishman: *Mirifici Logarithmorum Canonis Descriptio*, but had not had an opportunity of studying it. He had had to wait two years before obtaining greater insight into this Englishman's pioneering achievement. Kepler immediately saw clearly the significant simplification offered by the new logarithms for the many voluminous and time-consuming tasks of computation necessitated by the practice of astronomy. He thought that anyone who used his tables should have this excellent facility in a suitable form at his disposal. However, he was not content simply to accept the new mechanical aid as he found it. Napier, in his work, had simply presented the tables of numbers without stating how his logarithms were to be computed. So in the first instance his "wonderful canon" must have operated like a magic trick. In fact, in the beginning, mathematicians as serious as Maestlin mistrusted the new aid to calculation. Was it permissible for a rigorous mathematician to utilize numerical tables about whose construction he knew nothing? Was there not danger that employing them might lead to false conclusions, even if the calculation was proved to agree in many cases? When Kepler, during his visit in Württemberg in 1621, discussed these questions with Maestlin, the latter even ventured so far as to observe "it is not seemly for a professor of mathematics to be childishly pleased about any shortening of the calculations." Kepler differed. He wanted to prove and interpret the new aid to calculation by solid methods and subsequently calculate logarithms himself.

In the winter of 1621–1622 he carried this plan out and composed a book about the subject in which he again demonstrated his fine mathematical instinct. The work was an achievement completely independent of Napier's. Whereas the latter in establishing his logarithms, which were indeed not yet defined in today's sense as exponential powers, started from visible geometric notions, Kepler, in complete contradiction to his otherwise usual preference for geometry, derived his numbers from the theory of proportions, thus purely arithmetic, referring to the fifth book of Euclid. Certainly, it cannot be said that his arguments, in which considerations of infinitesimals and especially the mathematically important concept of limit again take part, are particularly clear. However, considering the novelty of his reflections, this is understandable.[1]

The printing of the logarithm book has an unusual history. Kepler sent the completed manuscript to Maestlin to have it printed in Tübingen. But Maestlin was not interested and postponed the matter. It took considerable effort on the part of Kepler's friend, William Schickard, to get the manuscript back from Maestlin. When this was finally successful in September, 1623, Kepler had just been requested by Landgrave Philip of Hesse-Butzbach to remove certain objections in the carrying out of logarithmic calculation. Therefore, he felt obliged to leave the printing of his work to this prince, to whom it is dedicated, and to leave it to him whether he wanted to order it printed in Tübingen under Schickard's guidance or whether he had

---

[1] AU. NOTE. The definition of the logarithm of a number x, to voice his thoughts in the modern style, is expressed by the formula: $\text{Log } x = \lim_{n \to \infty} 10^5 \cdot 2^n \left(1 - \sqrt[2^n]{\dfrac{x}{10^5}}\right)$. From this follows the equation for the Keplerian logarithm: $\text{Log } x = 10^5 \ln \dfrac{10^5}{x}$.

some fit person in Frankfort, who would correct carefully, "because there are lovely types there." Thereupon, Kepler heard nothing further about his opus until, to his great astonishment, he read in the catalogue of the 1624 autumn fair that it had appeared. The landgrave had had it printed in Marburg without getting in touch with Kepler again. Meanwhile, Kepler had worked at improving and expanding his logarithm tables and had put them in the form suitable for the astronomical tables.

Finally this work, too, was completed. "Video portum" (I see the port), Kepler, uttering a sigh of relief, could announce in December, 1623, to Edmund Gunter, the English astronomer, and a few months later he wrote to his friend Bernegger in Strasburg: "The *Rudolphine Tables*, which I received from Tycho Brahe as father, I have now carried and formed within me for twenty-two whole years, as little by little the foetus forms in the mother's body. Now the labor pains torment me." These were bad and long. The interference of Tycho Brahe's heirs, the money question, the choice of the place of printing retarded the publication. The solution of these questions turned out to be so much the more difficult because the political waves lashed higher and soon dashed against the walls of the Upper Austrian capital. Since the publication of the *Tables* belonged to the official tasks which devolved upon Kepler as imperial mathematician, and in them he had taken over a commission which was first bestowed on Tycho Brahe, he could not now do as he liked according to his own opinion, and in truth it was part of his nature's tough will, as well as of his intellect's good sense, to bring the completed work to a happy ending.

In spite of the long interval of time, the Brahe family was still jealous of Kepler. Admittedly, the most difficult of them, Tengnagel, had died in 1622. However, Tycho Brahe's son George, who henceforth appeared above all as representative of the family, also kept a careful watch that no danger befall his father's fame, although such a worry was completely unfounded as regards Kepler, to whom it never occurred to adorn himself with other people's feathers, and who at all times had unreservedly and thankfully acknowledged and proclaimed Tycho Brahe's merits. So it came about that now Kepler had to defend his right. Even without taking into account the previously mentioned agreement reached with Tengnagel in 1604, Kepler's hands were tied by a written declaration he had made to Tycho's heirs in 1612 to obtain their consent to his using the Tychonic register of observations in Linz. In it he had pledged himself to submit the completed manuscript of the tables for inspection before publication to the heirs or to an intelligent man agreeable to both sides and to pay regard to possible considerations as to the form, the publication and the location at which this was to take place. In the title and in the dedication he wanted to mention Tycho Brahe honorably as the first author and beginner and his observations as the correct foundation on which the tables were erected and to give himself no more credit than for what he had deduced from these observations. Accordingly Kepler forthwith communicated with the Brahe family. As mediator and arbitrator, in whom he placed his complete trust, he chose Severin Schato, the Bohemian physician, who had known Tycho Brahe personally. However, it soon became apparent that the transaction would not come off without friction. The heirs took the stand that the contract of 1604, despite the fact that it concerned only Tengnagel personally, possessed validity even after his death. Although lacking any knowledge of the field whatsoever, they posed as the patrons who wanted to repay Kepler for the trouble he had taken with the tables. Against this, the latter defended himself energetically. It was not the heirs who had hired him and, as directors, pushed the work, but rather the emperor was the "patron" and he his "appointed one." "It is quite sufficient that the heirs are next to and before me in the preface or dedication." In the learned world they did not have their father's prestige, and Kepler himself was the one who had to make a reputation for the work by what had been constructed out of the observations by his methods. Since not much was to be achieved by the agreed plan, Kepler immediately withdrew the authority conferred on his agent Schato.

The choice of a place of printing presented no fewer worries in those restless and uncertain times. Johannes Plank's Linz printing press was not sufficiently efficient for the great work and Plank, besides, was thinking about departing since Linz under the new conditions regarding the confession no longer pleased him. Kepler had to look about for another place. Whither should he turn, he asks friend Bernegger. The *Tables* cannot be printed in accord with his desire if he is not present. Consequently, either he needs a printing press in Austria or he must go away from home for a long time. In this case he would have to leave his family. If he were to take them along, they would be separated from their friends and forgotten by them. The household goods would be ruined by storage or would be endangered by the chaos of war. The whole need of the time and the desperate mood of the people is illuminated as though by lightning by the question: "Quis locus eligendus, vastatus an vastandus?" (Which place should I chose, one already laid waste, or one which has yet to be laid waste?) Bernegger would like to propose Strasburg, but he knows the greed of the printer there and the cheap attitude of those who would have to promote such an undertaking. The Frankfort people motioned him away with a reference to the "devastation of the German land." The city, however, to which Kepler above all turned his eyes was Ulm. In the course of war it seemed the first to offer him security, and Johann Baptist Hebenstreit, the rector of the Ulm gymnasium, a very learned humanist and for many years friendly with Kepler, did everything in order to smooth the paths to the realization of this plan. It must have been a joy for the astronomer to hear how very well disposed the Ulm senate was to him. The technical requirements for the printing also seemed to be best fulfilled there.

In considering the financing of his undertaking Kepler thought of drawing for this purpose upon Emperor Rudolph's old debts which were still in arrears. Indeed, the *Tables* were to bear his name. This arrangement appeared more wholesome and more appropriate to Kepler than to have recourse to the liberality of Emperor Ferdinand. There still remained 2,333 gulden in salary arrears and 3,966 gulden in yearly allowance from that period for him to claim. To obtain the first sum he had been referred to the Silesian treasury which, however, in all this time had paid but 100 gulden thereof. Nothing was to be expected from there. How could he get the sum due him? He thought, could not the city of Nuremberg, which at all times had encouraged astronomical observations and works and distinguished itself by printing them, by paying cash, and the cities of Memmingen and Kempten, by supplying paper, a special quality of which was to be obtained there, be responsible for it?

It was not possible to solve all these questions from Linz. So in October, 1624, Kepler set off for a rather long visit to the imperial court in Vienna. There he succeeded in promoting his affairs in various respects as he desired. His position relative to Brahe's heirs was fundamentally improved and strengthened by a bit of luck. The emperor arranged for two persons, the counsellors and physicians Gisbert Voss von Vossenburg and Wilhelm Rechperger, to be commissioners. Acting partly in an intermediary capacity, partly making regulations and decisions, they were to intervene in the negotiations between the two sides and conclude them. Thus was created a superior office which could reduce the claims of the heirs to their proper proportions and, considering the understanding which both gentlemen had for Kepler's distinguished accomplishment, would do justice to him. It was to be expected that their intervention would bring about a settlement between the two sides and that especially for the *accidentalia*, title, dedication and preface, a style would be found with which all could be satisfied. Since these *accidentalia* were not handled until the end of the printing, there was still time for further reflection. Kepler was intent on holding to the principle: "Fine candid German and straightforward; that cannot hurt anyone." He got nowhere with the heirs with his proposal to share the costs and the profit equally. They kept fighting in vain for the payment of the sum which they still claimed from Emperor Rudolph.

The solution of the financial question even appeared very hopeful at the Vienna conference. The emperor approved the proposal to impose the payment of the outstanding sums on the cities Nuremberg, Memmingen and Kempten. Kepler found an understanding promoter of his request in Abbot Anton Wolfradt von Kremsmünster, at that time president of the royal exchequer, who had previously as one of the Upper Austrian representatives shown him his good will. Thus in December, 1624, the office of the imperial treasurer in Augsburg received the imperial order to present Kepler with the yearly allowance due him in the amount of 3,966 gulden out of the city of Nuremberg's Frankish district allowance and his back court salary of 2,233 gulden out of the Swabian district allowance of the cities of Memmingen and Kempten. Of this latter, 1,297 would be charged to Memmingen and 936 to Kempten. Kepler obtained an appropriate note delivered to him which he had to hand over in Augsburg, where he then was supposed to receive the appropriate instructions to the three cities; at the same time notification of the order issued would be made to the cities themselves. Pleased by the success Kepler returned to Linz early in January, 1625.

In one point only had he had no success. The emperor did not accept his proposal to be permitted to print the *Tables* in Ulm. On the contrary, he insisted that the printing take place in Austria. The progress of events will show how, by this request, the completion of the work was to be postponed more than a year. Nothing further remained for Kepler than to submit. Since under the given circumstances only Linz was taken into consideration as the place of printing, he had to see how he could improve the printing press there enough so that it would suffice for the not inconsiderable technical requirements of the task at hand. He thought about the trip within the empire which he now had to undertake not only because of the money but also to look around for type material and suitable workmen. For this he placed special hope in Nuremberg.

The start of the journey, however, was delayed. Since from the beginning Kepler harbored the fear that Nuremberg might look for loopholes, he struck an agreement in Vienna with the Nuremberg representative to wait in Linz for the answer of the Nuremberg senate so as not to journey there in vain. The answer only arrived in March. It read as he had foreseen. The senate had decided to decline gently to pay the requested sum of a round 4,000 gulden out of the district treasury or on the anticipation of the imperial contribution, because such anticipations are forbidden during the recesses of the Imperial Diet, also because much had already been previously anticipated and the treasury in the course of this war and continual marching through of troops was very short of money. The city excused itself to the emperor for not being able to follow his order even though there was nothing it would rather do. But when Kepler had afterwards learned that a second imperial order would go to the people of Nuremberg and that the emperor would see to it that they would not offer resistance much longer, he cheerfully began his journey into Swabia on April 15, 1625.

Immediately on his arrival in Augsburg he met with a misadventure. As the officials of the office of the imperial treasurer broke open the writings which he had brought along from Vienna, it became apparent that in consequence of an unpunctuality of the expedition (or were intrigues hidden here as Kepler presumed) these papers were not in order, so that the office could not draw up the drafts on the three cities and a protest in Vienna became necessary. Because of the long distance and the wearisome course of business a longer postponement was to be counted on. Moreover, the court exchequer in this time of war also had other concerns than those over the printing of astronomical tables. Nonetheless Kepler repaired forthwith to Kempten and Memmingen to see what he could accomplish there. Both cities seemed inclined to follow the emperor's order which had reached them. So Kepler could negotiate and sign agreements with the manufacturers of paper there, although the drafts on the part of the office of the imperial treasurer were still not drawn. The latter settlement followed without friction.[1] To be able to wait such a long time for the answer from Vienna without expense, he went via Ulm to his sister Margarete in Rosswälden near Göppingen. Her husband, Pastor Binder, had been transferred there. The visit afforded him the opportunity of using the excellent acidulous waters in Göppingen to cure a very troublesome rash which he had caught in Vienna. When the wait became too long for him, he sought out the imperial vice-chancellor, Baron Ludwig of

---

[1] AU. NOTE. It is noteworthy that Kepler paid out of his own pocket for the paper furnished and let the sum paid by Kempten, which he rounded off to 2,000 gulden with the Memmingen money, stand at the city of Kempten. Later the obligation in question was still in his widow's estate. In 1660 and 1668 his son Ludwig, as well as two nephews, received interest on this debt which eventually was completely paid off.

Ulm, at Erbach castle near Ulm and asked him to intervene for a quicker settlement of his affair. The request was most willingly complied with. As a consequence, after a visit in Tübingen, to which he was repeatedly drawn, he could at last at the beginning of August carry out his journey to Nuremberg. But here he was greatly disappointed. His long wait had been for nothing. His arrival and the proposal of the imperial letter came at a fatal moment. For Colonel Aldringen, by order of Wallenstein, and Kepler came to Nuremberg simultaneously. Since that city was foreseen as the next troop quarters after Regensburg, it had offered the sum of 100,000 gulden as compensation if it should be spared this distinction and if it and its subjects should be assured security and freedom from further burdens at the passage of the imperial troops. Now Aldringen was supposed to unburden the rich Frankish city of one half of the promised payment. Under these circumstances it is understandable if the counsellor of the city also on this occasion refused with finality to pay the astronomer the requested sum.

When Kepler returned to Linz on August 22, it was a meager result that he brought back from his journey, which had cost him four months' time and "had already eaten 400 gulden." He had received a third of what he had counted on. Once more he turned to the emperor for further funds, but in vain. In order not to postpone the publication of his work still further, he now decided to draw on his own funds. With God's help he wanted to find his own way out and "not be held off from the completion of the work by any fear in regard to future maintenance with his wife and six children." Yes, if there were no other way, he would prefer to collect the money by begging from high and low and from the middle classes, rather than put off the printing any longer. That was magnanimously contemplated. He also followed it by fulfillment and took over the printing costs on his own purse. Nothing remained for Kepler other than to try out Plank, the printer. He, it is true, intended quitting Linz. With that possibility in mind, Kepler had already looked around in Nuremberg for a printing press. He also had found one there which was purchasable and a printer who had placed his services at Kepler's disposal and was ready to come to Linz and set up the printing press there. But Plank let himself be persuaded to stay. Kepler, to be sure, could not promise himself unchecked progress in the work.

Now before the work was in full swing, a blow struck Linz and all Austria. For months it created the greatest excitement and most severely impaired business life. Events resembled those which had previously happened in Graz, when Kepler had returned full of hope from his journey to Prague and then been surprised by the general banishment of Protestants. Similarly now, a few weeks after his return home, the main blow in the long active Counter Reformation was delivered by the religious patent published on October 10, 1625. The measures up to now had been only idly carried out in part. Opposition to them had, however, led the people to adopt political means, which in the close interlacing of religion and politics were not always consistent with the loyalty due the sovereign. Now a clear situation with but one meaning was supposed to be created. The expulsion of "preachers and un-Catholic schoolmasters," already previously arranged, was renewed and they were forbidden to remain in the land under threat "of heavy bodily and capital punishment." Holding clandestine conventicles, reading books of homilies, sermons, and instructing in matters of religion were severely forbidden, as well as going out to foreign non-Catholic preachers. Children of burghers, who were at foreign, non-Catholic schools, were to be brought back inside of six months; also lords might not send their children to such schools. All heretical books had to be delivered up inside of one month's time. All inhabitants would have to attend the Catholic divine service. Anyone who did not want to become Catholic could depart freely. Anyone disregarding the day appointed for this, Easter 1626, would be expelled and would have to leave behind a tenth of his fortune.

The city and land suffered heavily. Under such hard prescriptions is it not necessarily the best who suffer the most, they who cannot change their religious convictions the way one changes a shirt? Kepler was among these. But whereas before he had to share the common lot, this time an exception was made for him. The decree of banishment did not apply to him. He enjoyed, as he informs us, the privilege of a court official. Besides, no obstacle was placed in the way of his performing his work. Plank, the printer, who likewise clung to the Augsburg Confession, in consideration of the tables received a concession to remain. The fair way of thinking of Count von Herbersdorf, president of the Bavarian district, further succeeded in getting the Reformation commission to give Kepler permission to employ suitable people without regard to their creed until the completion of the printing. Notwithstanding, this does not mean that the measures of the Counter Reformation had not also interfered with his personal affairs. He had to have his little son, born in 1625, christened by a Catholic clergyman. His other children attended the Catholic divine service. Indeed, it is not wrong to assume that he himself likewise took part in this under some protest. It was an aggravating humiliation for him to be called to account by a member of the Reformation commission, a Jesuit.

The latter reproached him harshly for taking part in secret instruction. Kepler did not fail to reply. Even if he would offer to teach, he said, still no one would venture to listen to him. Also no one would want to listen to him, since he was excluded from his church. He hates attacks which are not open, even against the enemy. "I teach and test my children, even if they attend your houses of God; I teach them that which serves peace, that is what is common to both sides. So I act as father of the family." Kepler, however, also did not scorn to snap his fingers at the ruling lords. He let some "goodhearted country gentlemen" take his then nineteen-year-old son Ludwig out of the house, without saying where they were going, so that the father "could swear with good conscience that he does not know at what place his son is staying and consequently cannot require him to come home." By the kindness of Count Palatine August von Sulzbach, Ludwig was brought to the gymnasium there and a few months later through the intercession of the rector of the university he was accepted in Tübingen on the Fickler scholarship.

Although, in consideration of his position as a court official, Kepler was promised immunity, he could not prevent a truly agonizing measure. On December 31, 1625, his library was put under seal by the Reformation commission. Only a very few books, dealing with scientific practice, were left to him. The condition for getting the others again was that he should himself choose those to be handed over. So the mother dog herself should sacrifice one of her young, he says bitterly. "The mark of such a slavery forsooth burns!" He loved his books. There was almost none among them with which a part of the fruits of his studies would not be taken at the same time because of the marks which he had introduced and the notes which he had written in. Of his books of religious content, a Greek edition of the Bible in folio is of the greatest importance to him because of the various variants. A German Bible, in Luther's translation, because of its chronological notices is an indispensable aid to him in recollecting. He owns an old book of homilies by the reformer Brenz, which he loves because of its remarkable woodcuts. The burdensome measure, which hindered him in his studies, seems, however, to have been lifted again after a few weeks. The right spokesman was at hand to assist him in this, Paul Guldin, a Jesuit Father in Vienna, a well-known mathematician, who took great interest in his scholarly production. It was fortunate for Kepler that Guldin wanted to borrow some books which were among those which Guldin's fellow Jesuits had placed under seal.

Naturally, under the circumstances described, there could be no talk of speedy progress in the work of printing. Besides, having soldiers quartered in the house in which the printing press was situated was a great obstruction. As owner of the house, Plank had to assume this citizen's burden and bear the annoyance by the soldiers whom he was to board. To be sure, the false rumor reached even as far as Danzig that the soldiers had destroyed all the material in the printing press, cast bullets out of the letters, made cartridges out of the printed and written sheets. "If the works proceed, then we have every reason to thank God for this." How far the printing prospered and what in detail was undertaken, can no longer be said. Kepler's reports speak only generally of numerical tables. The matter certainly could not bring satisfaction. From the fact that Kepler had the paper purchased in Kempten conveyed from there right to Ulm, it is apparent that no more than a beginning was made. He had not given up the hope of being able to carry out the printing in this city.

14. Siege of Linz and Kepler's departure

Alarming new events were soon to push the whole undertaking on another track. The flames of war, which up to now had thrown their light here only from afar, suddenly flared up in the immediate vicinity.

Recently we first met the name of the man who in the following years exercised a decisive influence on the course of events and was even to enter into the narrower circle of Kepler's life as powerful patron: Wallenstein. Just then, in 1625, this ambitious man had succeeded in having the emperor charge him with the formation of an army of his own, in order to take part, on the side of the League's commander-in-chief, Tilly, in the military operations in preparation in Lower Saxony. His success against the followers of Frederick of the Palatinate had certainly strengthened the emperor's power, but consequently also alarmed the opponents of the house of Hapsburg. The war, under the influence of foreign princes, now threatened to take a different direction. Richelieu, who in 1624 had come to the helm in Paris, was able, by financial support, to get the conquest-happy Danish King Christian IV to intervene. Tilly and Wallenstein were supposed to meet this danger jointly and expand and secure the emperor's power in Protestant north Germany. Far-reaching plans were weighed and the first successes in the spring of 1626 revived the hopes which were entertained. At this juncture, in the rear of that army there emerged a menacing danger which put a severe damper on these hopes. In the same spring in Upper Austria a fruitful uprising of the peasants took place. Angered by the compulsory conversion required of them and embittered by the burdens imposed on them by the Bavarian garrisons, the peasants rose in great numbers. They gathered in large troops which traveled burning and plundering through the land. They destroyed castles and cloisters and spread terror and confusion everywhere. Under the determined leadership of Stephen Fadinger they achieved considerable success. The city of Wels was occupied and on June 24 the capital, Linz, blockaded. The siege lasted more than two months. It ended on August 29, at the approach of an imperial relief-column. However, that in no way ended the uprising. The bloody battles continued, and only toward the end of the year did Count Pappenheim succeed in suppressing the peasants, who had suffered horrible losses. The consequences of the uprisings were destructive for the entire country.

The siege brought the heaviest suffering to the city of Linz. Hunger and pestilence were hard on the inhabitants; numerous conflagrations kept them in continual fear and excitement. With all these horrors Kepler got off unscathed. "By the help of God and the protection of His angels I endured the siege unhurt," he informs a friend with relief. Also he had been one of the few who, without tasting horse meat, had not had to suffer hunger. But he experienced a great loss when as early as June 30, the peasants started a blaze to which seventy of the houses in the outskirts fell as sacrifice, and in which Plank's house and printing press also went up in flames. As much of what had previously been printed as Kepler did not have at home with him was destroyed. The report occasionally read, that a work by the astronomer was destroyed by this fire, is falsely interpreted. His manuscript, at any rate, suffered no harm. At that time he lived in the country house. It seemed like a blessing to him when a year before the commissioners granted him this dwelling. Now he saw the work of an evil spirit in it. The country house stood at the city wall. From his dwelling he had a free view of the graves and the outskirts from which the battles would be waged. A whole company of soldiers had quartered themselves in his house and occupied the ramparts. All doors had to be kept open for them. Their going to and fro disturbed his sleep by night and his studying by day, as Kepler recounts. Constantly, so he relates to us, his ear would be annoyed by the noise of guns, his nose by bad fumes, his eye by the light of fires. (A happy time, when one still complained about such annoyances in a bombardment!) It is surprising that in such a situation Kepler found quiet and energy to work. He was busy with questions of technical chronology. A book by Joseph Scaliger had stimulated him, but also incited him to contradict. Against this man he acted as the garrison had against the peasants, he wrote in a letter. He had fear and worry only for his books and his collection of notes.

Because the printing press was destroyed, a continuation of the printing of the tables could no longer be contemplated. As soon as the siege had given way in August at the approach of the imperial troops, he wrote to the emperor, requesting permission to move to Ulm. Now the emperor could no longer refuse; he agreed to the long-cherished plan. On October 8, 1626, the court chancery made out the permit. Now nothing, absolutely nothing, could detain Kepler any longer in Linz. Many of those who had been close to him had already emigrated or were about to do so. The departure could not have been hard for him, although he had been there for more than fourteen years. The city had become inhospitable to him in every respect. As had happened in Graz and in Prague, events in which he had not participated had developed in such a way that he could not remain any longer. So he packed up everything he possessed, wife, children, household furniture, books, manuscripts, and printing material, and in the cold season of the year, on November 20, traveled on a boat up the Danube toward Passau, inspired, driven, strengthened by the one thought: the *Tables* must be completed.

V

LAST YEARS IN ULM AND SAGAN
AND DEATH IN REGENSBURG

1626–1630

1. *Ulm and the* Tabulae Rudolphinae

WHEN Kepler left Linz he did not know whither the path he trod would lead him. Ulm was only a stop on the way through. The magistrate there granted him a permit to remain for a half year with the prospect of an extension, in case it proved necessary. The man who his whole life had pursued the goal of scientific research, which was to calculate what will be, and had taken endless pains to increase the certainty in the prediction of phenomena by reckoning, had to experience ever more severely the uncertainty of all human events. He who was overcome by astonished rapture at the contemplation of the wonderful order which God's creation offered him in the starry heavens was being drawn ever deeper into the disorder which man with his free will arranged round about. The world denied him the fulfillment of his longing for a place where, in the security of a home, he could dedicate himself, undisturbed and free of care, to his peaceful studies. He declined an attractive offer in order to preserve his freedom of conscience. So, with unbroken courage, he strode further on his way until, four years later, death put an end to his travels.

The journey to Ulm was difficult. Kepler tarried for several days in Regensburg and found accommodations for his wife and children in a modest dwelling situated in what is today Kepler Street. Since the Danube was frozen, he himself continued his journey in a wagon in which he also had stowed away his valuable letter type. In Ulm he took up residence with the city physician, Gregor Horst, with whom he had long been on friendly terms—a noted representative of medical science, who had previously been professor of medicine in Wittenberg and Giessen and physician-in-ordinary to Landgrave Ludwig of Hesse; in

Prague the two men had once contrived joint anatomic experiments. Kepler's quarters were in the "adjoining dwelling" to Horst's in the alley today called Raben near the Cathedral place.[1] It must be assumed that Kepler originally had intended to take his family along to Ulm. In a hitherto unpublished letter to Landgrave Georg of Hesse he writes to the effect that he had had to leave his wife and the three youngest children behind in Regensburg, "because my boat did not continue on account of ice." Also the Ulm magistrate had granted him the permit to remain with his household, and Horst, in the past spring, had expressly undertaken architectural changes in his adjoining house and connected it with his main house. So it was the inclemencies of the winter which compelled Kepler to forego living with his family.

The printer who had been under consideration for two years was Jonas Saur. He had been recommended to Kepler as a skilled, willing and cheap representative of the "black art." The paper purchased in Kempten lay ready. Kepler had ordered it to be sent direct from Kempten to Ulm, because he had never given up the hope of being able to print his work there. His Ulm friend, Johann Baptist Hebenstreit, rector of the gymnasium and a very learned humanist, had received it and taken charge of its storage. There were two bales of the better and two of the plain kind. The printer praised it. Sufficient letter type was on hand. Kepler himself had previously had astronomical symbols molded. The work began and was soon proceeding vigorously. Kepler himself spent many hours of the day in the printing plant, to supervise the type setting. How necessary that was is shown by the manuscript, which is the only one of all the manuscripts of his printed works which is preserved. It comprises 568 pages. Leafing through it and considering the proposals, mostly not clearly arranged— the sample examples, the proof sheets, the numberless corrections, the inserted calculations, the various values of the individual tables—one perceives the difficulties which the compositor had to overcome in order to achieve a neat form. This would have been impossible without the author's helping hand. Since the printing plant was situated directly across from his dwelling, it was easy and handy for him to inspect and intervene at any time. Wolfgang Bachmaier, at that time pastor in Möhringen, was an occasional co-worker of Kepler's and corrected a proof with him. And his friend Hebenstreit may also have helped by advice and deed.

Joy in the work's progress was soon very much checked by disappointment in the printer. In complete contrast to what he had previously heard about this man, Kepler came to know him as an

[1] AU. NOTE. The house was destroyed in December, 1944, by an aerial attack.

unpleasant, proud, extravagant and impetuous person. He even informs us that the printer, finding himself in financial difficulties, had tried to extort money from him by threat and tricky interpretation of the contract. It may, of course, also not always have been easy to remain patient in the face of all Kepler's desires and requests. The costs especially worried him. They proved to be twice as high as he had assumed in Linz. Sharp discussions resulted and Kepler feared he might have to make a report against the printer to the magistrate. That, it is true, he did not do, but he weighed the plan of taking the printing away from Saur and completing it in Tübingen. Therefore, as early as February, 1627, he put a series of questions to Schickard, his Tübingen friend. Besides, he wanted to become personally acquainted with the circumstances in that city and started for there. What is more, he went on foot, because sitting on a horse or in a wagon was very uncomfortable as a result of an abscess. He did not get far. His strength did not suffice for such a long journey, especially in February weather, so that he had to turn around again right away, in Blaubeuren. The plan to transfer the printing was not carried out. It had certainly arisen in a psychological depression, such as Kepler's choleric temperament repeatedly brought about.

Kepler appears to have provided himself with ample means for his stay in Ulm. Complaints relating to this are heard almost solely in connection with his son Ludwig, whose desire for study and conduct of life did not satisfy his father. Ludwig, who wanted to become a physician, was at that time still studying in the faculty of arts in Tübingen and, as is customary, needed more money than his father gave or wanted to give. The latter kept the purse closed. He had, as he tells us, given his son, as "expense money," the healthy knowledge that they were poor, and now let him know with some fatherly exaggeration that he was certainly approaching poverty step by step. Friend Schickard, who was taking care of the son on behalf of the father, had a better understanding of the student's needs and tried to help him out of his straits. This resulted in anger and distrust, and not only Ludwig but also his indulgent mentor had to put up with reproaches. "I am, I grant, provided for at present for a short time, for the printing of the work, which takes form under my hands," the father writes to Tübingen. "To take away only a little of it for my son and thereby delay the work, I consider almost as a sacrilege. The emperor appointed me because of my work, not to establish a son. The matter stands thus: the work will be concluded now or never."

When the printing was well along, Kepler decided it was time to remind the Brahe family of the dedication which they desired to add to the work. It was to be directed to Emperor Ferdinand II, the second successor of the monarch whose name the tables bore. The brothers Tycho and George Brahe therefore sent a title composed by them and a short dedication, which they had previously shown learned men living in Prague. Kepler showed these drafts to the imperial commissioners with the comment that he was not averse to the contents. Only he had composed the dedication somewhat "more formally" and also altered the title in such a way that he could hope the heirs would have nothing further to say. The concept of his own dedication, in which he especially wanted to set forth the causes for the long postponement, he likewise forwarded to the commissioners with the request that they see whether there was anything in it at which the heirs could take offense, and that they present the drafts of the title and the dedications to the emperor and strike out what perchance displeased the emperor. By this he believed he had done justice in this matter to all fair demands.

The Ulm magistrate appreciated the honor which devolved on the city for having been chosen as the place of printing for the great work. But he also drew profit from the presence of the imperial mathematician, who was no less educated in many technical questions than in pure science. So the opportunity was used and Kepler was asked to help in the regulation of weights and measures. Kepler had previously demonstrated, in his *Archimedes' Art of Measuring*, that he was quite at home in the confusion presented at that time by this field, so important for trade, since each country, each imperial city possessed its own units of measure. With the assistance of Johannes Faulhaber, the town's well-known constructor of fortresses, who by order of the magistrate had come to his aid and supplied the necessary foundations, Kepler solved the task set him. In a lengthy opinion he analyzed his principles, and accordingly had an attractive basin molded, from whose interior measurements of width, height, weight and capacity it should be possible to conclude, with an exactness adequate for trade, the units for the length, weight and capacity measures. In this manner, the size of foot, ell, hundredweight, pail, imi (an old corn measure) was established. The basin is still preserved in Ulm.

In the beginning of September, 1627, the printing of the tables was at last completed. The edition numbered one thousand copies. Since Kepler had had the work printed at his own expense, it must have been of consequence to him to sell it advantageously. It had been his custom to give writings, whose publication costs he supplied out of his own pocket, to Tampach, the bookdealer in Frankfort, on commission. He wanted to use the same procedure with the tables. So, bearing some of the copies, he set out on the journey to Frankfort on

September 15 in the company of some Ulm tradesmen. He arrived on September 22, at the time of the autumn fair. However, before the sale could be started, the price had to be fixed. Because he was tied up with the Brahe family, Kepler was not free in his decision. He had agreed with them that the first proceeds from the sale should belong to him as reimbursement for his expenditure in the printing; thereafter, any profit should be divided equally between him and the Brahe family. The imperial book commissar, who was present, called on specialists among the scholars and bookdealers for the setting of the price. The former, who were Jesuits, proposed 5 gulden, the latter who were asked only to look at the wares as such, 2. Thereupon the commissar fixed 3 gulden, and for the copies on the better paper 40 kreuzers more. This setting of the price did not occur until October 2, just as the fair ended. However, the delay thus caused in the sales was not unwelcome to Kepler. No matter how often the call for publication of the tables had resounded, the number of those interested, as the author surmised, was still not great. "There will be few purchasers, as is always the case with mathematical works, especially in the present chaos." He hoped now, by the spring fair, to obtain a larger number of orders, if in the meantime the information about the published work were to be properly spread. Particularly, Bernegger in Strasburg would be asked to aid in this and to busy himself with the sale in the foreign land lying closer to him. Behind Tampach's back Kepler sent his Strasburg friend a number of copies for the circle of interested people in the university there.

On November 9, Kepler returned to Ulm. The magistrate of Esslingen, where he stopped on the way and was honorably received, had put a horse from the home for the infirm and a young boy at his disposal for the last lap of the trip. In conclusion, there was one more addition to the long printing history. The drama ended in satire. The Brahe family made a rumpus. Contrary to the previous agreement, Kepler had neglected to show them a sample for examination before completion of the printing. Now they were indignant about individual phrases in the title and dedication, especially about Kepler's saying there, he had improved what Tycho Brahe had previously done. In the eyes of the heirs that was a disparagement of the dead master. And then Kepler's dedicatory writing to Emperor Ferdinand was longer drawn out than theirs. This would not do. Therefore, they saw to it that the title signature with some corrections and a longer dedication by their hand was reprinted a few months later in Prague (Kepler had left Ulm a long time before). Now Tampach, who had to exchange the signature, became angry. Since, besides, he found that the new printing of the title was bad, that there was "no rightly adjusted writing to be found in it, rather everything was piece work and patched together," months later in Ulm there followed afresh a third printing. The second signature, which contains a long poem by Hebenstreit, was also interchanged twice for similar reasons. So it comes about that in the extant copies the two first signatures occur in three different drafts.

The *Tabulae Rudolphinae* appears as a stately folio volume. What a deep satisfaction Kepler must have experienced at the completion of the work on which a good part of his life history hung! It had decisively influenced not only the external course of his life, but also the form of his creative ability. How many obstacles from all sides had stood in his way! But in spite of all these checks he had now reached the goal; he had overcome all difficulties, those which he had made for himself by the high demands he set for the work, and those which the outer circumstances had placed in his path. He designated the *Tables* as his principal astronomical work. The discovery of his planet laws and the working out of the completely new methods for the calculation of planet orbits based on them had been preparatory labors for this work. His theoretical hypotheses had to be tested by these tables. These hypotheses were the fruits which had grown on the tree of his new astronomical knowledge. The greatest observer of that period, Tycho Brahe, and the most gifted theoretician had both given their best to, had frankly seen their life's task in, creating this work. In that time when astronomy was still completely and almost exclusively directed toward the study of the path of planets, it was necessary that the palm of victory in the scholarly competition fall to him who succeeded in representing this path most exactly and in thereby solving the great age-old task of astronomy. Kepler himself was well aware that even now the goal was not completely attained. Just as in an earlier report had he spoken about so molding the calculation "that it should be valid for many hundreds, yes thousands, of years before and after," so now in the introduction to his *Tables* he points out that justice had indeed been done to the observations of his time by the calculation, but that the secular changes in the motions, which would have to be taken into account, were not yet quite certain; yes even in the mean motions of the planets everything is not yet in agreement and, as the eclipses showed, unexplained deviations in motion also appear with the sun and moon. In spite of these deficiencies, which no one recognized better than the author himself, the *Rudolphine Tables* formed the basis for all calculations in the solar system for more than a century. They were snatched at by astronomers who wanted to test their theories, horoscope casters who needed constellations at some one time, calendar makers who wanted to state the position of the wandering stars and, last but not least, seafarers who were compelled to make geographical place determinations. For them particular directions are given. Kepler's tables, then, also accompanied the men who at that time went to the far east and west on journeys of discovery, just as once Regiomontanus' tables had accompanied Columbus on his bold and successful undertaking. Without themselves having moved very far out in the world, these two Germans had performed important services in the conquest of the earth's globe.

Before the tables proper, which begin with logarithms and end with a catalogue of one thousand fixed stars, the book has a first part of the same size, which contains rules or instructions for the use of the tables. Originally Kepler had intended to add demonstrations also, that is to explain in detail how he had calculated the tables. But he gave up this plan. Indeed, it was now already partly fulfilled in the *Epitome* and, then, Kepler still thought of developing his long-planned *Hipparchus* into a new *Almagest*, which, however, he never did. In addition to the purely astronomical tables, there are also tables for technical chronology, as well as a voluminous catalogue of cities in the whole world with statements of their geographical latitudes and longitudes, with reference to the meridian of Hven.

To this catalogue of places Kepler wanted to add a world map, on the production of which he had already worked a couple of years before. Schickard in Tübingen was supposed to lend his skilled hand to it. Through the intervention of Philip Eckebrecht, a merchant in Nuremberg, who was an ardent follower of astronomy, and whose hospitality Kepler used to enjoy, he contacted the notable copper plate engraver, J. P. Walch. To Kepler's regret, there was repeated delay in the completion of the map. When, shortly before his death, he once more came to Nuremberg, he found that it was almost completed. However, in the situation created by his decease, the map was forgotten. As an inscription on it states, it was first brought to light again in the reign of Emperor Leopold (1658–1705). It is not known who found it and circulated it among the owners of the *Tables*. There is good reason to conjecture that Albert Curtius, the Jesuit Father in Dillingen, later in Nuremberg, was behind this. From Ulm, Kepler carried on an active correspondence with Curtius who would gladly have played a larger part in the field of astronomy.

The frontispiece, which Kepler himself designed and Georg Celer engraved, makes a pleasing adornment for the book. It shows a ten-sided temple, open all around, whose vaulted roof rests on ten columns,

some roughly, some skillfully worked. Within this space five persons
meet in a variety of attitudes as an academy: an old Chaldean, Hipparchus, Ptolemy, Copernicus, Tycho Brahe. On the roof stand six
allegorical figures, symbolizing the fields of astronomy, mathematics
and physics. In one of the side faces of the high base Kepler immor-
talized himself; he works by candlelight at his writing table and from
his study looks forth quite anxiously at the observer of the picture.
Above the roof the imperial eagle hovers as a symbol of imperial
liberality. It lets talers fall from its beak and a few of them land on the
astronomer's writing table. The picture was supposed to show the
accomplishments of previous astronomers in their relation to the work
of the *Tables*, and to illustrate in meaningful details all the factors from
whose combined action the work developed. In Hebenstreit's introductory poem entitled "Idyllion," the symbolic picture is significantly
interpreted in well-chosen words.

2. In search of a new dwelling place: Landgrave Philip of Hesse, Emperor Ferdinand

In the beginning of November, 1627, on Kepler's return to Ulm from
his Frankfort journey, he was faced with the difficult question
"whither," which once more demanded decision. The feeling of satisfaction and relief which he must have experienced after the completion
of his task was covered over or replaced by worry over the prospect
of the immediate future. The thought of the further arrangement of
his life tormented him some months previously. For that reason he
turned to Bernegger for advice. What he would want for himself, he
wrote to him, would be a place where he could deliver lectures about
the use of the tables to some group of listeners, if possible in Germany,
if not, in Italy, France, Belgium or England, provided only that a suitable
salary would be at the disposal of the foreigner and his family. Also he
expressed himself further about his concept of these lectures. If he
were to find a proper number of students, he would like to take refuge
with the nurse of astronomy, astrology. He would let individuals tell
him their birth hours and pledge to teach not only the procedure for
calculation of the positions of the planets but also the significance of
nature's signs. Thus would each one of his audience understand only
individually what was lectured about his affairs publicly. It could be
supposed that in this manner Kepler had wanted to sell himself to
astrology completely. But such a supposition would be entirely false,
if astrology is understood as the conventional interpretation of stars

with its prognostication of particulars. Indeed, in the introduction to his *Tables*, where he surveys similar works from earlier times, he had most definitely taken a position against this erroneous idea and stressed and praised all the efforts which aim at pure research in nature and at the highest goal of all scientific study, the knowledge of God. So he commends Reinhold, the author of the *Prutenic Tables*, the last and best which had appeared before the *Rudolphine*. Reinhold, too, it is true, had wanted to serve astrology with his work. But in this man's other performances there were to be found flowers from the innermost garden of philosophy, from which a wonderful, most fascinating exhalation escapes. And praise is given Tycho whose mind had been completely free of astrological superstition and followed the highest, purest goal of scientific study. He who at every opportunity scoffs at and curses the empty, unscientific, useless and squalid activity of astrologers meets with Kepler's full approval. But, and in this he knows himself in agreement with Tycho Brahe, Kepler affirms the belief in an "effectus generalis," a general influence of the heavenly phenomena on nature and man, and in a cosmic subjection of man which has nothing to do with the rules by which the Arabic art of prophecy pledges to predict the individual fates, the *eventus individui*, from the stars. To help get this thought appreciated was precisely what he contemplated with his lectures.

Friendly Bernegger willingly entered into Kepler's plan. He noted that Kepler would not be reluctant to come to Strasburg to take over a professorship. He spoke about it in an influential place but, unfortunately, without success. Yet with amiable words he invited his friend to be a guest in his home. "If it should please you to honor our city by choosing it as a dwelling place, then I do not doubt that you would be an extremely welcome guest with all." He offered Kepler and his family a dwelling in his house with assurance of great hospitality. As house-rent, nothing would be more valuable for him than the daily conversation with his friend. However, Kepler would not and could not consent to such an offer. No positive results were gained by the written recommendations with which Bernegger promised to turn to Basel and Georg Michael Lingelsheim[1] promised to turn to France. Consequently, Kepler gave up his original plan. The uncertainty of his future, the feeling of dependency on others, the pressure of circumstances in which he was wedged, kept him deeply despondent, and he had to resist this with his utmost strength. It is impossible not to be moved by the following passage from a letter in which he begs for

[1] An experienced diplomat with wide foreign connections in Strasburg. (Transferred from text.)

advice: "Reflect that I write this with a heart filled with worry and wounded by fear for the future, a heart which by a single correct word and by evidence of well-wishing opinion can be calmed and aroused to new hope." In what abyss of sadness must a soul be if it openly presents such a request to another.

Since the course pursued did not prove passable, another plan was considered. On his journey to Frankfort, Kepler had visited Landgrave Philip of Hesse from October 6–19. It was to him that he had dedicated his work on logarithms. The instruments and arrangement of the prince's observatory at Butzbach aroused Kepler's greatest interest. There the two astronomical friends observed sunspots with a tube fifty feet in length; a little opening (without the use of a lens) on the front end of the apparatus produced the picture of the sun on a white screen at the back. Now in his distress, Kepler placed his hope in the kindness and intercession of this prince. But the talk was no longer about lectures which he wanted to deliver, rather about the carrying out of another plan which he had long entertained. When making preparations for the printing of the tables he had thought of publishing, as a supplement, Tycho Brahe's observations which had been in his care since the latter's death and "in which posterity on behalf of this art had laid much store." When he purchased the paper in Kempten he had taken this plan into account. It would have been bad business if the paper for which he had paid 2,000 gulden had been for one work only, to be sold in an edition of one thousand copies at 3 gulden each. Now he hoped, by the support of the Landgrave, to find a way to carry out his plan. In the negotiations, the various difficulties were discussed. Kepler, of course, was not yet dismissed, either as district mathematician of the Upper Austrian representatives, nor as court mathematician of the emperor. Naturally, because of the salary, which he drew from both positions, it must have been important to him to hold them as long as was practicable. Therefore, he did not want to take any kind of step of his own accord to hasten his dismissal, at all events, not so long as he did not have another sure position. It goes without saying that in addition he was animated by a feeling of sincere reverence toward the House of Hapsburg to which he owed so much. On the other hand, he was certainly aware that both his positions were imperiled as a consequence of his attitude in the question of creed. Given the threatening aims of the emperor, which emerged from all his words and transactions, Kepler could not, so he supposed, hope to hold on for long. In fact, in the summer of 1627 imperial decrees had been promulgated implementing anew the discharge of all non-Catholic officials in the service of the Upper Austrian representatives.

Kepler then considered himself already discharged, as he explicitly confirms. According to title he was a court official, in reality an exile; the knife was already at his throat. Landgrave Philip showed full understanding for the distress of the astronomer whom he esteemed highly and was ready to help. Philip was willing to use his influence with his nephew, the ruling Landgrave Georg, on Kepler's behalf. Kepler thereupon presented a petition to this prince submitting the proposal: "If only I might see some means to last out during a printing for a year or two, together with wife and children, in such a way that I could earn my and my family's livelihood by working in a professional side line, then as soon as I came to the imperial court to present to his majesty my book dedicated to him I should like to inform him about the needs for the future and already commissioned printing and request gracious permission to leave." There was no doubt that this furlough would be easy to obtain. He was sure of it because of the "most recently proclaimed resolution of his majesty to get rid everywhere of hostile religious elements." In case this plan could not be carried out, Kepler would like to ask the landgrave to use his influence with the duke of Württemberg on his behalf. He had—a last relapse—knocked at the door of the councilors at the Stuttgart court. They, however, had partly been silent, partly replied that he should not have any false hopes. Landgrave Georg met the petitioner's wishes willingly. Most liberally he promised him a home of his own and the necessary means of support in case the payments of the imperial court failed to come because of his long absence. Marburg was agreed on as dwelling place. There, in its time, the logarithmic work had been printed. The circumstance that, in contrast to Landgrave William V of Hesse-Cassel, who was a Calvinist, Landgrave Georg remained loyal to the emperor, might have made Kepler's plan easier for him at the imperial court.

The immediate prospects, consequently, were not bad when Kepler took final leave of Ulm on November 25, 1627. After a two-day stop with the Jesuits in Dillingen he arrived in Regensburg on November 29. For a whole year he had, of necessity, been deprived of life with his family. But even in the short time now granted him with his beloved ones, he could not be idle. During his visit in Dillingen he had learned about a letter in which Johannes Terrentius, the Jesuit Father, who was active in China and highly respected there because of his astronomical knowledge, asked for information concerning the newest research results in astronomy. He had in mind especially the achievements of Kepler and Galileo. Kepler did not put off the answer long, but immediately, in Regensburg, composed a little treatise which he wanted to publish at a later opportunity. Even before the Christmas holidays he traveled further. On December 29 he arrived in Prague, where the emperor was then staying, and took lodgings in the "Walfisch bei der Brucken."[1]

The Bohemian capital was in an exalted mood. The reason for the move of the imperial court to Prague was the selection, as king of Bohemia, of Ferdinand's son, who later became Emperor Ferdinand III. The coronation had been celebrated with splendid magnificence. Besides, there were the military and political successes of the previous year. Much had happened while Kepler, in Ulm, gathered the harvest of his studies of many years. In 1626 the military operations had gone favorably for the armies of the League and of the emperor. In the battle at the Dessau bridge Ernst von Mansfeld had been defeated by Wallenstein's troops. Rebel Hungary had had to make peace. Tilly's great victory over King Christian at Lutter am Barenberg had caused a decisive change in the Danish war in favor of the emperor and the League. The uprising of the peasants in Upper Austria, which in the beginning appeared so menacing, had ended with the complete defeat of the insurgents. Death had carried off the old opponents of the imperial side, Ernst von Mansfeld and Christian von Braunschweig. Now in the year 1627 the clever strategy of the imperial commander-in-chief, Wallenstein, succeeded in rolling back the military and political front of the Danish king in North Germany from east to west. Brandenburg, Mecklenburg, Pomerania, Holstein, Schleswig and finally all Jutland to its northernmost point were occupied, so that the defeated king was only able to save himself by retreating to his islands.

The triumphant commander-in-chief, summoned by the emperor, had come to Prague a few weeks before Kepler and had set up his head-quarters in his pompous palace in the Small Town below the Hradschin. Now was the time to form plans in order to make secure what had been accomplished, to complete what had been begun. Not only some of the princes of the empire but a number of others opposed the new state of affairs. In foreign parts, too, the mighty strengthening of imperial power was looked at askance. France and Sweden prepared to interfere in Germany's fate. Wallenstein again used the favorable opportunity to follow his personal designs with clever calculations and tenacious resolution and to come closer to the far-off goals of his ambition. He had previously come into possession of the duchy of Friedland; now he succeeded in having the Silesian duchy of Sagan conferred on him as fief. His appointment as "General-colonel-commander-in-chief" and as "General of the Baltic and oceanic seas"

---

[1] ED. NOTE. The translation of the name of the hostelry would be "Whale at the Bridge."

exalted him and surrounded his person with the halo of power. But this was not all. By the transfer of the two Mecklenburg duchies, whose rulers, as followers of the Danish king, were declared to have forfeited their fiefs, he obtained possession of a German imperial principality. Not to be left behind, the leader of the League, Elector Maximilian of Bavaria, who followed Wallenstein's rise with suspicion, was also able to achieve a long striven for promotion for himself and his house. The emperor now formally pledged to him the transfer of the hereditary electoral dignity. Thus, during the first months of the year 1628 when these events took place, there was much activity in Prague. Worry about the political and military exigencies of the immediate future, agreement and variance of points of view and interests, personal aspirations and affairs, diplomatic deliberation in closed and wide circles, receptions and festivities kept all interested parties out of breath. Among the actors in this scene stood out in marked relief the figure of Wallenstein, in whose hand openly or secretly most of the threads converged.

When Kepler, who was not politically minded, entered this world, in order to present his tables to the emperor, he experienced a surprise. He was most graciously received by the monarch, was very fortunate and successful with him in his work, and contrary to expectation found many well-wishers and admirers among the men at the imperial court. His fear of being dismissed as court mathematician proved unfounded. From the distance he had misjudged the temperature of the court air and, swayed by a depressed disposition, such as was not unusual for him, had formed a false picture of the true situation. Now he, too, was seized and elevated by the waves of life flowing all about him. His mood became confident. He may have recalled his visit to Prague under Rudolph II. Now he saw himself once more transplanted into the agitation of court life, which did not fail to impress him even though as a man of science he took only a modest part in it. Of course, he must have noticed that much had changed in the intervening time and that his present imperial ruler pursued different goals from those of the emperor whose confidence he had enjoyed for twelve years. His old friends who had belonged to the Augsburg or Bohemian Confession had disappeared. They were either dead or banished. Ferdinand's Counter Reformation measures had transformed Prague into a purely Catholic city. In the place of arts and sciences the profession of arms was fostered; homage was paid to the god of war instead of to the muses.

In addition, Kepler was put in a good mood by the magnanimity with which the emperor rewarded the dedication of the tables.

Because of Kepler's expenditure on the work and in benevolence, the considerable sum of 4,000 gulden was granted him. The cities of Ulm and Nuremberg were each to be responsible for half of this. Could he, so benevolently received, quit the imperial service, as he had intended doing? Under the influence of the very friendly reception which he had found, he abandoned the agreement which he had made with the landgrave of Hesse. Besides, this could barely be carried out because in the meantime imperial and League troops had taken up winter quarters in Hesse, Franconia and Swabia, so that there could be no talk of undisturbed work in these territories. Other negotiations were entered into; new future plans emerged. Kepler's desire to remain in Hapsburg lands met the emperor's wish to keep him in his service. An enticing offer is made him. Only one condition is attached: he should become Catholic. Once again the question of the choice of creed intervenes in his life. Again he is faced with the decision: either, for the sake of external benefits, to abandon his conscience, which is most sacred to him, or to preserve moral freedom and sacrifice for it. We know his morals and how he will decide.

3. Disagreement with the Jesuits

We do not know what the offer was by which an attempt was made to win Kepler; it may have been a professorship in Prague or else a well-endowed position at some place in the emperor's lands where he could carry out the printing of the Tychonic observations undisturbed. We are better informed about the attempts to prevail on him to be converted to Catholicism. For this the imperial advisers and Kepler's well-wishers at the court used all their art of persuasion and even the emperor himself let his mathematician know how much he wanted such a switch. The Jesuits considered it their task to dispel the theological doubts of this man who resisted. In 1627 while Kepler was in Ulm, Father Albert Curtius of neighboring Dillingen, whom we have already mentioned, had repeatedly dealt with theological doctrines along with astronomical questions in the active correspondence which he carried on with Kepler and had tried to refute the objections which his correspondent put forth against the Catholic point of view. Now, also, Father Paul Guldin of Vienna, who himself came from Protestant parents and only as an adult had transferred to the Catholic church, broached the question of religion in a few lines of a letter. His short admonition caused Kepler to write a long reply in February, 1628, excitedly setting forth his position in regard to the request put to him and expressing everything which had accumulated in him concerning this matter more openly because he could value Guldin's friendly confidence highly. In his own environment, where caution was indicated, he surely would have expressed himself with greater discretion.

"My devoutness regarding God up to now would have been in a bad state," he commences his argument, "if I just now had to begin to become Catholic. But right on the threshold of life I was introduced to the Catholic church by my parents, sprinkled with holy baptismal water and thereby spiritually endowed as a child of God. Since then I have never left the Catholic church." On the strength of this clear avowal he prides himself in his membership in the Augsburg Confession, in which he had been instructed from childhood, and turns toward his admonisher with the explanation: "If you say the Church is the community of the people who are united under one chief in order to spread the wrong uses cast off by the Augsburgians and to command them as matters of conscience, then you specify an attribute which prevents me from accepting the Church, if it offered this alone." Here we meet Kepler's view of the Church as he had formed it since his theological studies. He understands the Catholic church to be the community of all baptized people. The Church is for him "one and the same at all times." Rome, Wittenberg, Geneva are for him parts of this one Catholic church. "Just as disputes arise among the citizens and parties of a state so, out of human weakness, errors arise among the members of the one church who are divided by time and place." The dogmatic differences of the various creeds accordingly rest on errors which have crept in as a consequence of human weakness. But how can Christ succeed in distinguishing truth and error? In what consists the rule of faith according to which it is possible to measure the purport of that which is to be believed? Whereas the Catholic church (so called in common speech) assumes one unerring mastership which establishes the purport of the faith, Kepler refers with emphasis to the anointment, the chrism, which he had received at baptism and which survives in him. Thanks to this anointment, so he fancies, he does not need to have anyone instruct him beyond what he reads in the scriptures and in the documents of faith. "As the anointment teaches me, *that* is true and it is no lie." Kepler overlooks the fact that this subjectivity must lead to ever new sects and can never lead to an elimination of the schism which he laments and would like to conquer. Thus his proposal, that in order to heal the wounds of the entire church the members of the Roman church should follow the laws and the Lutherans the spirit which best advises, is also a sign of his tolerant, peace-loving character but cannot achieve the goal of uniting the separate creeds.

After these fundamental statements about the church and the *regula fidei*, Kepler turns in his letter to a series of doctrines and institutions of the Roman church which he does not want to recognize, such as adoration of the Eucharistic God, the worship of images and saints, sacrifice of the mass and communion in one form. Against these he sets forth the reasons which he had absorbed from childhood on. In all these expressions of Catholic religious practice he sees the work of the devil, who in the course of time, according to Christ's predictions, has sown weeds among the wheat. He considers them inadmissible innovations which have falsified the faith of the early Christian. It had indeed also been the resistance to the innovations which had induced him to the condemnation of the ubiquity teaching of his own creed. Not out of pride, he states, does he deny obedience to the mother church of the west in this and other doctrines, but rather because he believes that one must obey Christ, its head, more and because strong exhortations of Christ make him fear it could stumble and err in that kind of innovation; it could fall asleep and be surprised by the intrigues of the devil, who secretly sows weeds. That is his conviction and his motive. In an emphatic explanation he now draws the conclusion: "Accordingly think thus of me, best friend: I remain in the Catholic church. Only I am ready, for the sake of rejecting what I do not acknowledge as apostolic nor as Catholic either, not only to abandon the recompenses which are now being offered to me and to which His Imperial Majesty has magnanimously and liberally consented, but also to abandon the Austrian lands, the whole kingdom and, what weighs much heavier than all this, astronomy itself." The man who, for the sake of his faith had once, in Graz, accepted banishment, endured exclusion from his church and buried the hope of a position in his native land, could have given no other answer. He remained true to himself. In a period when thousands, for the sake of temporal advantages, changed their confession at the wish or command of their prince, he proved an example of a lofty character, which in honest and earnest contest about the perception of religious truth places freedom of conscience above all that the world can offer in possessions.

After such an avowal, could he hope to be able to remain in the service of the emperor? No matter how positive and decisive his answer was, nevertheless on his part he did not want to be the one to close the door. With this in mind, he added an explanation to his letter telling how he would behave should he be granted a working place in the imperial lands. At the same time that he directed the letter to Guldin he received a greatly delayed written communication from the representatives in Linz, in which he, like other public servants, was ordered by the command of the Reformation commission to declare whether he was ready to "accommodate" himself. At that time he surely, though perhaps only remotely, must have thought of returning to Linz, however unsatisfactory such a solution of his situation must have been for him. The words he wrote to Guldin indicate how he conceived the "accommodation." "I hold fast to the Catholic church. Even when it is enraged and strikes, I remain connected to it with a heart full of love, as far as human weakness can do so. Should I be tolerated with those few reservations, then I am ready in silence and patience to carry on and to perfect my science under the followers of the prevailing party; I will refrain from all slander, mockery, hatred, exaggerations, calumny and disparagement of those who are of good will. The sermons I shall take to heart in each case to the extent that I see in them the light of divine grace; I shall shun processions and similar affairs so as not to give offense to anyone, not because I want to pass sentence on those who take part, but because it is not the same when two people do the same. Yes, I can also attend mass and unite my prayers with the prayers of the other faithful under a certain condition, namely, if my protest and that of all my family is accepted, in so far as we do not agree with what according to our convictions is an error, but only with the general and final, sacred and Catholic purpose of the mass: to present to God our prayers and the sacrifice of our praise and our good works in consideration of that unique sacrifice consummated on the altar of the cross; to turn this sacrifice to good account for us, to teach the church by these visible acts about this use and about the memory of the death of the Lord."

Guldin answered Kepler's letter with a long theological treatise refuting his opinions with a great show of learning. He had had this treatise composed by a brother in the order. Since Kepler did not learn the brother's name, his distrust was aroused; he did not want his confidential letters circulating from hand to hand in the order. Consequently, in his reply, he confined himself to stating the reasons for his attitude, but refrained from sending off a theological refutation for which he had put pen to paper. The dialogue was silenced. It had been conducted on both sides in dignified, frank, even cordial form. Considering the customs of that period and the trusted relationship that existed between Kepler and Guldin, it would be wrong to characterize the procedure of the Jesuits as officious. Even Kepler himself did not take the attempt at conversion as officiousness. A man who is completely imbued with the truth and strength of his religious conviction will strive the harder to win another over to it the more he respects and esteems the other's personality. Guldin and Curtius were sincere in assuring Kepler of their respect and love and expressing concern for the salvation of his soul. Kepler understood this well. Had he not at one time been aware of the same tones from his Tübingen friends? On this side and that, men trembled for his eternal salvation; they attempted to pull him to one side or to the other. He stood in between. It pained him that he stood alone. He, too, thought about the salvation of his soul. "I assure you solemnly that I am furthest from toying with my salvation," he wrote to Guldin and assured the ecclesiastical admonisher he would have the greatest confidence if, on the day of judgment, God would put him face to face with those who tried to convert him.

Kepler's avowal of the "Catholic" church naturally was not sufficient for the emperor. At the crossway where he stood, he had to pursue the direction which pointed to foreign lands.

4. Kepler and Wallenstein

In these circumstances, it was propitious that the most powerful man at the imperial court, namely Wallenstein, was free of confessional thoughts. Just as he, of late, for political reasons stood up for a free and peaceful side by side existence of the various creeds in the individual lands (a goal for which the Peace of Westphalia first made the hypothesis), so Kepler, for religious considerations pledged himself to this position. Both rose thereby above the narrow and, for permanency, untenable church-political principles which although not permanently tenable had been, up to now, put into practice by both the Catholic and the Protestant side. Now fate achieved a great conjunction when it brought together Wallenstein, who had in his nativity Saturn as *dominus geniturae*, as ruler of his birth, and the more jovial Kepler in whose nativity the planet Jupiter occupied an important position, the middle of the heavens. This great conjunction molded the last years of the astronomer's life. The conjunction was in essence great because it united the brightest stars in the Prague heaven of that time. Different though the two men were in what they practiced and produced, both were imbued with burning passion for the goals which they pursued, out of which alone do great deeds grow; both, according to their contemporaries, were surrounded by mystery and were gifted with extraordinary powers. Also both felt themselves most closely united with the starry heavens, to be sure, in a completely different way. Wallenstein, caught in astrological delusion, consulted the stars regarding all his political and martial decisions and transactions; one must certainly call on the astrological considerations on which he was dependent when one seeks the reasons for his often puzzling behavior, since he sometimes hesitated, sometimes acted impetuously, more so than is usually the case. Kepler, on the contrary, shaped the form of his own life by permitting his sensitive and receptive mind to be overwhelmed and carried away by the unspeakably effective beauty and order in the heavenly regions.

The two men's spheres of life had touched before. In 1608 Kepler had been visited in Prague by a physician named Stromair who, on behalf of an unnamed noble lord, asked him to cast his nativity. As Kepler relates, already then, and later, he held to the principle of complying with such requests only if he could be assured by those who made them on their own behalf or for others "that my work [is] for one who understands philosophy and is infected with none of that contradictory superstition that an astronomer should have been able to predict coming particular things and future contingencies from the heaven." Since Dr. Stromair was known to him as a learned physician, he considered this assumption as fulfilled and gratified the request.[1] He calculated the position of the stars for the birth hour told to him as well as the corresponding directions for a series of years. From the beginning, in the interpretation of the constellations he protested against the conventional procedure of astronomers which he considered "superstitious, prophetic and a supplement to Arabic fortune telling," whereas he perceived therein the correspondence between heaven and man, namely, that the hidden powers of the soul are much inclined toward the heavenly configurations, are inspired and, indeed, at the birth of the person, instructed and formed by them. With these principles, he copied out of the heavenly conformation the picture of a person in such a way that he stands plastic before us in his uniqueness. Since this person was, indeed, Wallenstein and in the figure which Kepler sketched, drawing and truth agreed to a great extent, the wording of his explanation might be repeated as an example of how he was able, thanks to his ability to fill in and his psychologic art of presentation, to discharge one of the astrological commissions which came to him.

"I might truthfully describe this man as one who is alert, quick, industrious, of restless disposition, with a passionate longing for various innovations, who dislikes common human nature and affairs, but who rather strives for new, untried or otherwise unusual means, yet has much more in his thoughts than he lets be seen and perceived

[1] ED. NOTE. Kepler ,*Opera* (Frisch, ed.), I, 386–90.

outwardly, [for] Saturn in ascendency makes deep, melancholic, always wakeful thoughts, brings inclination for alchemy, magic, sorcery, communion with the spirits, scorn and lack of respect of human law and custom, also of all religions, makes everything suspect and to be distrusted which God or humans do, as though it all were pure fraud and there was much more hidden behind it than was generally assumed.

"And because the moon stands in abject position, its nature would cause considerable disadvantage and contempt among those with whom he has dealings, so that he would be considered a solitary brute, shunning the light. He would also be formed unmerciful, without brotherly or conjugal love, esteeming no one, surrendering only to himself and his lusts, hard on his subjects, grasping, avaricious, deceptive, inconsistent in behavior, usually silent, often impetuous, also belligerent, intrepid, because the sun and Mars are together, although Saturn spoils his disposition so that he often fears for no good reason.

"But the best feature of the positions of the heavenly bodies at his birth is that Jupiter follows, bringing hope that with ripe age most of his faults would disappear and thus his unusual nature would become capable of accomplishing important deeds.

"For with him can also be seen great thirst for glory and striving for temporal honors and power, by which he would make many great, dangerous, public and concealed enemies for himself but also he would mostly overcome and conquer these. It can also be seen that this nativity has much in common with that of the former Polish chancellor, the English queen, and other similar people, who also have many planets standing around the horizon in positions of rising and setting, for which reason there is no doubt, provided he only would pay attention to the course of the world, that he would acquire high honors, wealth, and after making a court connection for himself also a high-ranking lady as his wife.

"And because Mercury is so exactly in opposition to Jupiter, it almost looks as though he might yield to wild schemes and by means of these attract a great many people to him, or perhaps at some time be raised by a malcontent mob to a leader or ringleader."

The question, whether Kepler when painting this character picture knew whom it concerned, has already been variously discussed. For this, two later (hitherto unpublished) letters from Gerhard von Taxis, an officer in Wallenstein's service, who had sent the go-between, Stromair, to Kepler, are very enlightening.[1] The letter writer speaks of the above-mentioned horoscope of Wallenstein in such a manner

---

[1] ED. NOTE. *Johannes Kepler Gesammelte Werke*, XVII, 131–2, 144 (letters numbered 704 and 717).

that it must be assumed that this man who gave the commission although he kept in the background was well known to the astronomer. Besides, since Kepler, in the horoscope diagram still extant in the original, recorded right on the sketch the name of Wallenstein in the cipher he sometimes employed, it cannot be doubted that Kepler knew for whom he was supposed to cast the nativity. In criticizing his character delineation one should, of course, bear in mind that Wallenstein, then only twenty-five years old, was not yet the universally known great man of later times.

To his astonishment sixteen years later, in the autumn of 1624, Kepler received his manuscript back, once more through the mediation of Gerhard von Taxis, who was now Captain-General of the duchy of Friedland. In the latter's covering letter, in which this time the name of the man who gave the commission was again not spoken,[1] the astronomer was requested to explain his earlier *Judicium* in more detail and at greater length and also to extend it for future years.[2] Wallenstein, in his own hand, had supplied the earlier horoscope with marginal notes defining his attitude to the predictions. Kepler, it is true, drew little satisfaction from this strange demand but fulfilled it gladly because he wanted to retain the favor of his powerful client and an ample honorarium was offered him; but still he did not fail to warn the high lord, whose name he himself does not express in the document, to abandon his "quite visibly erroneous delusion." Should he, although he thought nothing of the rules of astrology "let himself be used like an entertainer, actor or else a mountebank?" "There are many young astrologers who have the inclination and belief for such a game; he who wants to be cheated with open eyes may enjoy their efforts and entertainment. Philosophy and thus also true astrology are a testimony of God's work and thus sacred and not at all a frivolous thing, and I for one do not want to dishonor them." There were strange questions about which Wallenstein wanted information from the astronomer: whether he would die of a stroke, or away from his native country, or if he would attain positions and wealth outside his native land, how long he should continue his military career, in which lands he would have to perform military duties, whether good or bad luck was to be expected in them, whether he would have enemies and

---

[1] ED. NOTE. The German edition reads "offen genannt" which Miss Martha List, Dr. Caspar's collaborator, says should be replaced by "wieder nicht genannt" and it is from the latter that the translation was made.

[2] ED. NOTE. The exchange of letters can be found in *Johannes Kepler Gesammelte Werke*, XVIII (published in 1959), 217–18 (letters numbered 999 and 1000). These and the augmented and extended nativity can be found in Kepler, *Opera omnia* (Frisch, ed.), VIII, pt. 1, Frankfort: Heyder and Zimmer, 1870, 343–58.

who, and beneath which heavenly sign they would be located. The answer to these crazy questions Kepler introduced with an explanation which could not be misunderstood: "Whatever person, educated or uneducated, astrologer or philosopher, in the discussion of these questions turns his eyes from the individual's own temperament, or otherwise does not consider his behavior and qualities in relation to the political circumstances, and wants to derive the answers solely from the heavens, be it because of an inner compulsion or only from inclination and disposition, he truly never really went to school, and never yet properly cleaned the light of reason which God ignited for him." The individual answers were given in accordance with this belief. Kepler held unconditionally to his conviction of the impossibility of solving such particular questions from the stars and attempted to satisfy his client by telling at which conclusions the believers in the stars would arrive in the above case by the use of their rules. His explanation, after mentioning various constellations in a particular year, is characteristic: "A ruler who thought as much of astrology as the person in question, and knew all this, would without doubt send such a commander with such an impressive constellation against his current foreign enemies, provided that he was sure of the man's faithfulness." What Kepler would be ready to admit is the possibility of drawing conclusions about the mutual sympathy or antipathy of two people by comparing their nativities. In this connection he points to the nativity of the young Bohemian King Ferdinand and observes something which was very significant to Wallenstein, namely "that between the two men there are to be expected, not particularly affection and inclination toward each other, but rather various frictions." In figuring the revolutions for the coming years Kepler stops with the year 1634, prophesying "horrible disorder" for March of this year. As everyone knows, Wallenstein was murdered on February 25, 1634.

After the two eagles had thus circled about each other from the distance, they now had to meet, since they stayed in the same place, where Wallenstein used every means to raise and strengthen his personal ascendency and Kepler was in search of a favorable spot for his scientific activity. Several years before, the general, through Gerhard von Taxis, as thanks for the horoscope had offered "to show" the astronomer "attractive favors, when the gentleman will feel that his princely Honor can be of advantage to the gentleman." Now such an opportunity had come. Wallenstein had just received the duchy Sagan from the emperor as fief. As always, he immediately took measures to raise the position of his new possession. Kepler gladly looked toward Silesia, because here the Protestants were not yet as restricted in their religious practices as in the other Hapsburg lands, although this territory still belonged to the emperor's sovereign domain. In order to preserve his position and his claims as court mathematician, he did not want to quit the empire. So in February, 1628, conferences regarding a removal to Sagan were started. Kepler himself makes the following comprehensive statement about his settlement with Wallenstein: "I obtained the favor of Duke Albert of Friedland and Sagan, the chief general of the imperial troops, a favor which previously I had looked at with suspicion. The duke is an exceedingly valiant hero as well as an ardent admirer of mathematics and thanks to these double characteristics, so to speak, a second Hercules, to whom the emperor is much devoted. Therefore, with the very first meeting, he easily spared me the necessity of looking for a place of residence outside the imperial hereditary lands and for that reason of presenting a petition to the emperor. He most graciously assigned me a quiet place in Sagan, settled on me an annual stipend in keeping with his otherwise brilliant appearance and also promised me a printing press, with full consent of the imperial circle." However, the decision was not reached as simply and quickly as one might believe from this account. Kepler had misgivings regarding the unheard of rapid rise of the commander-in-chief. Would his good fortune, which was dependent on military chances and looked upon by many with suspicion or jealousy, last? Bernegger strengthened this mistrust and, worrying for his friend, asked whether he had the courage to entrust himself to the wagon of that Phaethon. How strong were Kepler's hesitations and how cautiously he went to work on his decision, is made clear by his letter of the end of February, 1628, to Elector Johann Georg of Saxony. In it he strove hard to obtain reinsurance, by asking the elector to grant him refuge, in case it should be necessary once more to leave his contemplated place of residence, Sagan. It reads as follows: "Since the present difficult times also cause, in the above-mentioned place, sundry inconveniences *per consequentiam*, and since I together with my family might be driven out overnight and be obliged to move on still further in a region which is still entirely unknown to me, so in addition my most submissive request reaches your electoral highness: may it please you to permit me in this my always onward continuing wanderings and the anxiety these bring to grant me so much solace as to condescend to permit me and my wife and children, if this is necessary, to have safe refuge and an undisturbed domicile in your highness' lands so that, for the benefit of the profession, I may at some time also teach orally in your honored universities, Wittenberg or Leipzig, that which I now publish in print."

Yet, in the embarrassing situation in which Kepler found himself, all these hesitancies had to yield. Even if his move to Sagan would not be able to promise any lasting solution, he had to act and come to a decision. By April 15, nothing definite had yet been agreed upon. But a few days thereafter the decision was reached. On April 26, Wallenstein informed Grabes von Nechern, the Captain-General of Sagan, that he had consented to Kepler's request to live in Sagan, since he was "a man qualified and highly skilled in mathematics and astronomy." Wallenstein ordered that he be provided with a comfortable home for a small payment and that otherwise a helping hand be extended to him in everything. Two days later Wallenstein instructed Gerhard von Taxis, the Captain-General of Friedland, to pay the newcomer an annual salary of 1,000 gulden. Kepler thought that the reason for the large scale reception by the Friedlander was that he was the one who three years before had prevented the payment of the sum of 4,000 gulden which the emperor had imposed on the people of Nuremberg. Moreover, on May 10, the emperor requested his commander-in-chief to order that Kepler be paid his outstanding debt of 11,817 gulden which the imperial treasury still owed him. The emperor indicated that he had no doubt that Their Honors would gladly be helpful in satisfying the astronomer. Their Honors, to be sure, were in no rush to comply with this imperial request. Yet for the time being everything appeared to be arranged for the best. The path to the near future was open for the troubled wanderer.

Kepler's relationship to Wallenstein should not be interpreted as his having become court astrologer to the warrior, who had faith in the stars. Kepler had not lent himself to that. Surely, he saw very clearly the great dangers concealed in the combination of astrology with state affairs. Since he knew the astrological inclinations of his new patron just as well as the difficult decisions with which he was faced, he had to be so much the more cautious of those dangers. Certainly Wallenstein himself, after the advice which the astronomer had offered in the horoscope, was not entirely satisfied by what the former was able to offer in accordance with his scientific conviction. Wallenstein wanted to discover more from the stars than Kepler was able to read from them and consequently also repeatedly obtained information from other astrologers. Zeno, the Genoese, is known; Schiller introduces him on the stage as Seni, an insignificant man, who later entered the general's services; he was never mentioned by Kepler. But what Wallenstein especially requested of his new servant were astronomical calculations of the positions of the planets. He knew that no one could work these out as exactly as the author of the *Rudolphine Tables*. With the results of these calculations he then turned to his obliging astrologers in order to learn from them what he requested. So he obtained, as the available evidence shows, written and oral advice and information from the court mathematician. It is surprising that in this reference is repeatedly made to the Bohemian King Ferdinand whose rivalry the General-issimo openly feared for the future. After all that, there must be seen in Kepler's summons by Wallenstein the magnanimous gesture of the grand seigneur, whom it pleased to provide the then already famous astronomer with a place for his activity and who also knew that he was gaining honor for himself when he supported science in this manner. This interpretation completely agrees with what Kepler himself wrote to Bernegger at that time: Wallenstein is trying to gain fame for himself by furthering science without regard to religious creed.

5. Sagan

The two men departed from the Bohemian capital almost at the same time in May. Wallenstein went to Mecklenburg, where his general, Hans Georg von Arnim, besieged Stralsund in vain, and where he himself was to meet with his first failure. Kepler went first of all to his family in Regensburg and in the following month to Linz to wind up his affairs there. In fact, he still held the position of Upper Austrian district mathematician. For obvious reasons he had assured himself of this position until another arrangement of his future was found. As late as February he had written from Prague to Linz that he was ready to adapt himself in case the emperor should want to assign Linz or another place for him as his further residence; the maintenance of his appointment as district mathematician, wherever he should settle, would, as previously, be presented to him "for special consolation and continuation of mathematical studies." He was now most cordially received in Linz by the representatives, just as he had been by the emperor in Prague. He reported on his activity in the years of his absence and about his arrangements with Wallenstein. His plan was approved. The representatives had 200 gulden paid out to him by the collector's office for the presentation of the tables and for his traveling expenses and, on July 3, 1628, granted him the requested release. Thus was the Linz period of Kepler's life finally closed. The former district mathematician departed from the place of his fruitful activity of many years, full of happiness that his old patrons retained their friendliness for him. He met his family again in Prague. They had come from

Regensburg with their household possessions.[1] Together they soon proceeded north. They arrived in Sagan on July 20. A not inconsiderable part of his household furnishings, and in addition valuable objects, books, globes, and instruments had been left behind in Regensburg with friends, of whom he had many in that city. This is confirmation that Kepler did not view his move to Sagan as a permanent solution. Yet could there be a permanent solution for this man in those times?

He found it difficult to get used to the new situation. This man from Swabia had felt very much at home in Austria; he got on with difficulty with those who had the north-German manner. "I am guest and stranger, almost completely unknown, and barely understand the dialect, while I myself am considered a barbarian," he still reports one year after his arrival. In Linz he had deplored the lack of friendly intercourse with men of similar aspirations, who were interested in his research. In Sagan he must have missed this intercourse to a much greater extent. In Linz he had at least found a number of intellectually active friends with whom he, a communicative man, could interchange thoughts. But here the feeling of loneliness cast a net over his soul. Only by letter is he connected with the world to which he belongs. He conducts the most ardent correspondence with his old friend Bernegger, who participated in and understood all his requests. "I cannot say at all," he writes to him, "how much, in my loneliness and disquiet here, you comfort me by your endeavor to prove kind to me. For it is the loneliness which oppresses me, away from the great cities of the kingdom, where letters come and go only slowly and are costly." In addition to Bernegger, Philip Müller, the Leipzig professor, physician, mathematician and astronomer is the one with whom he carries on an intimate correspondence; Johannes Seussius, the Saxon adviser, who had long been friendly with Kepler, had brought about the acquaintance between the two. Besides the feeling of isolation, the uncertainty of his position keeps worrying the lonesome man. He knows himself dependent on a man who himself is dependent on the fortunes of war. And these are changeable. Will he be able to remain in Sagan? This he immediately asks himself, worrying about

[1] AU. NOTE. The extant documents are puzzling in regard to the number of Kepler's children by his second marriage. While it is certainly established that in December, 1626, on his trip to Ulm, Kepler arrived in Regensburg with three small children, in many of his letters in February, 1627, he speaks of four children. This would lead to the conclusion that Frau Susanna in the meanwhile was delivered of a child in Regensburg. Yet in the baptismal register of this city there is no entry about this. Since four, for the number of children for the time before the move to Sagan, is also testified to in documents from the period after Kepler's death, it is necessary to stick to this number, although it is most surprising that Kepler, who always carefully noted down family events in his papers, nowhere gave information about the birth or the name of the fourth child.

his future. So long as things go well with Wallenstein, he is cared for. During this time, with the high salary which he draws and which is generally punctually paid him, there is little talk of need for money.

But what Kepler waited for in vain was the payment of his claim on the imperial treasury, which Ferdinand had entrusted to his generalissimo. Wallenstein needed his money for other purposes. Once, it is true, he instructed the imperial commissar, Johann von Oberkamp, to see how the mathematician could be satisfied with some fief which had fallen to the imperial exchequer. The estate, Görlachsheim, near Friedland was mentioned; also an endeavor was made to seize a certain small fief in little Halberstadt. However, these steps did not yield results. Another way to stabilize Kepler's situation and satisfy his claims had already become apparent when, in the spring of 1629, Dr. Thomas Lindemann, rector of the University of Rostock, on the order of Wallenstein, called Kepler to the university there. Kepler did not refuse but set conditions which he knew could scarcely be fulfilled. He asked that Wallenstein himself obtain the emperor's approval of the acceptance of the call. In addition he should now fulfill the disbursements which he had promised from the duchy of Mecklenburg, namely, the payment of all Kepler's claims on the court in the amount of nearly 12,000 gulden. In reality Kepler did not have the slightest desire to go to Rostock. This city lay still further north than Sagan and, what is more, in a land that just then lay at the center of military arrangements. Wallenstein had shortly before received this land Mecklenburg as fief after the expulsion of the rightful dukes. With good political instinct Kepler did not trust this situation and had doubts about its duration. Even in Rostock he would have been dependent on Wallenstein and would be able to keep his position only so long as the latter should be in power. "If there is peace in the Baltic, then the duke is compelled to draw further away from there with his troops. But if there is no peace, as will be almost universally assumed, then he will have there as enemies the Swedes, the Danes, and the Dutch fleet." So, far from the court, he would also get into straits with his salary as mathematician. The plan necessarily came to nothing.

Leaving aside the question of these obstacles, still another dark shadow, which troubled his mood, lay over Kepler's life in Sagan. As though he was never to be granted the opportunity of living quietly in his creed, just at the time of his arrival in Sagan the confessional war broke out in this city, too. In the duchy of Sagan nearly a hundred years before the Reformation had already spread itself out in wide circles. Now in the entire domain only a very few among the lordly and knightly rank were Catholics; the people mostly followed the Lutheran or Calvinist doctrine. Albeit Wallenstein, as we know, took a rather free position in the question of the confession, nevertheless, for political reasons he immediately set to work to lead the subjects of his newly acquired duchy back again to the Catholic religion. Grabes von Nechern, his captain-general, although himself of Protestant origin, did everything to fulfill the plans of his lord. In November, 1628, in the name of the duke, he ordered the citizens of Sagan to become Catholic or to leave. With inner reluctance, they gave in to the pressure exerted on them and swore to comply with the command. At the same time, Jesuits were summoned to support and promote the work of conversion. They were supposed to set up a gymnasium and bring up the young in the Catholic spirit. Everything had to go according to Catholic ritual for confession and communion. Heretical books had to be handed over. Taking part in the Corpus Christi procession was considered a serious duty. The citizens were forbidden to accompany the body of a non-Catholic to the grave. The disbursements which the city had previously offered the Lutheran preachers had to be paid to the Jesuit college. It is clear that the newly converted, whose repugnance had to be suppressed with force, were only waiting for an opportunity to revert to their earlier creed. Kepler himself with the unique position which he occupied was not concerned with these regulations. But still he felt himself secretly watched over and obstructed in his relations with his fellow citizens, who had to keep away from him. He reports to Bernegger about sad examples and pictures, how acquaintances, friends, people from his close neighborhood were ruined, and talking with the frightened ones was cut off by fear. He took it bitterly that he was denied a place in the Lord's house.

To keep from suffocating in this atmosphere, Kepler had just one remedy, ceaseless work. In fact his delight in creating was unbroken. In all seriousness something like high spirits is heard about when Kepler writes to Philip Müller: "In the midst of the collapse of cities, provinces and states, of old and new families, in the midst of fear of barbaric invasion and violent disturbance of home and hearth, I see myself, a disciple of Mars, even if not a youthful one, compelled to hire printers and publish, or rather start, the edition of the Tychonic observations, without feeling any fear whatsoever. With God's help I also wish to truly complete this work, and indeed in military manner because today I deal with commands, fume, rage but leave the worry about my burial for the morrow." His pure ethos of work is evident from the words: "When the storm rages and the shipwreck of the state threatens, we can do nothing more worthy than to sink the anchor of our peaceful studies into the ground of eternity."

However, before Kepler could begin with the productive work, he had to deal with a series of time-consuming preparations. First it was necessary to lay a new foundation in this city, which did not possess any learned tradition. The most important matter for him was the establishment of a press, which Wallenstein had promised him. But it took almost a year and a half for this goal to be reached. As early as the fall of 1628 he had discovered, in Frankfort-on-the-Oder, a press which could be purchased for 80 Reichstaler. Since, however, he wanted to get a printer at the same time and there was none such to be found, because at that time "the Reformation activity was going on" and as a consequence foreign non-Catholics took good care not to go to Sagan, the purchase was not made then. In the following February Kepler turned once more to his lord in order to achieve his goal. In case he had not achieved his aim by Easter, he would be unable to complete in the current year the printed works he had undertaken unless Wallenstein should give him the opportunity of going elsewhere for half a year, to a place which had a printing press, perhaps Frankfort-on-the-Oder, Görlitz, Prague, or Leipzig. In October he was finally able to send two people to Leipzig. They were a compositor and a printer whom he had meanwhile hired. By interceding with Wallenstein he had procured the protection which they needed because they were not Catholics. In Leipzig, through Philip Müller's good offices, they were to purchase a press. In addition, there were negotiations about the most reasonable possible buying of paper, which he ordered from Prague, Friedland, and Leipzig. Besides, it was necessary to procure letters in Leipzig or Wittenberg to complete the set of number type which he had brought along. Arranging for the payment of the necessary expenditures for all these things, for which Kepler assumed responsibility, involved much writing and made travel necessary. Presently the restless man, because of this and other business, was in Görlitz, Friedland or Gitschin, where Wallenstein had his residence. In Görlitz he had a few leaves entitled *Sportula* printed. They formed an appendix to the *Rudolphine Tables* and gave directions for the astrological use of these tables; the elderly mother (astrology) was not supposed to permit herself to complain of becoming abandoned and scorned by her thankless daughter (astronomy). From June 1, 1629, on, Wallenstein granted twenty bales of paper annually and 20 gulden for printing expenses weekly. In March, 1630, to be sure, the order was issued not to remit anything further to Kepler for his printing work beyond his annual stipend. In December, 1629, the press was at last set up in Sagan, undoubtedly in the house in which Kepler lived. He himself acted as printer. Everything was ready, the printing could commence.

But what were first taken in hand were not the Tychonic observations, albeit these were placed in the foreground in the advices. Indeed, astronomers everywhere waited for this publication, too, because they wanted to employ that very valuable material for their theoretical efforts. Kepler knew that. He was, however, also aware that by the discovery of his planet laws and the publication of his tables he had already separated from this ore the noble metal therein. Besides, the printing of the voluminous observations meant a prodigious effort, and it is understandable that this man, burdened with his own thoughts and plans, must have found it hard to tackle this colossal, purely editorial undertaking. (Not until our day was it completed by J. L. E. Dreyer in the complete edition of Brahe's works, in which the observations fill no less than four stately quarto volumes.) It was a different task which forced itself upon Kepler and which he now wanted to carry out. We recall that ten years before as the first part of a comprehensive work he had published *Ephemerides* for the years 1617–1620. Now he wanted to continue this work. He saw that, after publication of the long-awaited *Rudolphine Tables*, astronomers everywhere would want to use this work for the purpose for which it was created and would want to publish ephemerides. To assure himself of the fruits growing on the tree planted by him, it was necessary that he himself get to work quickly so that others would not precede him. Indeed, along with his work on the tables he had already done extensive preparatory work for the continuation of his ephemerides. The harvest had to be brought under cover. Work for the whole of 1629 had been scheduled precisely, so far as other affairs left time for it. The answer, already composed two years before, to the work by Johannes Terrentius, the Jesuit, who was visiting in China, was printed with the new press in order that by this little first production Kepler could dedicate a sign of thanks to his high patron, Wallenstein. Thereafter, in the first months of 1630, the ephemerides went to press, in fact, first the third part for the years 1629–1639, then the second part for the years 1621–1628. In September the whole thick ephemeris volume was complete. The parts for the years 1621–1629 contain, as had previously that for the year 1617, Kepler's day to day weather observations. The second part is dedicated to the Upper Austrian representatives to whom the author considered himself tied by gratitude. The third part is dedicated to Wallenstein. As noted on the title page,

Kepler had printed the work at his own expense. Tampach, the book-dealer in Frankfort, took over the sale.

6. Somnium

During the pauses in the printing of the ephemerides caused by trips and other circumstances, still another work went to press. It is entitled *Somnium seu Astronomia Lunari* (*Dream or Astronomy of the Moon*). Its beginnings are far back in Kepler's life. When a student in Tübingen, as an inspired follower of Copernicus, he had written down thoughts concerning how the various heavenly motions would appear to an inhabitant of the moon, who directly comprehends by his senses the proper motion of his dwelling place just as little as we inhabitants of the earth do that of ours. A few years later he was especially stimulated by a study of Plutarch's work about the face of the moon. In the summer of 1609 in learned conversation with his friend Wackher von Wackenfels he discussed the subject at length, so that he was happy to realize his plan to write a "moon-geography" and composed the *Somnium* to please his friend. In 1620 when he had returned from his long visit in Württemberg, he once more brought out his manuscript, the subject of which often whirled round in his head. It was apparent to him that his little writing contains just as many problems as lines, problems to be solved in an astronomical, physical or historical way. Since, however, so he supposed, people do not like such difficulties and want entertainment of the kind which has soft arms to throw around their necks, he decided to solve these problems himself by the addition of notes. He packed so many discussions into these notes that they are far more extensive than the main work. Observations made in the meanwhile with the telescope especially excited his fancy. Were not cities with circular ramparts seen there? He asked himself should he not have written a "moon-state," the way his contemporary, Campanella, composed a "sun-state"? Would it not be a good idea to describe the Cyclopean customs of his time in lifelike colors but, for the sake of caution, quitting the earth and going to the moon to do so? Unfortunately, he did not carry out this plan. He wanted to keep his hand out of the stickiness of politics and instead remain in the pleasant pasture of scientific contemplation. That appeared to him to be less dangerous, especially since Thomas More with his *Utopia* and Erasmus with his *In Praise of Folly* had not been safe and had had to defend themselves.

Now Kepler thought the time had come to bring into the open the child of his mind. "Since we shall be driven out of the earth, the book will be a useful provision for our journey to the moon."

The title *Somnium* (*Dream*) points to the fanciful raiment in which he expresses his selenographic thought. In addition to the powerful discipline of thought which he showed in many works, he was gifted with a fancy which brought him flowering ideas from all sides. In the dream, so goes the tale he tells, he sees himself as a youth, who grows up in Thule under the guidance of his mother, Fiolxhilde. (This name he chose because on an old map of Europe, which hung in his abode at the home of Bachazek, the rector, in Prague, the name Fiolx appeared instead of Iceland.) The mother is on friendly terms with wise spirits, one of whom transports her to other lands whenever she wants and brings her information about them. After the youth has met with some adventures, one of which led him to Tycho on the island of Hven, and has grown up, he can once induce his old mother to conjure up this spirit on a bright night by some secret signs. The spirit now relates, as though about a distant land, Levania, and its inhabitants, all that which Kepler wants to impart to the reader about the moon and the heavenly phenomena to be observed from it. There is a contrast between this poetic frame of his account and the no less fanciful descriptions which he, never at a loss, is able to give of the nature of the moon's inhabitants, their dwellings and habits, of the seas and swamps, of the plant and animal world, of the encircling walls, which he interprets as skilled construction. It is often difficult to distinguish what is supposed to be poetry and what scientific conviction. The very exact description which he gives of the firmament as seen from the moon and which is the main interest of his book is all the more strikingly contrasted with the previous passages. Here he attends to all the phenomena which are presented by the sun, the earth, the planets as regards their motions, their light and their sizes for the dwellers on the moon, both on the side turned toward us and on that turned away. The alternation of day and night, the length of periods of time, seasons, the alternation of heat and cold—all these he includes in his consideration. His manner of rigorously carrying through in all detail this not so simple experiment in thought shows the master aware of the scientific goals he pursues: "to make an argument for the motion of the earth taking the moon as an example." The scientific importance of the book is enhanced by the contents of the appended notes with their numerous astronomical, physical and geographic digressions. These bear witness to the author's great book-learning as well as to his intellectual agility. This agility is always throwing new ideas to the reader, drawing him across and through all spheres of Kepler's wealth of knowledge and continually confronting him with new considerations which are in no way always simple. To his own work Kepler also added the Latin translation he had made of Plutarch's *Face of the Moon*. He even supplied passages appropriate to the context which were missing in the traditional Greek version.

7. Jakob Bartsch

For the laborious reckoning of the ephemerides and for the later printed works Kepler found an excellent helper in the person of Jakob Bartsch, a young scholar from not far distant Lauban in Lusatia. Bartsch was a pupil of Philip Müller, had studied astronomy and medicine in Strasburg, and had met Kepler for a short time in 1625 in Ulm, when on a journey to Padua. After the *Rudolphine Tables* appeared, he had at once tackled the calculations of the ephemeris for the year 1629 and published the result of this work. Since he could not learn where Kepler was staying at that time and would gladly have been in closer touch with the highly esteemed master, he appealed to him in a public letter in this printed work, offering him his co-operation. Since Kepler was most desirous of the aid of a younger, trustworthy strength, the two men joined in common work in the autumn of 1628. Bartsch took over the calculation of the ephemerides for the years 1629–1636, and zealously carried them out in the course of the year 1629 in Lusatia, from where he repeatedly came to Sagan for discussion in person.

The association between the two men was soon to become even closer. During all his trouble over the course of the heavenly planets, Kepler still did not forget the part which Venus plays in the course of earthly matters. Could he desire a better husband for his daughter Susanna than Jakob Bartsch, the master and candidate in medicine, his clever and industrious assistant? This was the thought of the anxious father as soon as he had tested the eagerness and character of his new collaborator. However, Susanna was not in his home; for some time past she was "one of the woman servants in the princely Margravian household in Durlach." Apparently Bartsch also secretly harbored similar wishes. The affair had to be set in motion. Friend Bernegger had to help in that. He knew the candidates and could come in contact with Susanna in nearby Durlach more easily than could Kepler in far-off Sagan. So Kepler inquired of him how Bartsch lived in Strasburg, what his habits were, and how much money he spent. Being certain of a favorable reply, he asked his friend to prod this bashful young man into speech. For the sake of emphasis it should also be pointed out to the latter that Kepler's intercession in his application for the Strasburg mathematics professorship, still held by old Malleolus but necessarily soon to be vacated, would be very useful to him. Only one thing in the contemplated son-in-law would not please Kepler, namely that he anchored his studies in astrology. Bernegger gladly supported the plan. Yet several months passed before it came to the point. Bartsch, indeed, had not even seen his future bride, and the latter also had a word to add concerning the matter. At last the timid suitor summoned up courage and properly sued Kepler for Susanna's hand, even before he had met her. The father, not at all surprised, said "yes" provided Susanna also gave her consent. This occurred, so that nothing further stood in the way of the union. Father Kepler and Bernegger, who commended the maiden's modesty, piety and unusual good sense, were Bartsch's guarantors of his future happiness.

Now it was necessary to prepare for the wedding. Where should it take place, in Leonberg, Regensburg, Sagan or Strasburg? The confusions of war were making travel difficult. Strasburg was selected. There the bridegroom hoped soon to establish himself, and Bernegger and his wife were happy to act the part of the distant parents of the bride. On March 12, 1630, the marriage was performed with great ceremony, after the bridegroom had been decorated with the hood of a medical doctor that very morning. The rector of the university, a number of professors, the leading men and women of the city took part in the festivities. All the streets through which the bridal procession moved were full of onlookers. The name Kepler was in every mouth and in veneration for him the people pointed out for each other with their fingers those of the great astronomer's relatives who were present. The magistrate had donated two buckets of rich wine for the banquet. Music was the only missing element which might have added to the gaiety. In consideration of the very serious times, its silencing had been commanded. Kepler had to content himself at home with a detailed account of the celebration in writing. But it was not just the great distance which kept him from taking part in the celebration. His wife was about to be confined and on April 18 gave birth to a little daughter, who was baptized Anna Maria. When soon thereafter the young couple came to Sagan, there were many happy faces in the astronomer's house. That also Bartsch, by formal decree of the senate, had meanwhile been designated substitute or, in accordance with the circumstances, successor to the chair of mathematics at Strasburg University promised well for the future. The fortunate success of the marriage plan, which occupies much space in contemporary correspondence, was Kepler's last great happiness in his sorrowful life. His days were numbered.

8. Regensburg and Kepler's death

During his residence in Sagan there were political events whose outcome was a reversal of the emperor's hitherto favorable position. It was so much the more in order for the imperial mathematician in Wallenstein's sphere of rule to keep an eye on these events since his personal fate was ever entangled with political change. When Kepler left Prague, the emperor stood at the peak of his power. In view of the endeavor which throughout his life fixed the line of Ferdinand's politics, it was inevitable that he utilize his fortunate situation for restricting Protestantism and strengthening the Catholic cause. On March 6, 1629, he published the so-called Edict of Restitution, which ordered restitution, that is the introduction of the Catholic Reformation in all ecclesiastical possessions immediately subject to the emperor that had been seized since the Passau contract of 1552 and in all chapters of which Protestants had taken possession since the Religious Peace of Augsburg of 1555. The proviso concerned, above all, the archbishoprics of Magdeburg and Bremen, as well as a series of bishoprics. Since the duke of Württemberg's territory was also involved, the prophecy, made by Kepler to Maestlin and the Tübingen theologians many years before, was now being fulfilled. (As the war progressed, his native land was to become even more severely afflicted.) One can imagine the unrest called forth by these measures and picture how much stronger the already existing tensions became. Although the Peace of Lübeck in May of the same year put an end to the Danish war, a universal peace was still not to be thought of. Wallenstein, who objected to the Edict of Restitution, nevertheless felt compelled, in his position, to dedicate his entire attention to the disputes growing out of it.

As for the rest, the years 1629 and 1630 were taken up less by military affairs than by political intrigues of individuals and powers, whose interests opposed each other, or who believed they could derive advantage for themselves out of the confused situation. Among the latter was King Gustavus Adolphus of Sweden. In earlier fights with Russia and Poland he had conquered the coastal lands of the Baltic Sea from Riga almost to Danzig. Now he considered the opportunity favorable for subduing the German Baltic coasts also and for both carrying out his plan of founding a powerful northern kingdom and at the same time helping to make the evangelical matter which was of such great moment for him victorious over the emperor in Germany. In June, 1630, the conquest-happy prince trod on German soil in Pomerania. Few recognized, in its entirety, the danger which threatened here; the emperor likewise underestimated it. The man with the keenest sight was Wallenstein. But just at this critical moment the sword was to be taken from his hand. In the same June the emperor had summoned a congress of electors to Regensburg. His main purpose was to assure his son Ferdinand of the succession to the imperial throne. The crown of Hungary and Bohemia was already entrusted to him. General displeasure with the heavy taxes imposed by Wallenstein on all the lands he entered, fear of his power and strength, mistrust for his obscure plans, the jealousy of the Bavarian elector, Maximilian, whose aspirations were being supported secretly by France, all worked together so that at this congress of electors the emperor consented to the discharge of that generalissimo who had been proved in so many battles. In August Wallenstein, who had gone to Memmingen with his splendid retinue, was informed of this step. He was not surprised.

On October 8, while the electoral congress was still assembled, Kepler started from Sagan for Regensburg. Why did he take this, his last, journey? It is repeatedly stated that at the Reichstag he wanted to obtain payment of the sum owed him by the emperor. However, in this form the case is misrepresented. The electoral congress as such had nothing to do with claims which the emperor had not yet settled, and one cannot suppose that the court mathematician would have wanted to seek his rights by a path which his master would necessarily have considered insulting. Nor is there anywhere, in the extant deeds and letters, talk of such a purpose. The main goal of Kepler's journey was Linz. He wanted to go there the year before but his affairs did not permit. He owned two 6 per cent bonds, a 2,000 gulden one and a 1,500 gulden one, payable to the district Austria above the Enns. He had great trouble obtaining the interest. In the autumn of 1629 he had written to Linz about it and in the spring of 1630 sent a messenger there at considerable cost to himself. Wallenstein, himself, at Kepler's request, sent a reminder to the representatives. In August, the reminder was renewed. There was never any money in the district treasury. Kepler's fear, he could suffer serious damage to his property, rested on the fact that the regents had changed in the district above the Enns, and the men now there did not know him and his conditions. Finally, he was told to present himself on St. Martin's day (November 11); there would be a chance of giving him satisfaction on that day. So, he arranged the appointed time for his journey accordingly. It was convenient that in September the ephemerides volume was finished. Presenting this work to Wallenstein, who as Kepler had been informed would be in Nuremberg or Memmingen, was a second reason for this journey.

In carrying out these objectives he came to Regensburg of his own accord, so to speak. It was a fortunate circumstance that the congress of electors, which broke up in the first days of November, was still assembled at this time. There were present the emperor and many other men whom he knew and with whom he could discuss his situation. It can well be imagined that prominent in these discussions was the question of his big claim on the imperial treasury after his previous ill success. Wallenstein's dismissal, so he had to suppose, could not fail to affect him. Would not the general now be even less ready to comply with the decree to help him to procure his money? Indeed, could his double position as mathematician to His Majesty and to the duke of Friedland, as he signed himself, be permanent, after a tension had arisen between the two lords? Conditions being as described, one might guess that Kepler intended to look around in Regensburg for another residence. For here the presence of many persons of authority offered the best opportunity to do so. At any rate, what he took on this journey from Sagan is astounding. He took along not only books and clothes but also the written documents which "contained all his wealth"; that was all sent ahead with Leipzig carters. Besides, he must have been disquieted by the thought that Silesia, too, could become a theater of war. In fact it was not very long before, in the new phase of the war begun by the entrance of Sweden, Sagan was drawn into the whirlpool of the battle confusion. Bartsch later wrote that his father-in-law had departed from Sagan without any hope, in such a frame of mind that his wife and children would have sooner expected the Day of Judgment than his return. This report cannot be interpreted as though Kepler had had the intention of settling somewhere else and letting his family follow. Clearly, this time the taking leave of his beloved ones was especially hard for the departing Pater Familias, as though he had had a premonition that he would not return. This premonition was to be realized in a sad way. Had the stars confirmed him in this?[1]

[1] AU. NOTE. For the beginning of each of the years of his life, Kepler had calculated horoscope figures, that is revolutions, a large number of which are preserved in the papers he left. Now, for the beginning of the sixtieth year of his life, which he now approached, there is to be found, surprisingly, a note to the corresponding figure, whereas the figures for the immediately preceding and following years lack such explanative additions. He observes that for that period the planets occupy all the same positions as in his birth figure.

Kepler traveled first from Sagan to Leipzig, where he spent several days with Philip Müller. The earnest mood which here also cast a shadow over his disposition is apparent from his last letter from there to Bernegger; it is after all the last letter of his which we possess. In it he returns without any preliminaries to his Strasburg friend's earlier invitation and explains that he would gladly accept his hospitality, since in the current uncertainty of the general situation no opportunity of finding a place of refuge, however distant it might be, should be refused. The misery of his native land touches his heart. In the last sentence of this letter, his last, he asks his friend to pray for him. "Hold fast with me to the only anchor of the church, prayer to God for it and for me." In Nuremberg he visited Eckebrecht, who had the commission for the land map for the *Tables*. His hope of being able to take the finished map along and present it to the emperor was dashed, even though the work was nearly finished. On November 2 he rode, tired, on a skinny nag, over The Stone Bridge into Regensburg. He took up quarters in Hillebrand Billj's house in the street now named after him. This acquaintance was a tradesman and later an innkeeper.

Only a few days after his arrival Kepler came down with an acute illness. His body was weakened by much night study, by constant worry, and also by the long journey at a bad time of year. In the beginning he attributed no significance to his being taken ill. He had often before suffered from attacks of fever. He believed that his fever originated from "sacer ignis," fire-pustules. As the illness became worse, an attempt was made to help him by bleeding. But soon he began to lose consciousness and became delirious. Several pastors visited him and "refreshed him with the vitalizing water of consolation." It is not said anywhere that holy communion was afforded him. In the throes of death Pastor Christoph Sigmund Donauer rendered him aid. When, almost in the last moment of his life, he was asked on what he pinned his hope of salvation, he answered full of confidence: only and alone on the services of Jesus Christ; in Him is based, as he wanted to testify firmly and resolutely, all refuge, all his solace and welfare. At noon on November 15 this pious man breathed his last. His mortal remains were laid to rest in the Protestant cemetery of St. Peter on November 17, as the register of deaths of the New Parish indicates, or on November 18 as Bartsch reports, referring to the lunar eclipse on the day after the interment. A numerous retinue from among his many friends and acquaintances, as well as from the illustrious society still in the city because of the congress of electors, paid the last tribute to the dead man. The memorial service for the late court mathematician was arranged "by order of the reigning chief chamberlain." Pastor Donauer gave the funeral oration, based on the text of Luke 11, 28: "Blessed are they who hear and preserve the word of God." The inscription which was put on the tombstone praised the deceased as a man who is known in all Christendom through his published works and who will be counted the prince of astronomy by all scholars. It ends with a distich which Kepler himself had composed and, as Bartsch reports, had communicated to him only a few months before (a confirmation of his premonition of death). It reads:

> Mensus eram coelos, nunc terrae metior umbras.
> Mens coelestis erat, corporis umbra jacet.

> *I used to measure the heavens,*
> > *now I shall measure the shadows of the earth.*
> *Although my soul was from heaven,*
> > *the shadow of my body lies here.*

Kepler's soul had entered its eternal home. Enlightened by a ray from God's countenance, he had explored the magnificence of the visible heaven and ventured to capture and to express the harmonic order of creation in terms of numbers as the adequate medium. The body lay stretched out in the grave unsubstantial like a shadow. The earth had received what belonged to it. The soul returned whence it had come.

The foregoing report about Kepler's illness, death and burial is based on statements contained in letters by Jakob Fischer, the Regensburg preacher, and Stephan Lansius, a younger friend, in whose album the deceased had written a verse ten years before in Tübingen. Since both authorities were present during those critical days, their testimony may be considered dependable. The same cannot be said of communications by two Regensburg chroniclers, since from the beginning their statements show a dislike and confessional hate for the astronomer who had been expelled from the ecclesiastical community. Daniel Tanner, the clerical chronicler, reports: "In his illness Kepler was somewhat confused in his head, said nothing, but pointed with his index finger sometimes to his forehead, sometimes above him to the heaven. He had wanted to bring into accord the evangelical and Papist religions: *sed frustra, Christus enim et Belial numquam concordabunt.*" (Yet in vain since Christ and Belial can never be brought into harmony.) Was not what this man brings forward as a reproach really a title of glory for the departed one? The report of Kepler's death by Plato Wild, the chronicler, demonstrates the same narrow-mindedness: "This man as he doubted in religion, also died in doubt, so that one could not converse with him *de capitibus fidei.*" Tanner is also the source for the information given by a few biographers, that on the evening of the burial day fiery balls fell from heaven, as were seen not only in Regensburg but also in other places.

Considering all that we know about his last days, it appears very questionable whether Kepler had an opportunity to undertake negotiations about the desires which led him to Regensburg. The emperor departed very soon after the arrival of his mathematician. Lansius is able to report about a demonstration of kindness on the part of the prince. He relates that the emperor, when he had already mounted the ship and was about to pull off, heard of Kepler's illness and thereupon had ordered gentlemen of his retinue to visit him and to bring him 30 or 25 Hungarian ducats to aid in his recovery.

Some biographers want to construe the fact, that Kepler as a Protestant found his resting place outside of the city walls, as an act of intolerance. This is an erroneous interpretation. The citizenry of Regensburg which had joined the Lutheran creed early belonged at that time to the by far predominating portion of the evangelical creed. Consequently, already ninety years before, right after the introduction of the Reformation, the churchyard in front of the Peter's Gate had been opened for the followers of the new dogma. The fact is that Kepler, although he was excluded from the communion of his church, nevertheless was interred in the churchyard of his congregation.

The place of burial and the entire churchyard soon suffered a very sad fate. War swept everything away. When, scarcely a year and a half after Kepler's death, Gustavus Adolphus with his forces passed southwards through Bavaria, the strategically important Regensburg was most hurriedly made ready for defense and thus the destruction of the churchyard was begun. The Swedish king, it is true, passed the city by, since he turned toward Munich. The misfortune, however, was only postponed. In the following year Bernard of Weimar, who had taken over the command after Gustavus Adolphus' death, advanced, and besieged and captured the city. At that time the destruction of the churchyard was completed, by the actions either of the defenders or of the besiegers or both. Whatever remained of the sacred place of peace was demolished and ploughed up when, once more a year later in 1634, the Bavarian and imperial troops reconquered the much overrun city. In their great misery, which was further increased by the outbreak of the plague, those who had to live no longer bothered themselves about those whom death had taken and even the fame of the great astronomer did not suffice to wrest his grave from oblivion.

Thus, but a few years after his death, the resting place of his bones was no longer known. Tycho Brahe's grave is in the Tyn church in Prague, Galileo is buried in the venerable church of Sante Croce in Florence, Newton rests among the great dead in Westminster Abbey. Veneration for genius erected these worthy monuments. But no tombstone covers the place where the no less gifted Kepler was interred. It is as though the fate, which in life gave him no peace, continued to pursue him even after death.

9. The fate of Kepler's family and manuscripts

The decease of the master, whose name and work were famed in the learned world not only of Germany but also of Europe, aroused sorrow and sympathy, wherever news of it penetrated. Friends expressed in moving lamentations their pain over the gaps which the death had torn. Scholars sorrowed over the fact that the sun of astronomical science had set, and deeply lamented the loss that astronomy had suffered. They had waited for further works which Kepler's gifts and diligence would still give them, above all for the *Hipparchus* which he had announced long before and for the Tychonic observations, which he was going to make accessible for public use. It was known that no one of his time had been master of the whole domain of astronomy in so perfect a fashion, no one had been able to enrich and transform astronomy with such fruitful new ideas, as the imperial mathematician who had now passed away. The world will marvel whenever it contemplates the Herculean accomplishments and the incomparable genius of this man, said Pierre Gassendi, the renowned French philosopher and astronomer, who honored the memory of the dead man in an especially warm letter.

Most severely and grievously stricken, however, was Kepler's family, when in the first days of December a special messenger brought the bad news to Sagan. Susanna Kepler, who had always felt that there she was in a foreign country, now stood alone with her troop of children, the youngest of whom was scarcely more than half a year old. Need and worry towered over her when she thought of the future. Yet in her stepdaughter's husband she found a faithful helper. Bartsch, who himself felt torn from the busy peace of joint creativity with his master, cautiously took charge of the forlorn ones; just as previously he had shared good fortune with the Kepler family, so now he helped it to bear the misfortune which was also his own. With the pain of the soul was combined the need occasioned by a lack of pecuniary resources. Kepler had taken almost all of the ready cash on his journey and besides had borrowed 50 gulden in Leipzig from a Sagan merchant. Immediately, in December, Bartsch went to Gitschin to learn from Wallenstein what in the present state of affairs was to be expected for the widow and her children and what was to be done for him with the printing. He reported on the works which were incomplete in the press and about the plans which his late father-in-law had further intended to carry out, and which he courageously proposed to take in hand himself. For fourteen days he was kept in suspense in Gitschin. The upshot of his journey was piteous and crushing. Wallenstein wanted to hear nothing more about the printing; he refused the costs for the future. Bartsch could not even obtain from the ducal chamber the still unpaid remainder of the dead mathematician's annual salary; hopes for later payment were held out to him. In the beginning of January the widow again turned to the counsellors of the ducal chamber regarding this matter. She described her need in moving words. The merchant from whom Kepler had borrowed money was very desirous of being reimbursed. The printers demanded their wages. Because of paying for mourning clothes she was severely pressed. The debts incurred by Bartsch during his stay in Gitschin, the great length of which was unforeseen, had to be met. Yet her entreaty was in vain. The imperial treasury did not pay the outstanding salary.

The sale of the books, chiefly the *Ephemerides*, brought some money into the house. Bartsch completed almost the entire printing of the *Somnium* and an improved logarithm table which were already in the press. He thought about moving with his family to Strasburg at Easter. But Bernegger suggested he should not. Malleolus, the old incumbent of the mathematics professorship which was promised him, had no intention of resigning and was very touchy on this point. In addition to everything there was the constantly growing public uncertainty and stress of war. After Gustavus Adolphus had captured Frankfort-on-the-Oder in April, 1631, Silesia lay open before him. He could be in Sagan in a few days. In the excitement many fled the city. Even Bartsch and his wife went on foot to Lauban. The widow could not join them because on the very day of the flight measles broke out among her children. The danger of war passed for the moment because the Swedish king turned westward. The agitation remained. In May, Magdeburg met its frightful fate. In June, Frau Kepler and her children moved to Bartsch in Lauban. She did not yet know whether she could remain here for the future. Sagan had been left forever. Besides it was not very long before the city was drawn into the military events. Saxon, imperial and Swedish troops alternately took possession of it in the years immediately following. Had Kepler remained alive, he too would soon have had to strike his tent in Sagan and proceed once more on his travels.

Had Kepler's heirs been able to cash in the obligations which they had in their hands, there would have been an end to all material need. However, the descendants, also, could not settle the difficulties which Kepler himself had not been able to surmount. On September 3, 1631, Bartsch and Frau Kepler set out to present a petition personally in Linz and Vienna, in order to be paid the sum due them. They had planned the trip for a long time, but because of the turmoil of war had not been able to carry it out. Their first goal was Prague where they achieved this—Wallenstein on September 21 gave his captain-general of Sagan the order to pay out the back salary owed in the amount of 250 gulden. This hitherto unknown document refutes the reproach often made of Wallenstein that he did not discharge his liabilities to Kepler's heirs. He did not remain in debt to them, but neither did he make them any present. From Prague the journey led to Regensburg, where they both visited the grave of the dear departed one and received the property he had left. Soon after his death an inventory had been made of this. Here they also met Kepler's son Ludwig. Meanwhile Ludwig had dedicated himself to the study of medicine at Basel. In his presence the legacy was divided.

In the beginning of 1632 Susanna Kepler went to Linz with Bartsch and Ludwig to demand payment of their due. Apparently they did not have much success in this. No one knows whether the obligations for 2,000 and 1,500 gulden were ever discharged. Ludwig undertook the trip to Vienna alone. Frau Kepler and Bartsch returned to Lauban. Ludwig remained in Vienna for a year but was not successful in his efforts to gain possession of the money that the imperial treasury owed the heirs. The result of his negotiations was solely a statement of the court bookkeeping about the size of the debt due Kepler. In capital and interest an indebtedness of 12,694 gulden was shown on April 27, 1633. This debt was never discharged. As late as 1717, William Hilbrand of Königsberg, the husband of a granddaughter of Ludwig Kepler, made a vain attempt to get paid. Hilbrand personally showed the promissory note in Vienna, but had to accept the decision that the commission for the liquidation of debts had decided no longer to recognize debts older than for 1680. As consolation, 75 gulden were tendered to the petitioner for travel expenses. Ludwig went from Vienna to Geneva as the traveling companion of August von Sintzendorff, the Austrian baron.

Kepler's widow earned a scanty livelihood in Lauban. Her lot became still more oppressive when, at the end of 1633, Bartsch was carried off by the plague. He never came into the promised professorship, and with his death all his plans for continuing the tasks, left behind by his father-in-law, also dissolved. In the autumn of 1634 Frau Kepler took up residence in Frankfort-on-the-Main, where her stepson Ludwig, who had returned from his journey, awaited her. She hoped he would be able to help her. But he himself required outside help and support. Both now awaited an improvement in their situation by the sale of the *Somnium*. Frau Kepler had brought along the incomplete copies of it. Ludwig had the still lacking title-signature printed; he dedicated the work to Landgrave Philip of Hesse. Whereas Ludwig soon thereafter departed again and went to Danzig as companion of an English legate, Frau Kepler first remained in Frankfort, leading a life of greatest poverty. Later, in the late autumn of 1635, she moved to Regensburg, which the previous year had been conclusively repopulated by the emperor's followers. Here among her husband's many acquaintances and friends she found more friendly and more familiar surroundings. Of her children, only the two girls were still with her; the two boys she had lost. Ten-year-old Hildebert had been buried on October 18, 1635, in Wertheim-on-the-Main, presumably an offer to the plague raging there. Fridmar, the younger son, likewise died in the neighborhood of Frankfort. Nothing further is known about his death. Frau Kepler was not destined to bear the burden of her wretched earthly existence much longer. At the beginning of September, 1638, at the age of forty-seven, she followed her husband in death, in the same city in which he had met his end. Dr. Marchtrenker, the Regensburg burgher, took charge of Cordula and Anna Maria, the deserted orphans. Until now it was thought the two girls had also died young. Nothing further is known about the younger, Anna Maria; in most recent times, however, information could be produced that Cordula married in Vienna and that her marriage was blessed with children. How far this hitherto unknown Kepler family continues is reserved for further research.[1]

Jakob and Susanna Bartsch had two children. A couple of years after her husband's death, Susanna entered into a second marriage with Martin Hiller in Lauban. There are still descendants of this daughter of Kepler. Ludwig, who did not resemble his father much but gladly lived off his fame, saw much of the world before settling down. From Danzig he went to Königsberg where, as early as 1636, he was received as a practicing physician. But he did not stay there long. He was in

[1] AU. NOTE. Thanks are due Miss Martha List for this information, furnished on the basis of studies of the archives. It has not hitherto been published.

Vienna the following year. Here there were long negotiations about the journal of Tychonic observations which Ludwig had taken over from his father's estate. They were not Kepler's property. The emperor now demanded that the valuable manuscripts be given up. Ludwig refused to accede to the demand as long as his claims on the imperial treasury were not fulfilled. Albert Curtius and Christoph Scheiner, two Jesuits who had long been wanting to get them for themselves, were mixed up in the agonizing attempt to tear the volumes away by craft. Whereas, in this none too pleasant affair, copies already produced in Brahe's lifetime reached Vienna, Ludwig could hold on to the original records. Later he sold them to the king of Denmark in Copenhagen where they are still to be found. In after days he went to Italy and in Padua obtained the degree of Doctor of Medicine. He did not disdain to implore Galileo's recommendation to the duke of Tuscany, in order to obtain money. After a rather short period of activity as physician in the Hungarian city Oldenburg, he finally settled permanently in Königsberg as practicing physician, city physicist, personal physician to the elector of Brandenburg and to the king of Sweden. Here he died in 1663. He was twice married and had numerous children. Of these one son and three daughters survived. It is not known whether his stock procreated in the male line but this appears questionable. Descendants of his daughter, Susanna Elisabeth, are still living today.

Something must still be said about Kepler's manuscript legacy. It consisted of thousands of sheets with sketches and fragments from all fields in which he had engaged—notices, numerous astronomical calculations and tables, preliminary studies for his works and, also, especially hundreds of letters which he had received or copies he had made of his own letters. Many of these letters are lengthy scholarly treatises, because in those days when there were not yet scholarly journals, knowledge and understanding were exchanged in this way. Ludwig Kepler took this entire valuable legacy with him to Königsberg. He repeatedly promised publication from this rich material, but never kept his word. Neither did he redeem his pledge to put the Tychonic observations into print and to write a biography of his father. For such works he lacked not only the time but also, from the outset, the necessary scientific hypothesis. After Ludwig's death that legacy came into the possession of Johannes Hevelius, the well-known Danzig astronomer who, after much trouble, bought it from the heirs for a considerable sum. Hevelius, also, never published them, as he had intended, but made the learned world, especially in England, aware of the great treasure which he possessed. As by a miracle the legacy

escaped complete destruction in 1679 when a great fire, set by an un-
faithful servant, enveloped Hevelius' dwelling and observatory. The
library and many other manuscripts were destroyed. The treasure
passed from Hevelius to his son-in-law, Ernst Lange. The change of
ownership did not disturb the peace which befell the manuscripts after
Hevelius' death. In the year 1707 Michael Gottlieb Hansch, who was
born in Danzig and domiciled in Leipzig, acquired the collection, with
the intention of finally making all the material here piled up accessible
to the learned world through publication. In his labors he enjoyed the
experienced counsel of Leibniz, who took an active interest in the
plans. The fruit of Hansch's labors, a large folio volume containing a
portion of the letters from the legacy, appeared in 1718, with the support of Emperor Charles VI. Hansch also had the entire collection
bound in vellum in twenty-two volumes. Various circumstances, such
as the absence of further support from the emperor and difficulties
inherent in the material, prevented Hansch from continuing his task.
In 1721 financial need even compelled him to pawn eighteen volumes
of the manuscripts for a sum of 828 gulden. The volumes of letters,
whose contents he had published, reached the court library in Vienna.
In spite of all exertions the impoverished scholar during his lifetime
never succeeded in redeeming the pawned property. The estate for
many years returned to oblivion.

About 1765 Christoph Gottlieb von Murr, a man very familiar with
manuscript literature and a great friend of learning in Nuremberg,
discovered the valuable treasure in a trunk at the home of the wife of
Trümmer, the mint councilor. The owner was prepared to hand it
over for 1,000 taler. In his *Encouragement of Germans to put Kepler's
writings into print* in the appendix of his work *Notes about Mr. Lessing's
Laocoön*[1] Murr pointed out to the scholarly world this opportunity of
saving the treasure from destruction. In numerous letters he personally
turned everywhere, to Göttingen to the well-known mathematician
and poet Abraham Gotthelf Kästner, to Heidelberg, to Vienna, to
Tübingen to the physicist Johannes Kies, to Zürich, to Prague, to
Berlin to Johann Bernoulli, the director of the observatory. But from
all sides he received refusals. The value of the manuscript collection
was appraised only according to its positive yield for the current state
of science. Scholars had no sense of its significance in the history of
science, nor of the obligation of reverence owed the discoverer of
the planet laws. Characteristic of this is the observation of the very

---

[1] ED. NOTE. *Anmerkungen über Herrn Lessings Laokoon, nebst einigen Nachrichten, die
deutsche Litteratur betreffend...*Erlangen: Walther, 1769, pp. 47–60, *"Ermunterung an die
Deutschen, Keplers Schriften zum Drucke zu befördern."*

renowned Johann Heinrich Lambert who remarked to Bernoulli: he did not believe that anyone would be found who would pay 1,000 gulden for the manuscripts; fifty years ago something could have been made out of them, when the Kepler tables had been in vogue; today the manuscripts could be desired only out of sheer curiosity.

Nevertheless, Murr did not cease his efforts. A last step was to lead to the goal. It was Leonard Euler, the great Basel mathematician, who solved the problem. At that time he was living in St. Petersburg. On his advice the Russian empress, Catherine II, a German princess of the Anhalt-Zerbst house, bought Kepler's estate in 1773. The purchase price was paid in jewels. Catherine presented the collection to the Russian Academy of Science which seventy years later turned it over to the library of the newly founded Pulkova observatory in St. Petersburg. There it is still preserved as a valuable treasure.[1]

A hundred years later the drama had a further sequel. In the middle of the nineteenth century Emma and Auguste Schnieber, two descendants of Kepler's daughter Susanna, lived in Lauban. They reverently guarded a few pieces which they had received from their grandfather, three little pictures painted on copper of the youthful Johannes Kepler, of his first wife, and of Jakob Bartsch, Susanna's prayer book with various handwritten insertions as well as some of Susanna's trinkets and useful objects. Since the aged owners had no legitimate heirs and were in financial straits, they thought of selling these momentos during their lifetime to someone who valued them. They turned to Johann Gottfried Galle, the well-known astronomer, at that time director of the observatory in Breslau. The best advice he could give was to sell these pieces where the manuscripts were. This was done. The pictures, the prayer book, the headband, the little thread basket, the brooch, and the little plate traveled to Russia in 1876 and since then, to the shame of the Germans who had let themselves be surpassed in reverence by the Russians, they rest in Leningrad in the place where high science is cultivated. Letting oneself be reminded of cherished people by things which they once owned, even if the things have little value, is indicative of a fine and deep requirement of the heart. Of course, it must be noted that the then director of the Pulkova observatory, Otto Struve, was a German.

Thus ended the fate of Johannes Kepler so far as concerns what he had left behind.

[1] These manuscripts have been transferred to Leningrad, USSR, Academy of Sciences, Archives. See *Trudy arkhivi Akademii nauk SSSR*, 1946, vol. II, no. 5, 287–312.

REVIEW AND EVALUATION

1. Kepler's physical constitution and his character

"O CURAS hominum, o quantum est in rebus inane." (O the cares of man, how much of everything is futile.) This verse by Persius, the ancient poet of satires, forms Kepler's motto. He used it repeatedly, supplemented by a line or so of his own, in his entries in the family album. We still have a great number of these. An insignificant verse. Yet in Kepler's mouth it could not be solely a resigned sigh about the inadequacy and wretchedness of earthly activity and striving, not only a moralizing complaint about the lowliness, shallowness, and pettiness, so frequently displayed in the endeavors of men as soon as one examines their busy zeal more closely, but rather, in those words there lay for Kepler a summons not to let himself be inundated by everyday cares, to set his goal higher, to direct his faculties and aspirations to lasting values. We have seen how he was often up to his neck in the misery of life, how he had to defend himself against the common needs of existence, how he always had to struggle against misunderstanding and hatred in order to hold his own. We have followed him on his life's path and learned how he was evicted, dislodged, pushed off, from Graz, from Prague, from Linz and how in Sagan, too, he would have had to retreat had not death relieved him of the burden of his travels. Whenever he believed he had found a home somewhere, circumstances, through no fault of his own, became distorted in such a way that he could no longer remain. The period in which he lived loaded him down with its restlessness, its brokenness, its torn state. But we have also seen how he was able to push through against all reverses which he encountered because he remained true to himself and, following that challenge, reached and won the high goals which his genius had set for him. He distinguished each one of the above-mentioned cities by the composition of one of his astronomical masterpieces, Graz by the *Mysterium Cosmographicum*, Prague by the *Astronomia Nova* and Linz by the *Harmonice Mundi*. Tirelessly, indefatigably, unbroken, he held out to the end in the knowledge of a special mission, to which he knew himself summoned.

The detailed description of Kepler's life and work given here shows the originality of his thought and the basic features of his character.

But it might still be wise to take from the mosaic of our portrait and assemble in this final chapter that which is essential, and to complete, in a few strokes, the picture of this rare and strange man. Likewise it seems indicated that, putting them together briefly, we evaluate his many-sided and significant achievements with which we became acquainted in connection with the course of his life, exhibit the picture of the world in which and out of which he lived and describe for ourselves the main substance of his thought and research.

"I cannot marvel sufficiently that such a great mass of solid learning, so many treasures of knowledge about the most profound secrets can be locked and concealed in one such small body." That was the impression made by Kepler, by personal acquaintance, on an ardent foreign admirer. Fortunately, intellectual gifts are not usually proportional to material dimensions. For all that, Kepler himself laughingly writes about his constitution to a stranger who had asked him for a horoscope: "If you are not more corpulent and of more powerful appearance than I, you will never become burgomaster." But even if we do not marvel that his great intellect lived in a body which Kepler, himself, once described as small, active, well proportioned, another time, to be sure, as gnarled (*nodosum*), we must, however, be surprised that he completed such a colossal life's work, although so frequently visited by illness. Kepler belongs to those gifted people whose intellect shines forth from a truly fragile body, and who overcome bodily weakness with phenomenal resiliency. That in him, an astronomer, precisely that organ which is especially important for such a person, namely the eye, was weakened, was a drawback which he often bemoaned; he was nearsighted and suffered from multiple vision in one eye. Attacks of fever of various kinds often tormented him. He suffered much from ailments of stomach and gall. Every error in diet led to an attack. Besides he often speaks of boils and violent eruptions in various parts of his body, especially on the shoulder. It is difficult for him to sit for a long time; he must keep moving back and forth. And, indeed, his habits are not always conducive to his health. He could keep no order. Then, he is water-shy and wants to have nothing to do with baths and washings. He says he finds great pleasure in gnawing bones, eating dry bread, tasting bitter and sharp things, and considers it a festive pastime to walk over rough paths, up heights, through a thicket. His spiritual talent corresponds to his body. His soul is fainthearted and hides in scholarly corners; it is suspicious, timid and gladly roams lingeringly through difficult and knotty subjects. He knows no means of seasoning life outside of the sciences; nor does he desire any, and he rejects those offered him. Another time he compares his nature with that of a little house dog. Like the latter, he drinks little, snaps at everything which comes within eye-range, but is satisfied with the simplest thing. He tries to be liked by his masters, is dependent on others in everything and renders service to them, is not up in arms if he should be reproved, and takes great pains to make things smoother. He rummages through everything, is always busy and copies everything he sees others do. He has no use for conversation and greets people who come to his house as a dog does. If anyone takes the least thing from him, he snarls and gets angry. Also he is snappish and always ready with scorching sarcasm. Consequently people kept out of his way, but his lords liked him.

It is not exactly a great soul that he paints here. Yet it must be borne in mind that he made these private notes during the early Graz period. He had held a mirror before him and underlined somewhat too strongly the image with which he was pleased, spinning out, in his naïve frankness, specific weaknesses and idiosyncrasies which doubtless clung to his nature. Insight into impediments under which one suffers is the first step toward overcoming them. Then, in contrast, the later Kepler appears much freer, larger, surer, superior. In particular the Prague period was very beneficial in his character development. At a young age it drew him out not only from the narrowness of the Graz situation, but also from the narrowness of his soul. Rich experiences in life had broadened his view, intercourse with the great world had freed and relaxed him, the realization of successful achievements had increased his self-assurance. When he says people got out of his way, he exaggerates for the Graz period and even in later years exactly the opposite occurred. The best sought his company. His frankness, his strength of character, his amiability, his faithfulness, his warmth of feeling, the purity of his way of thinking, no less than his wealth of knowledge, the ready wit of his judgment and his sociability in conversation attracted all who had taste and understanding for human goodness and greatness. Much evidence for this is contained in the letters which he preserved. And when he says he copied everything which he had seen others do, he refutes himself, for we observe how in life and research he went his own paths, paths in which the majority could not follow him, which he trod and had to tread alone, because his moral conscience and his scientific thought pointed them out to him.

Nevertheless, the key to some difficulties which Kepler encountered in his life is still found in his youthful self-characterization. He never completely got rid of a certain feeling of dependence, of being subordinate. He had to resist it constantly or give himself a jerk in order to free himself from it. We have seen how difficult it was for him not to be able to conform to his surroundings, how he suffered when those who stood above him and to whom he considered himself bound were not in accord with him. Here also belong his efforts to justify himself not only in religious but also in scientific matters against re-proofs of having a passion for innovation. He did not want to be an innovator, nor to separate himself from others, and found it painful to have to go his own way. In the dedication of the *Epitome* he says, "I like to be on the side of the majority. Consequently I take pains to instruct many in the details of a given matter and I find great pleasure when I can agree with the party of the majority."[1] He did not have the imperious nature, the defiance, the feeling of superiority with which a Tycho Brahe or a Galileo assumed himself, as a matter of course, to be right against everybody and in everything. He had to justify him-self before others and before himself and wrest an attitude for himself. If he once advised someone, seeking his help, "One must carry one's head high and discard the low opinion of oneself" (which statement he immediately remarked held only for the relationship to the outside but not for that to God), he only expressed what he himself wanted to believe. But what he had recognized as right, to that he held un-shakably and accepted lonesomeness, need, privation, accusations, distrust rather than surrender one tiny speck of his conviction. "I have not learned to play the hypocrite," he announced to both sides, the Catholic as well as the Protestant, to these who expected the im-possible of him. That feeling of dependency and need for companion-ship has its foundation in part in a certain innate delicacy of feeling, in the vivid consciousness of his duty to be thankful to those above and in the respectful love he had for the simple man. It can also be explained as the result of influences he had experienced in his upbringing in his parents' home and in the seminary. The traces left by the oppressive atmosphere in the former and the demand for strong subordination in the latter must have been so much the deeper the more responsive and delicate of feeling was his disposition and the more zealously his efforts had been directed always to do a good job. It must have been difficult for him, with his talents as opposed to these influences, to carry through on that path on which a higher hand was ready to lead him. He was, as regards his station, not born a lord. Between his genius and his humanness there remained a latent gap. Although the idea of harmony kept his thoughts busy, he was not harmonic, not adjusted in his

---

[1] ED. NOTE. "Enimverô mihi cum multis sentire volupe est, quoties non errat multi-tudo; eoque id operam do, vt quod in re inest, quamplurimis persuadeam, eaque ratione cum magnâ multitudine sentiens, jucunditate perfruar majore." *Johannes Kepler Gesam-melte Werke*, VII, 8.

nature. He was a restless soul, fluctuating repeatedly between exhilaration and depression.

Concomitant with this spiritual attitude was the fact that all pride and conceit remained foreign to our astronomer. Sentiment changed into action. Assuredly he was aware of the great significance of his achievements; he must have been aware of it in order to be able to advocate them emphatically. It was clear to him that he was "almost the only architect and renovator of astronomy after master Tycho." He knew, so he says, that in his decade-long occupation with the science of the heavens he had come further than would have been possible for anyone else. This understanding, however, did not make him conceited, nor did it prevent him from recognizing publicly and gladly the merits of others, wherever he met them. He himself best characterized his attitude with the words: "I have always observed the custom of praising what in my opinion others have done well, of condemning what they have done badly. Never do I scorn or conceal other people's knowledge when my own is lacking. Nor do I ever feel subordinate to others nor do I forget myself when, by my own strength, I have made something better or discovered something earlier." His own fame is not at stake. Accordingly, he always declines the praise expended on him. Thus he writes to an Italian admirer: "I beg and beseech you, if you truly and sincerely want to show me your favor and devotion, to refrain from such pompously chosen language and from so much praise, in order not to set a bad example among philosophers. You must say to yourself that, if I fall short of such high words of praise, I must feel very hurt, but if I deserve them in the slightest degree I still have to fear them as a danger to my piety and also to my sincerity in the exploration of the truth." He could not have expressed the purity of his way of thinking better and more beautifully. Likewise, he turned to an Italian friend, who was spreading his fame in that land, with urgent exhortation: "Think no higher of me nor inculcate others with a higher opinion of me, than I can justify by achievements." Since it was never his own honor but the recognition of the truth which was at stake, he never, as has been observed of some, including great representatives of science, held back with his ideas out of fear that others could snatch them and precede him with new discoveries. "An exceedingly powerful craving for astronomy drives me and I cannot restrain myself from sharing my thoughts with the masters of science, so that I immediately progress in our divine art by their hints." When he once heard from Italy that Galileo was passing off some of his ideas as his own, he answered that he in no way restrained Galileo from engaging in his matters for himself. That makes no difference for one who has set truth and the honor of God as the highest goal, not his own fame. "Let the Garamantes and Indians learn these and other divine secrets, let my enemies make them known, let also my name perish, if only the name of God, the Father of the spirits, is thereby elevated." The truth and only the truth is the light in which he wants to walk. In the introduction to one of his writings he appeals to God in a prayer to favor him by such a pure service: "I ask God to make my spirit strong so that I direct my glance at the pure truth, from whichever side it should be presented, and do not let myself be misled, as so often happens today, by the admiration or contempt of persons or sides."

Freedom from pride and conceit also made it possible for Kepler to find the correct relationship to those who could not follow the flight of his thoughts and on whom life had imposed simpler tasks. Accustomed since boyhood to simple relationships, even at the height of his scientific success he preserved understanding for the thoughts and worries of simple men and evaluated their actions and struggles without superciliousness. He liked to converse with ordinary people and, as is shown by some of his German writings, especially the calendars, knew how to strike the right note with them. In the introduction to his master work, the *Astronomia Nova*, he addresses himself to these with the lovely words: "To him who is too dull to understand astronomical science ... I give the advice that he should quit the school of the science of the heavens ... and dedicate himself to his affairs. He should stay away from our wanderings through the world, return home and there cultivate his little field. But he should raise his eyes, with which alone he sees, to the visible heaven and give himself up with a full heart entirely to the thanks and praise of God the Creator, convinced that he shows no less reverence to God than does the astronomer to whom God gave the talent of seeing more sharply with the mind's eye ..."[1] Association with simple people was so much easier for Kepler as he in all liberal-mindedness still shared many wonderful concepts which were in vogue with them. There are many examples of this. Once when information about a pair of joined twins was spread abroad, Kepler believed he could explain this misbirth by the

[1] ED. NOTE. "Qui vero hebetior est, quam ut Astronomicam scientiam capere possit, [vel infirmior, quam ut inoffensa pietate COPERNICO credit:] ei suadeo, ut missa Schola Astronomica, [damnatis etiam si placet Philosophorum quibuscunque placitis,] suas res agat, et ab hac peregrinatione mundana desistens, domum ad agellum suum excolendum se recipiat, oculisque, quibus solis videt, in hoc aspectabile coelum sublatis, toto pectore in gratiarum actionem et laudes Dei Conditoris effundatur: certus, se non minorem Deo cultum praestare, quam Astronomum; cui Deus hoc dedit, ut mentis oculo, perspicacius videat, [quaeque invenit, super iis Deum suum et ipse celebrare possit et velit.]" *Johannes Kepler Gesammelte Werke*, III, 33.

mother's "seeing wrong" during pregnancy, since she had often watched her husband, a cabinetmaker, skillfully joining two boards. The phenomenon even made him think more about a union of the separated creeds.

Out of all the evidence adduced, and which we must restrain ourselves from increasing, the foundation on which Kepler's moral attitude grew—namely religion—is already recognizable. God is truth, and service to truth proceeds from Him and leads to Him. God is the beginning and the end of scientific research and striving. Therein lies the keynote of Kepler's thought, the basic motive of his purpose, the life-giving soil of his feeling. His deep religiousness expresses itself not only in occasional bents and passions of a pious soul; it feeds not only on reminiscences from the time of his theological studies. It permeates his entire creativity and spreads out over all the works he left behind. It is this feeling for religion which above all lends them the special warmth which we experience with such pleasure when reading them. All of its own accord at every opportunity the name of God crosses his lips; to Him he turns now with a request, now with praise and thanks; before Him he examines his deeds and omissions, his thoughts and words, to discover whether they can pass the test and are directed toward the proper goal. When he leads us over the wide open fields of his thinking, it is as though over them there lay a refreshing dew in which the rays of the sun shine in various colors. As the bird is created to sing, so, according to his conviction, is man created for his pleasure both in contemplating the magnificence of nature and in inquiring into her secrets, not for the purpose of extracting practical uses but rather to arrive at a deeper knowledge of the Creator. The entire work of Creation which God has prepared for man is for Kepler a richly laid table. And just as nature sees to it, that the living creature never lacks food, so, he supposes, we can also say that the diversity in the phenomena of nature is so great only so that the human intellect never lacks fresh nourishment nor experiences weariness of mind in age nor comes to rest, but far rather, that in this world, a workshop for the exercise of man's mind always remains open. Thus he can say that in contemplating the universe he immediately takes hold of God. And when in the intimacy of his soul he adds to this knowledge the question whether indeed he could also find Him in himself, he answers, in a thousand places, that so it is. There can be no better proof of it than the instance in Graz when after his discovery of the *Mysterium Cosmographicum* he breaks into tears and, feeling unworthy of this sign of God's grace, remembers the words which Peter spoke to the Master: "Withdraw from me, for I am a sinful person." Where is there another example of a natural philosopher who would make such a speech? But he who believes that in this sacred emotion only youthful exuberance is expressed should read over again the incomparably lovely prayers with which, at the height of his maturity, Kepler concludes his *World Harmony*, addressing the Father of light in devotion and humility: "O Thou, who by the light of nature increases in us the desire for the light of Thy mercy in order to be led by this to Thy glory, to Thee I offer thanks, Creator, God, because Thou hast given me pleasure in what Thou hast created and I rejoice in Thy handiwork. See, I have now completed that work to which I was summoned. In doing so I have utilized all those powers of my mind which Thou hast loaned me. I have shown man the glory of Thy works, as much of their unending wealth as my feeble intellect was able to grasp. My mind has been ready to correct the path and be punctilious about true research. If I have let myself be led astray by the astounding beauty of Thy work and become audacious, or if I have found pleasure in my own fame among men because of the successful progress of my work, which is destined for Thy fame, forgive me in Thy kindness and mercy.—Let my soul praise the Lord, Thy creator, as long as I live. Out of Him and by Him and in Him is all. That which is grasped with soul as well as that which is known in the mind. That which is still completely unknown to us as well as that which we know and which amounts to only a small fraction of the other; for still more lies beyond."[1]

His piety, the need—in the lonesomeness of his soul, in his family, in the community—to give a sure theological foundation to his religious activity, was also that which gave rise to and extended Kepler's conflict with the creeds which fought each other. For nothing would be more wrong than to believe that his religiousness was supported by feeling only, and people who, in connection with him, speak of dogma-less Christianity completely misunderstand him and are badly advised. We have followed this conflict in all its phases and backgrounds in such detail that we do not need to go into it further here. Certainly, he has carried on his intensive theological studies for himself alone, in order to arrive at clarity in his own affairs. His place was not in the pulpit, as he had supposed in his youth. He had a different calling. "I wanted to become a theologian," he wrote to Maestlin in the second year of his stay in Graz; "for a long time I was restless. But now see how by my pains God is being celebrated in astronomy also." A few years later he designated himself as a priest of God in the book of nature. In the dedication of the *Epitome* he repeats this avowal; as the priest of God

[1] ED. NOTE. *Johannes Kepler Gesammelte Werke*, VI, 362–3.

in the book of nature he wanted to offer that work to man as a hymn to the Creator. By this he characterized the ethos of his scientific research most strikingly. He has confused service to the word with service to the work. The priestly way of thinking is the same. Science is not its own goal, nor should it be used for advantages in this world. Its goal is to lead man to God. In this interpretation the scholar gives a consecration to his activity; in this consecration lie the nobility and the greatness of his calling. That was the conviction under which Kepler completed the prodigious work with which he was confronted; that was the view he held when he announced what he had to offer to others in understanding and discovery out of the rich treasure of his knowledge. He was a good servant in the work of creation, a faithful priest, for whom nothing was too much, who himself was filled with enthusiasm for his mission and who, by his announcement, was able to carry others away, and to awake in them love for that which was close to his heart. Astronomy, his chief sphere, is for him the delight of the human race. Heavenly speculations, he is convinced, quench the thirst of minds and impress on custom a certain similarity to the divine works. Secretly they bend the wills of mankind, tame his disorderly cupidity so that, because he is accustomed to the lovely order in geometrical and astronomical things, thereafter he also "gains a love for justice, moderation, decency and graciousness." With ever new words and monitions he therefore seeks to stimulate people to contemplate the loveliness of the work of creation. He wants to draw them away from the "barbaric neighing" which fills the land in the turmoil of war, and bring the blessings of peace closer to them. It is like a vision springing out of the enthusiasm of his overflowing heart when he writes in a letter: "If the intellect has agreed to contemplate what God had made, it also agrees to do what God has bid. Should this be attained by all, then there would be nothing further to desire for the human race than that all people in the whole globe should live together in one city and, already in this world far from every strife, have pleasure in one another, as we hope of the future [world]."

2. Kepler's view of the world and his doctrine of knowledge

Now what, more exactly, was it which invested Kepler with such great rapture and let him hope for such a great moral effect among people? It was, as we have seen throughout this biography, the passionate experience of order which he encountered in the universe, an experience which even in his youth had seized him most vehemently and which influenced him throughout his whole life so that, as his son-in-law Bartsch said after his death, his devotion to the contemplation of heavenly things was almost like a miracle. *Forma mundi*, the shape of the world, formed the great theme of his life's work. In it the idea, form, does not have the pale meaning of today's usage. It concerns the principles of order and configuration, that which makes the chaotic material into a cosmos, and also the epitome of the idea of the lovely, made real in the world. Copernicus, in the dedication of his work to Pope Paul III, had already advanced, as that which especially strengthened him in his conviction of the truth of his world picture and consequently as the goal of his investigation, the "form of the world and the symmetry of its parts" resulting from his concept. And Rheticus, who by talking with him had best come to know the thinking of the Frauenburg Magister, had uttered the thought that God had so arranged the world that a heavenly harmony in which each planet would have a particular place would be perfected by the six movable spheres. This thought had ignited in Kepler's head a half century after it had been uttered. It fashioned his life's program. What Copernicus had expressed only in one single sentence, Kepler wanted to develop unto the last conclusion and exhibit—the *forma mundi* as harmony. His enthusiasm had blazed up at this idea. It was to form the subject of his announcement to the world.

What supported and gave wings to Kepler in the execution of his program was a refreshing optimism about cognition. "Man, stretch thy reason hither, so that thou mayest comprehend these things"; that was the call which rang out for him from the material world. While he accepted this call with open ear, with the complete and unreserved readiness of a young mind, he believed in the reality of the things outside us and in the possibility of being able to comprehend them in their essence, order and meaning. What the eye brought him was that which he saw, and was in reality as he saw it; and the mind repeated the thoughts which God had materialized in His Creation. He did not start with doubt, as another[1] soon did, but with an unquestioned faith in *ratio*. He did not limit himself to the framework of immanent thought, but became intoxicated with the contemplation of a transcendental truth. He had not yet fallen into the abyss of relativity, but was deeply convinced that there is an absolute truth. Admittedly our mind never can completely grasp this truth, but it is the noble task of scientific and philosophical research to draw nearer to it. Thus to our searcher after truth that call signifies a moral obligation imposed on

[1] ED. NOTE. Doubtless, reference here is to Descartes (1596–1650).

mankind in the observation of nature giving research its direction and goal.

In his youth Kepler had expressed the basic thoughts of his outlook on the world in two memorable sentences: "Mundus est imago Dei corporea" and "Animus est imago Dei incorporea." "The world is the corporeal image of God" and "The soul is the incorporeal image of God." God, World, Human—prototype, copy, likeness: the circle of his thoughts is shut in this trinity. These ideas hold together and fasten tightly everything which presents itself to him, when he looks inward and reflects about his own soul or glances outward and inspects the world of phenomena or when, in veneration, he speculates on the original cause of all existence, out of which everything has come. Prototype, copy, image: these ideas give to his picture of the world that completeness and inspiration in which he found his happiness, his satisfaction, his peace and his delight.

"The world is the corporeal image of God." How can the corporeal be a copy of the absolute soul? According to Kepler this contradiction is reconciled by the idea of quantity which has its origin in the divine being. Quantities can be compared with each other; they form relationships. Now God, by certain selection in creating the world has, so to speak, taken such relationships out of Himself; He has made order out of chaos, given form to matter, in accordance with the word of the Bible that everything is regulated by number, size and weight. So to Kepler's contemplating eye the cosmos seems constructed like an ancient temple, a pyramid, a gothic cathedral, in the building of which the architect measured off the size according to aesthetic norms. And just as God himself proceeded in this work of creation, so he endowed everything to which he gave life with a power of creation which moves and works by the same laws. "As God the Creator played / so He also taught nature, as His image, to play / the very game / which He had played before her."[1]

Our mathematical mystic goes into still further detail in his view of the world as reflecting the divine image. The sphere is the most excellent of all geometrical images, the bearers of quantities. Therefore the sphere had to provide the prototype of the universe. The spherical shape of the real world was constructed for Kepler just as, indeed, our sensual-intellectual nature is so created in order that, in contemplating the whole, the rays of sight reach out from the eye equally in all directions into space. Kepler has the sphere formed so that an equal

---

[1] ED. NOTE. This is from section 126 of *Tertius Interveniens* and can be found in *Johannes Kepler Gesammelte Werke*, IV, 246, and is translated into English, p. 172, in *The Influence of Archetypal Ideas on the Scientific Theories of Kepler*, cited above.

flowing out (*effluxus*) takes place in all directions from one point. Therefore, aroused by similar speculations in the works of Nicholas of Cusa and others, he sees in it a symbolic copy of the Holy Trinity, a thought which he expresses in various ways in many places in his works and which even furnishes him with an argument for the correctness of the Copernican doctrine. The center point denotes God the Father, the surface God the Son, the space between, the Holy Ghost. From the center point of the sphere, as the origin, the surface emerges by means of radiation, whereby the surrounding equal intervening space is produced of its own accord. "The little dot in the middle gives birth to and forms the circumference, as soon as the point moves around in circles." The mere point would be invisible without spherical expansion; it can only make itself evident in the shape of the outflowing surface of the sphere. All three—centerpoint, surface, intervening space—stand in most intimate relationship, in loveliest accord, in the best proportioned ratio to each other. Together they form a unity so that not once can one of them be thought of as missing without the whole being annihilated. So the secret, unfathomable existence of the divine Trinity mirrors itself for Kepler in the visible world. The world is a sphere, the sphere a picture of the Holy Trinity; consequently the world is the corporeal image of God. This is not the ultimate thought on whose account he embraced the universe with such great fervor. For he loves the world of phenomena, because it is God's and wears God's features. God, playing, created in the sphere a picture of His Trinity, so worthy of adoration (*lusit imaginem*). And when Kepler sees children amusing themselves with soap bubbles he thinks of them as playing the role of the Creator because they blow up the drops into a sphere. In this he is only disturbed by the fact that a droplet continues to hang beneath the soap bubble instead of marking the center.

"The soul is the incorporeal image of God." This thought, confirmed by the Bible, which says that God created man in His own image, had been developed by Christian philosophers and theologians in profound speculations. Kepler took it up and embraced it with the complete accord of heart and soul. He expressed it in numerous places in his works in ever new words and established it as the foundation stone of his doctrine of cognition. Yet in doing so he follows different paths from those of the theologians, such as Augustine, who in contemplating his own soul sought therein a mirror image of the triune God and found the similarity to God based on the trinity, memory, will and cognition. Kepler does not speak of the trinitarian nature of God, as he had done with the picture of the universe. His speculation circles about the ideas of quantity and harmony. Just as God had fashioned these ideas out of himself at the creation of the world, so he also communicated them to man as his image when he breathed life into him and caused his countenance to shine down upon him. It is the reflection of divinity which becomes known to us in the quantities, their relationships and connections, in the structure of geometric figures, in the laws to which these are subject. "God wanted to have us recognize these laws when He created us in His image, so that we should share in His own thoughts. For what remains in the minds of humans other than numbers and sizes? These alone do we grasp in the proper manner and, what is more, if piety permits one to say so, in doing so our knowledge is of the same kind as the divine, as far as we, at least in this mortal life, are able to comprehend something about these." These thoughts are completely expressed in the sentence: "Geometry is unique and eternal, a reflection from the mind of God. That mankind shares in it is because man is an image of God."

In the variety of quantities Kepler beholds "a wonderful and positively divine state." They express the divine and the human symbolically in the same manner. It is they which establish the structure of order in the visible world. It is the deepest desire of the human mind to comprehend this order, to agree with it, to become assimilated with it. "Even if an order should once be produced by chance, still the minds fly together there; therein lies their pleasure, their life." If an order appears anywhere, there always arises in the mind a great trust, a strong confidence. He even seeks the reason for this "in the deepest original cause of geometry." The part which he assigns to this science at the same time illustrates his interpretation of the existence of mathematical things. These have their foundation in the divine being himself, and man meets them in his mind by virtue of his being in the image of God. What from outside meets him by means of the senses merely arouses him to become clearly cognizant of that which is already contained in them and belongs to his nature. That there are only five regular solids, that the diagonal of a square is in irrational proportion to the side, that the seven-sided figure cannot be constructed with ruler and compass—these statements Kepler, anointed with Pythagorean oil, considers metaphysically given. Therefore he also rejects Aristotle most positively when the latter compares the soul with a *tabula rasa*, a blank slate, which is first written upon by sense experiences. He follows Plato's doctrine of ideas and the views of Proclus, the Neo-Platonist, whom he highly esteems. "Geometry, being part of the divine mind from time immemorial, from before the origin of things, being God Himself, has supplied God with the models for the creation of the world and has been transferred to man together with the image of God. Geometry was not received inside through the eyes."[1] This conception also makes the close connection which Kepler establishes between mathematics and philosophy comprehensible. He is convinced "that the whole of philosophy arose out of mathematical things, exists therein mixed among them and so closely related that whoever proceeds without it in studies merely beats the air and fights with a shadow: nor may he in eternity be called a philosopher with honor."

So, for Kepler, understanding nature signifies nothing else than to think in accordance with the thoughts of God, who always pursues geometry. The fact that the world and man's mind are images of God in their manner, makes knowledge possible, a knowledge which not only is certain but also carries sense and value in itself.

The path he followed in his research as well as the ethos of his work is expressed by Kepler in the loveliest manner in a place in the *Epitome*, where he suddenly interrupts his scientific research and writes: "With a pure mind I pray that we may be able to speak about the secrets of His plans according to the gracious will of the omniscient Creator, with the consent and according to the bidding of His intellect. I consider it a right, yes a duty, to search in cautious manner for the numbers, sizes and weights, the norms for everything He has created. For He Himself has let man take part in the knowledge of these things and thus not in a small measure has set up His image in man. Since He recognized as very good this image which He made, He will so much more readily recognize our efforts with the light of this image also to push into the light of knowledge the utilization of the numbers, weights and sizes which He marked out at creation. For these secrets are not of the kind whose research should be forbidden; rather they are set before our eyes like a mirror so that by examining them we observe to some extent the goodness and wisdom of the Creator."

That Kepler here accounts to himself on the question of the admissibility of his questions of nature and thereby announces, as he also does elsewhere in various places, that he has questions which have not been permitted, is a fine thought. He knows that there is a limit between that which is accessible to our knowledge and that which is impenetrable, and that man must be prepared to accept this limit reverently. He is completely free of the sovereignty which seeks the measure of all things in mankind, and of the promethean stubbornness which, in reliance on its own knowledge and ability, challenges the Divinity.

[1] Ed. Note. *Harmonice Mundi*, Book IV, chapter 1. See above, p. 271.

First of all we broke down the mathematical-metaphysical side of Kepler's world of thought. That was necessary because, in the conventional expositions of his lifework, that part consistently is treated too briefly. In our expositions of his activity, however, we met in the discussion of his works an entirely different Kepler, the exact astronomer, the dispassionate mathematician, the tireless computor, the realistic physicist, the rigorous logician, the clear methodician, the experienced empiricist. We have come to know him as the discoverer of the planet laws and in doing so we followed him on the very tortuous path which he broke for himself through the denseness of his numbers until he reached the magnificent summit, rich in prospects. We recall his tireless effort in this, the mastery with which he handled the task, the ingenious strategy with which he laid his plans, the tactical skill with which he inserted the observations, and the unerring instinct for facts which excluded their ever being shrouded. And what incredible application, solicitude and patience did he spend in working out the tables until they reached the final point which he had set as his goal! We have evaluated his inestimable service to optics, which he could only attain by a strong scholarly (in the modern sense) combination of theory and practice. It was he who first successfully introduced physical contemplation of the heavenly motions into astronomy and thereby laid the foundation for celestial mechanics. He set entirely new tasks for mathematics and by his accomplishments in a significant manner paved the way for the infinitesimal calculus. Besides, in his chronological researches, by the critical examination and elaboration of extensive basic material he also proved himself a first rate philologist and historian.

The gap between his metaphysical speculations and his exact research shows the striking polarity in Kepler's intellectual giftedness. His innermost nature drives him to an aesthetic-artistic consideration of the universe, whose geometrical structure he wants to ascertain. Everywhere he looks for symmetry, for analogies, for a well-proportioned equalizing of the parts in accordance with a static order, in the contemplation of which he goes into the most extreme rapture. And yet it was he who established the dynamic explanation of the heavenly motions, who tracked the tensions between the sun and the planets and could not do enough in investigating the processes of motion. "The bodies would not be beautiful if they did not move," we hear him say. Motivated by metaphysical expectations he tried to interpret the meaning of the phenomena in all aspects, large and small. At every number which he established, at every connection which he ferreted out, the question always immediately flew in his head: why is that so? And still is there any scholar who would have followed up the given facts more diligently and more open-mindedly? However hard he was driven to encompass the whole, just so true and patient was he in inquiring into the parts. We see him now floating in the highest heights into which his faculty for enthusiasm carries him, now standing with both feet on the ground, indefatigably turning clod after clod. In contrast to his ever wakeful fantasy, which continually brought new ideas to him, was his strong ability to concentrate, which the solution of his difficult problems demanded. His exuberance hindered him no more than his rare nimbleness of mind from following a logical line of thought. The more one becomes absorbed in his main works, so much higher mounts one's admiration for the powerful logic disclosed in the planning and detailed carrying out. Yes, in all the bubbling delight in speculation, which inspired him, the mathematical-logical stamp of his thinking stood out so much the more conspicuously. He is equally familiar with deductive and inductive trains of thought. Even though he felt it his duty to explain the world by *a priori* principles, he was, nevertheless, the first who, making use of the inductive method, today self-evident to everyone, directed questions to nature and immediately applied this method with remarkable skill in the discovery of his two first planet laws.

Even if Kepler with his aesthetic world view was imprisoned by the Renaissance, still on the other hand his unusual genius opened up entirely new paths for scientific research. On the one hand he supplied his astounding intellectual capacity with special preference for morphological and teleological conceptions and deductions, but on the other hand he cleared the path for the causal explanation of nature. Alongside of the prototype, he helped the cause to obtain its right. He was suspended between an animistic and a mechanistic view of nature. The enthusiastic thinker, which he was, so easily talked and tunneled himself into an idea as to make his way to its root and to track down its last consequence. He penetrated deeply into the province of the psychic. He felt or divined psychic powers back of all visible and tangible things. For him the world is not an aggregate of dead bodies. Everywhere he finds life as an expression of a psychic principle; everywhere he suspects psychic influences. We know his view about the earth soul and its geometric instinct. In eloquent words he repeatedly expresses his belief in spontaneous generation; he lets the little souls of the plants be ignited at the earth soul. His astrology is supported entirely on conceptions from the domain of the psychic. And yet it was he who founded the mechanistic explanation of the heavenly motions. He disposed of the medieval concept which, referring to

Aristotle, had the rotation of the planet spheres taken care of by spirit beings or angels and showed that in this there is in operation only a force similar to that which draws the stone down from above to the earth. It is very fascinating to follow in him the transition from the animistic to the mechanistic explanation of the models of the motions. How can a planet find its path around the sun when this is in no way indicated by marks? How far is it possible to explain the individual phenomena in the planetary motions by a material compulsion (*necessitas materiae*)? On the other hand, to what extent is the acceptance of a psychic principle necessary for this? Those are questions which he investigates and decides with the greatest care. A later era has raised the completely mechanistic explanation of the models of nature to a principle and, with a remarkable shyness of everything which is called soul, required, in the name of science, the weeding out of every psychic power. Certainly, such a point of view would have been an abomination for Kepler. He would never have been able to conceive the universe, the animal, the human as machine. With all the dualism of his being, his view was more open, freer, deeper.

3. Kepler's cosmography

Now, what is the appearance of the astronomical world picture which Kepler, following the various tendencies of his thinking, developed on the foundation of the Copernican conception? Indeed, at the beginning of Kepler's research stood Copernicus; he was an inspirer. Kepler made serving him his life's task. "I deem it my duty and task," so he explains at an early age, "to advocate outwardly also with all the powers of my intellect the Copernican theory, which I in my innermost have recognized as true and whose loveliness fills me with unbelievable rapture when I contemplate it."

When speaking of Copernicus' world picture it is necessary to guard against introducing conceptions and knowledge which only arose in the course of the development for which he laid the foundation. It is customary to think of the countless number of stars swimming in an infinite space.[1] Our sun moving in this swarm is a fixed star like others and our earth a diminutive companion of it so that, in the immense party of dancers of milliards of fire balls which unite into systems and systems of systems, it appears as a completely insignificant member of the universe. Such concepts were wholly foreign to the Frauenburg

[1] ED. NOTE. There are many scientists in the twentieth century who question this view.

astronomer. He put the sun at rest in the absolute center of the world. If there is a center, there can be no infinite space. So he assumed the fixed stars as lights on a very large sphere. He had the earth with the five other planets circle the sun. Since this motion showed no displacement of the fixed stars, he had to assume that the diameter of the sphere of the fixed stars is so great that in comparison with it the diameter of the earth's orbit is of no account. He does not express himself about how the orbits of the planets will be formed. There is basis for the assumption that for each of them he assumed a solid sphere to which it is fastened. We know from our previous arguments that in his theoretical presentation of the planet orbits he saw his triumph in reducing them to superimposed similar circular motions. Kepler did not alter much in the over-all picture of this world concept. For him, too, the sun stands in the absolute center of the world. To be sure, he lets it rotate on an axis, but to assume a motion in space lies completely outside his range of conception. Correspondingly he expressly and decisively rejects the assumption of a real infinite space. Assuredly, he already knew that meanwhile, in bold speculation, Giordano Bruno had advanced the theory that the fixed stars are nothing else but suns, like ours, and distributed in infinite numbers in infinite space. Yet Kepler argued against these revolutionary theories with sharp words. He shuddered, so he says, at these concepts. He reproaches the Italian philosopher for misusing Copernicus' view and consequently all astronomy. Indeed, the thesis of Giordano Bruno, who understood nothing about astronomy and did not want to understand anything, did not grow in the soil of astronomical research but originated from theological speculations and a pantheistic interpretation of nature. The idea of an infinite space is not rooted in experience but in metaphysics. Indeed, several decades ago science again gave it up.

With the Greeks Kepler seeks the perfect in the moderate regulated finite. For him as for Copernicus the world is a real sphere. He assumes that the fixed stars, about the nature of which he does not express himself further, are distributed in a shell-like finite space, spherical inside and out. As a whole they form, according to his conception, a kind of wall or a vault and create the space in whose center stands the immovable sun which lights the whole space, the sun which is the heart of the world, the source of light. Thus, for him, the whole world is comparable to a great lantern or a concave mirror. On the inside of this ball lies the system of the six wandering stars. In composition and motion, this is exceedingly artistic. These planets or wandering stars— Mercury, Venus, Earth, Mars, Jupiter and Saturn—form the brilliant princely household of the queen, the sun. The measurements for the distances are furnished by the five regular solids which, after the sphere, are the most perfect and loveliest geometrical creations. These measurements are not entirely exact but in close approximation. Between these static sizes and the sizes of the motions there is a remarkable relationship governed by law: the squares of the periods of revolution are to each other as the cubes of the mean distances from the sun. And in relation to these distances, how great is the radius of the sphere of fixed stars? Since Kepler cannot ascertain it empirically, an analogy must help him. He fancies that his analogy fits extremely well if he assumes that the radius of the sphere of the fixed stars is to the distance from the sun of the outermost planet, Saturn, as this distance is to the radius of the sun's sphere. And how do the motions of the wandering stars take place? In the sun there is a moving force, which emits its rays as does light. The sun is the source of the motion. Since it rotates, it pulls the planets around with these rays of force. Since the effect of this force is so much the greater, the closer the planet is to the sun, the motion is rapid at perihelion, slow at aphelion. The radius vector describes equal areas in equal times. The circular form of the orbits is given up. The old axiom of similar shaped circular motion is annulled. By the operation of the solar power on the planet bodies, which are thought of as polarized, orbits arise. They have the form of ellipses with the sun in one focus. But the eccentricities of these ellipses are no more arbitrary and without rule than any other measures. No, in this fine construction the highly artistic formative hand of the Creator is shown in a very special way. Since the eccentricities determine the rates of the planets at aphelion and perihelion, they have been so measured by the Creator that between them appear the harmonic proportions which are to be presented by geometry and which are the foundation of music. So a divine sound fills the whole world. To be sure, sensual hearing is unable to perceive the wonderful harmony. But the spiritual ear perceives it, just as it is also the spiritual eye with which we see the loveliness of the sizes.

And the earth? Was it humiliated by being pushed out of the center of the world? By no means. "Our little hut" has retained a favored position between the planets. Two are inside, three outside its orbit. By its motion around the sun its inhabitants will be enabled to ascertain the size of the world. The unchanging inclination of the earth's axis takes care of the change of seasons and brings about an equitable distribution of the sunshine on the inhabitants of the various zones. By the earth's position midway between the other planets the spectacle, which her five sisters with their dance produce on the world stage for her, acquires a lovely variety. On this trip around the stationary sun, man can observe with understanding the wonder of the world in its diversity of phenomena. For everything is there because of man. That was the spectacle which stood before Kepler's eyes in his astronomical researches. His earnest endeavor was to paint this picture in all its details. The contemplation of this picture lifted him above all earthly misery and provided him peace, solace and happiness. Here there were no storms, no persecutions, no disputes, no wars. Here he found his never barred refuge, his home. He rejoiced because he knew himself to be inside of the sphere. For it is the image of God, of the Father, of the Son and of the Holy Ghost. And everything which happens in this sphere bears witness to the wisdom and kindness of the Creator. That gave the restless pilgrim on earth the feeling of blissful security. With curiosity he looked on the trip around the sun. Enraptured, he listened to the harmonies which resounded in his direction. This judgment flowed over him: everything is good which has been created here.

That was and is Johannes Kepler. For he continues to live among us through the work he created and the example he set us. As long as men reach for the stars in yearning and desire for knowledge, as long as they preserve the respect for spiritual and material greatness and their strength to draw themselves up before examples remains, his name will not perish. Besides, in the three hundred years since his death many have shown the highest admiration for him and extolled him in words full of appreciation and veneration. The loudest voices were raised abroad. Immediately after Kepler's death Horrox, the young English astronomer, enthusiastically called upon the poets to sing his praises and on the philosophers to make him known; for, so he says, "he who has Kepler, has everything." Bailly, the French astronomer and historian of astronomy in the second half of the eighteenth century, enrolls him among the greatest men who have lived on earth. Leibniz bows in appreciation before the greatness of the "incomparable man." Goethe, Hölderlin, Mörike considered themselves fortunate and exalted in their contact with the great genius. Since Kepler, science has built further on the foundation which he laid and has erected an amazing structure. It was Leibniz who used the same words about him which Kepler once had uttered about Copernicus: he did not know how rich he was. Certainly, with the progress of knowledge, some of Kepler's views have proved erroneous. His view of nature was conditioned by the time. But the ethos by which he performed his work has also changed and this is a pity. This change went so far that in a time which looked upon the task of science purely as the collection of facts all understanding for his aesthetic-metaphysical view of nature was lost and it was branded as "chimeric speculation." Laplace who pronounced this judgment even viewed it as "distressing for the human spirit" to have to see how ecstatically Kepler pursued the idea of his world harmony. The number of those who accepted the opinion of the great French astronomer is not small. Novalis, a great contemporary of that Frenchman, gave the right answer to all of them when he said: "To you I return, noble Kepler. Your noble mind created a spiritual moral universe. Instead of that, in our times it is considered wisdom to kill everything, to lower the high instead of raising the low and even to bend the mind of man under mechanistic laws." Novalis, too, has many followers. I hope the reader of this book will be among them.

THE PORTRAITS

T HERE are four extant pictures of Kepler made during his lifetime. They are as follows:

1. A picture made during his youth. It is the picture mentioned on p. 367 and now preserved at the Pulkova observatory. The original is a little oval medallion (7 × 5 cm.) painted in oil on copper. It dates from the Graz period, probably soon after his marriage to Barbara Müller, of whom a similar medallion picture is preserved at Pulkova.

2. An oil painting from the Linz period. In the autumn of 1620, Kepler sent this picture to his friend Bernegger in Strasburg, by his assistant Gringalletus. In 1627, as the inscription on the picture states, Bernegger gave it to the library in Strasburg. When Kepler heard this he asked his friend to take it away from this public place, especially because it was a poor likeness. Today the picture is in the Thomas seminary in Strasburg. The portrait shows Kepler at his height. Even if the unknown artist was certainly no master in his profession, still his picture is the only one that shows us Kepler's appearance at maturity.

3. A copperplate by Jacob von Heyden. In 1620 Bernegger had this picture made from the above-mentioned oil painting. Its general impression departs considerably from its model and consequently it found little approbation among the friends who knew Kepler well.

4. A small picture of Kepler in full length. It is from the frontispiece of the *Rudolphine Tables* which is sufficiently described on pp. 327-8. Since these tables appeared in 1627, we see here how Kepler looked in the last years of his life. It may be assumed, although there is no-where any mention of this, that he approved of this representation. In expression and composition the picture makes a pleasant impression. The words "Commentaria Martis" on the shield, stand for the *Astronomia Nova*. It is to be remarked that the *Harmonice Mundi* is not mentioned.

In addition to these pictures there is also a picture at the observatory of the Kremsmünster Benedictine seminary supposedly representing

Kepler which is frequently reproduced. But since there is good reason to doubt its genuineness it is not included here. A picture belonging to the Historical Society of Upper Palatinate and Regensburg which according to an inscription supposedly portrays Kepler and which served as model for the bust erected in Walhalla represents, as was later proved, Duke Ludwig X of Bavaria-Landshut. All the other pictures which are shown and which are mostly poor are based on the above representations 1 and 2 or arose from the free fantasy of the artist.

Two public monuments should be mentioned. The first was erected to our astronomer in 1808 in Regensburg on a square in the neighborhood of the cemetery where Kepler was buried. It is a circular temple with eight columns, open all around, in which the marble bust of Kepler rises on a pedestal. The energetic support of the prince-bishop, Karl Theodor von Dalberg, made the execution of the lovely plan possible. The design of the monument originated with the princely architect Emanuel d'Herigoyen. The bust was made by the sculptor F. W. E. Döll in Gotha after a model which was in Gotha but is no longer known today. The pedestal is adorned by a relief with an allegorical representation by the hand of J. H. von Dannecker.

In 1870 Kepler's birthplace, Weil der Stadt, erected a monument to her great son in the town market place. Cast in bronze, Kepler sits on a sandstone pedestal richly ornamented with figural decor. The stately work is by A. von Kreling, the Nuremberg artist.

In 1940, thanks to the generous assistance of Kommerzienrat Dr. Paul Reusch, it was possible to convert Kepler's birthplace and open it as a museum. An expressive bronze bust of Kepler, the work of the sculptor G. A. Bredow of Stuttgart, adorns the main room.

BIBLIOGRAPHICAL
REFERENCES

T HE list that follows replaces the one in the 1959 English edition, and reflects the standard
works listed there as well as more recent scholarship.

Bibliographia Kepleriana, edited by Max Caspar and revised by Martha List (Munich,
1968), systematically describes all of Kepler's works and gives information about the location
of extant copies in numerous German libraries. This bibliography also lists later editions and
posthumously printed works. The title page of each of the early editions is reproduced in
facsimile. In an appendix the most important books and articles about Kepler are arranged
chronologically. Martha List has given an extension of this latter bibliography to the year
1974 in *Vistas in Astronomy*, xviii (1975), 959-1003 and to 1978 in *Vistas in Astronomy*, xxii
(1978), 1-18. For another listing of Kepler's works, see Gerhard Dünnhaupt, *Personalbib-
liographien zu den Drucken des Barock*, iv (second edition, Stuttgart, 1991), pp. 2269-2308.

Joannis Kepleri astronomi opera omnia, edited in eight volumes by Christian Frisch
(Frankfurt and Erlangen, 1858-71), includes all the major printed works and also extensive
excerpts from Kepler's correspondence, copious editorial notes, a 361-page Latin *vita*, and
an as yet unsurpassed index. It remains a key source for the documents of the witchcraft trial
of Kepler's mother, and, as the bibliographical citations in the next section of this volume
show, of many biographical details of Kepler's life.

Johannes Kepler Gesammelte Werke (Munich, 1937-) is the monumental twentieth-
century edition of Kepler's works and correspondence. It was planned by Walther von Dyck
and Max Caspar and carried out under the auspices of the Deutsche Forschungsgemeinschaft
and the Bavarian Academy of Sciences. Von Dyck died in 1934 before any volumes were
published; Caspar, Franz Hammer, and Volker Bialas have served as successive editors. An
abridged table of contents, with short titles and dates of original publication, is given here
for reference:

I   *Mysterium cosmographicum* (1596), *De stella nova* (1606);

II   *Astronomiae pars optica* (1604);

III *Astronomia nova* (1609);

IV   *Kleinere Schriften: Phaenomenon singulare* (1609), *Tertius interveniens* (1610), *Strena
     seu De nive sexangula* (1611), *Dissertatio cum Nuncio sidereo* (1610), *Dioptrice* (1611);

V   *Chronologische Schriften: De vero anno* (1614), *Bericht vom Geburtsjahr Christi* (1613),
     Eclogae chronicae (1615), *Kanones pueriles* (1620);

VI   *Harmonice mundi* (1619), *Apologia pro opere Harmonices mundi* (1622);

VII *Epitome astronomiae Copernicanae* (1618-21);

VIII *Mysterium cosmographicum* (second edition, 1621), *De cometis libelli tres* (1619),
     Hyperaspistes (1625);

IX   *Mathematische Schriften: Nova stereometria doliorum* (1615), *Messekunst Archimedis*
     (1616), *Chilias logarithmorum* (1624), *Supplementum Chiliadis* (1625);

Bibliographical References

X   *Tabulae Rudolphinae* (1627);

XI,1   *Ephemerides* (1617-19, 1630);

XI,2   *Calendaria et Prognostica, Astronomia minora, Somnium* (1630);

XII   *Theologica* (1617-23), witchcraft trial (1621), Tacitus translation (1625), poems;

XIII-XVIII   *Briefe* (correspondence to and from Kepler and a few third-party letters, in chronological order);

XIX   miscellaneous biographical documents;

XX,1   manuscripts including *Apologia Tychonis contra Nicolaum Ursum, De motu terrae, Hipparchus, Lunaria, Consideratio observationum Regiomontani et Waltheri*;

XX,2   *Manuscripta astronomica (II)*, including *Commentaria in theoriam Martis*;

XXI   *Manuscripta varia;*

XXII   *Register* (index).

A series of important Keplerian documents not previously available, as well as Keplerian researches, have been printed in the *Nova Kepleriana* series of the Bavarian Academy of Sciences. Much of the material in the original series has now been (or is about to be) incorporated into *Johannes Kepler Gesammelte Werke*. The new series includes V. Bialas, *Die Rudolphinischen Tafeln, Nova Kepleriana*, n.s. 2 (1969), M. List and V. Bialas, *Die Coss von Jost Bürgi, Nova Kepleriana*, n.s. 5 (1973), and V. Bialas and E. Papadimitriou, *Materialien zu den Ephemeriden von Johannes Kepler, Nova Kepleriana*, n.s. 7 (1980).

Principal English Translations

Mysterium Cosmographicum - Secret of the Universe, translated by A. M. Duncan, with introduction and commentary by E. J. Aiton (New York, 1981).

The New Astronomy, translated by William Donahue (Cambridge, 1992).

The New Astronomy: Introduction, translated by Owen Gingerich in *Great Ideas Today 1983* (Chicago, 1983), pp. 309-23.

Five Books of the Harmony of the World, translated and with introduction and commentary by E. J. Aiton, A. M. Duncan, and J. V. Field (Philadelphia, 1993).

The Harmonies of the World: Book V, translated by Charles Glenn Wallis, in *Great Books of the Western World,* vol. 16 (Chicago, 1952), pp. 1005-1085.

Epitome of Copernican Astronomy: Books IV and V, translated by Charles Glenn Wallis, in *Great Books of the Western World,* vol. 16 (Chicago, 1952), pp. 839-1004.

The Birth of History and Philosophy of Science: Kepler's A Defence of Tycho against Ursus *with Essays on its Provenance and Significance*, by Nicholas Jardine (Cambridge, 1984).

Kepler's Conversation with Galileo's Sidereal Messenger, translated by Edward Rosen (New York, 1965).

The Six-Cornered Snowflake, translated by Colin Hardie (Oxford, 1966).

Rudolphine Tables: Introduction, translated by Owen Gingerich and William Walderman in *Quarterly Journal of the Royal Astronomical Society,* xiii (1972), 360-73.

Bibliographical References

Kepler's Somnium, translated by Edward Rosen (Madison, 1967).

Kepler's Dream, translated by Patricia Frueh Kirkwood, edited by John Lear (Berkeley, 1965).

"Johannes Kepler's *On the More Certain Fundamentals of Astrology*, Prague, 1601," translated by J. B. Brackenridge and M. A. Rossi in *Proceedings of the American Philosophical Society*, cxxiii (1979), 85-116. A translation by J. V. Field is found in *Archive for History of Exact Sciences*, xxxi (1984), 225-68.

Johannes Kepler: Life and Letters, by Carola Baumgardt (New York, 1951).

Principal German Translations

Das Weltgeheimnis (Mysterium cosmographicum), translated by Max Caspar (Augsburg, 1923, Munich and Berlin, 1936).

Neue Astronomie (Astronomia nova), translated by Max Caspar (Munich and Berlin, 1929).

Weltharmonik (Harmonice mundi), translated by Max Caspar (Munich and Berlin, 1939).

Grundlagen der geometrischen Optik (selections from *Astronomiae pars optica*), translated by F. Plehn (*Ostwalds Klassiker der exakten Wissenschaften* Nr. 198) (Leipzig, 1922).

Dioptrik (Dioptrice), translated by Ferdinand Plehn (*Ostwalds Klassiker der exakten Wissenschaften* Nr. 144) (Leipzig, 1904).

Neue Stereometrie der Fässer (selections from *Stereometria nova*), translated by R. Klug (*Ostwalds Klassiker der exakten Wissenschaften* Nr. 165) (Leipzig, 1908).

Neujahrgabe oder vom sechseckigen Schnee (Strena seu De nive sexangula), translated by Fritz Rossmann (Berlin, 1943).

Traum vom Mond (Somnium), translated by Ludwig Gunther (Leipzig, 1898).

Johannes Kepler in seinen Briefen (a large, two-volume selection of Kepler's letters), translated by Max Caspar and Walther von Dyck (Munich and Berlin, 1930).

Johannes Kepler Selbstzeugnisse, chosen by F. Hammer, translated by E. Hammer, commentary by F. Seck (Stuttgart-Bad Cannstatt, 1971) (includes Kepler's so-called *Selbstcharacteristik* and a group of letters).

Other Translations

Le Secret du Monde (Mysterium cosmographicum), translated by Alain Segonds (Paris, 1984).

L'étrenne ou la neige sexangulaire (Strena), translated by R. Halleux (Paris, 1975).

Discussione col Nunzio Sidereo e Relazione sui Quattro Satelliti di Giove (Dissertatio cum Nuncio sidereo and *Narratio de Jovis satellitibus)*, translated by E. Pasoli and G. Tabarroni (Turin, 1972).

Bibliographical References

General Biography

Arthur Koestler, *The Watershed* (Garden City, NY, 1960), being a section of his *The Sleepwalkers* (London and New York, 1959).

Owen Gingerich, "Kepler, Johannes," pp. 289-312 in *Dictionary of Scientific Biography* (New York, 1973), vol. 7; this article contains an extensive annotated bibliography.

Edward Rosen, *Three Imperial Mathematicians: Kepler Trapped between Tycho Brahe and Ursus* (New York, 1986).

Berthold Sutter, *Johannes Kepler und Graz* (Graz, 1975).

Collections of Articles

Arthur Beer and Peter Beer, *Kepler, Vistas in Astronomy*, xviii (1975). This massive volume contains many of the important papers from the 1971 Kepler quadricentennial.

Wilhelm Freh (editor), *Johannes Kepler Werke und Leistung* (Linz, 1971).

Fritz Krafft, Karl Meyer, and Bernhard Sticker (editors), *Internationales Kepler-Symposium: Weil der Stadt 1971* (Hildesheim, 1973).

Ekkehard Preuss (editor), *Kepler Festschrift* (Regensburg, 1971).

Universität Graz, *Johannes Kepler 1571-1971 Gedenkschrift* (Graz, 1975).

Kepler's Philosophical and Theological Outlook

M. W. Burke-Gaffney, *Kepler and the Jesuits* (Milwaukee, 1944).

Bruce Eastwood, "Kepler as Historian of Science," in *Proceedings of the American Philosophical Society*, cxxvi (1982), 367-94.

Gerald Holton, "Johannes Kepler's Universe: Its Physics and Metaphysics," in *American Journal of Physics*, xxiv (1956), 340-51, reprinted as pp. 53-74 in Holton's *Thematic Origins of Scientific Thought* (revised edition, Cambridge, MA, 1988).

Jürgen Hübner, *Die Theologie Johannes Keplers zwischen Orthodoxie und Naturwissenschaft* (Tübingen, 1975).

Jürgen Hübner, "Johannes Kepler als Geograph im Kontext des theologischen Denkens seiner Zeit," in *Wissenschaftsgeschichte um Wilhelm Schickard*, edited by Friedrich Seck (*Contubernium*, xxvi, Tübingen, 1981), 99-114.

Nicholas Jardine, "The Forging of Modern Realism: Clavius and Kepler against the Sceptics," *Studies in History and Philosophy of Science*, x (1979), 141-73.

Wolfgang Pauli, "The Influence of Archetypal Ideas on the Scientific Theories of Kepler," in *The Interpretation of Nature and The Psyche* (New York, 1955), 147-240.

Robert S. Westman, "Kepler's Theory of Hypothesis and the 'Realist Dilemma,'" in *Studies in History and Philosophy of Science*, iii (1972), 233-64, reprinted from Krafft, Meyer, and Sticker (cited above).

Bibliographical References

Kepler's Astronomy

Eric J. Aiton, "Kepler's Second Law of Planetary Motion," in *Isis,* lx (1969), 75-90.

Eric J. Aiton, "Johannes Kepler and the Astronomy without Hypotheses," in *Japanese Studies in the History of Science,* xiv (1975), 49-71.

Eric J. Aiton, "Johannes Kepler and the Mysterium Cosmographicum," in *Sudhoffs Archiv,* lxi (1977), 173-94.

Volker Bialas, "Keplers komplizierter Weg zur Wahrheit: Von neuen Schwierigkeiten, die 'Astronomia Nova' zu lesen," *Berichte zur Wissenschaftsgeschichte,* xiii (1990), 167-76.

J. Bruce Brackenridge, "Kepler, Elliptical Orbits, Celestial Circularity," in *Annals of Science,* xxxix (1982), 117-43, 265-95.

A. E. L. Davis, "Kepler's Resolution of Individual Planetary Motion," [a group of four articles] in *Centaurus,* xxxv (1992), 97-191.

William H. Donahue, "Kepler's Fabricated Figures: Covering Up the Mess in the *New Astronomy,*" in *Journal for the History of Astronomy,* xix (1988), 217-37.

William H. Donahue, "Kepler's Transition to the Oval," in *Journal for the History of Astronomy,* xxiv (1993).

Owen Gingerich, "Johannes Kepler and His Rudolphine Tables," in *Sky and Telescope,* xlii (1971), 328-33; reprinted in *The Great Copernicus Chase and Other Adventures in Astronomical History* (Cambridge, 1992).

Owen Gingerich, "Johannes Kepler," chapter 5 in *General History of Astronomy,* vol 2A, edited by René Taton and Curtis Wilson (Cambridge, 1989).

Owen Gingerich, *The Eye of Heaven: Ptolemy, Copernicus, Kepler* (New York, 1993) — contains essays published earlier including "Kepler and the New Astronomy," *Quarterly Journal of the Royal Astronomical Society,* xiii (1972), 346-73, "The Computer Versus Kepler Revisited," from Krafft, Meyer, and Sticker (cited above, p. 394), and "The Origins of Kepler's Third Law," from Beer and Beer (cited above, p. 394).

Giora Hon, "On Kepler's Awareness of the Problem of Experimental Error," in *Annals of Science,* xliv (1987), 545-91.

Alexandre Koyré, *The Astronomical Revolution,* translated by R. E. W. Maddison (Paris, London, and Ithaca, 1973).

Jürgen Mittelstrass, "Methodological Elements of Keplerian Astronomy," in *Studies in History and Philosophy of Science,* iii (1972), 203-32, an English translation of the article in Krafft, Meyer, and Sticker (cited above, p. 394).

J. L. Russell, "Kepler's Laws of Planetary Motion: 1609-1666," in *British Journal for the History of Science,* ii, no. 5 (1964), 1-24.

Robert Small, *An Account of the Astronomical Discoveries of Kepler* (London, 1804; reprint Madison, 1963).

C. Bruce Stephenson, *Kepler's Physical Astronomy* (New York and Berlin, 1987).

Derek T. Whiteside, "Kepler's Eggs Laid and Unlaid," *Journal for the History of Astronomy*, v (1974), 1-21.

Curtis Wilson, "Kepler's Derivation of the Elliptical Path," *Isis*, lix (1968), 5-25, reprinted in his *Astronomy from Kepler to Newton* (London, 1989).

Curtis Wilson, "From Kepler's Laws, So-called, to Universal Gravitation: Empirical Factors," in *Archive for History of Exact Sciences*, vi (1970), 89-170, reprinted in his *Astronomy from Kepler to Newton* (London, 1989).

Kepler's Harmony and his Astrology

Michael Dickreiter, *Der Musiktheoretiker Johannes Kepler* (Bern, 1973).

J. V. Field, "A Lutheran Astrologer: Johannes Kepler," *Archive for History of Exact Sciences*, xxxi (1984), 189-271.

J. V. Field, *Kepler's Geometrical Cosmology* (Chicago, 1988).

J. V. Field, "Kepler's Rejection of Numerology," in *Occult and Scientific Mentalities in the Renaissance,* edited by Brian Vickers (Cambridge, 1984), pp. 273-96.

Owen Gingerich, "Kepler, Galilei, and the Harmony of the World," in *The Eye of Heaven: Ptolemy, Copernicus, Kepler* (New York, 1992).

Franz Hammer, "Die Astrologie des Johannes Kepler," in *Sudhoffs Archiv,* lv (1971), 113-35.

G. Simon, *Kepler astronome astrologue* (Paris, 1979).

D. P. Walker, "Kepler's Celestial Music," in *Studies in Musical Science in the Late Renaissance,* (Leiden, 1978), reprinted from *Journal of the Warburg and Courtauld Institutes,* xxx (1967), 228-50.

Eric Werner, "The Last Pythagorean Musician: Johannes Kepler," in *Aspects of Mediaeval and Renaissance Music,* edited by Jan La Rue (New York, 1956), pp. 867-82.

Kepler's Mathematics and Optics

Eric Aiton, "Infinitesimals and the Area Law," pp. 285-305 in Krafft, Meyer, and Sticker (cited above, p. 394).

J. A. Belyi and D. Trifunovic, "Zur Geschichte der Logarithmentafeln Keplers," in *NTM-Schriftenreihe für Geschichte der Naturwissenschaften, Technik und Medizin,* ix (1972), 5-20.

Bibliographical References

Gerd Buchdahl, "Methodological Aspects of Kepler's Theory of Refraction," in *Studies in History and Philosophy of Science,* iii (1972), 265-98, reprinted from Krafft, Meyer, and Sticker (cited above, p. 394).

J. V. Field, "Kepler's Star Polyhedra," in *Vistas in Astronomy,* xxiii (1979), 109-41.

J. V. Field, "Two Mathematical Inventions in Kepler's *'Ad Vitellionem paralipomena,'*" *Studies in History and Philosophy of Science,* xviii (1986), 449-68.

Kuno Fladt, "Das Keplerische Ei," in *Elemente der Mathematik,* xvii (1962), 73-78.

David C. Lindberg, *Theories of Vision from al-Kindi to Kepler* (Chicago, 1976).

David C. Lindberg, "The Genesis of Kepler's Theory of Light: Light Metaphysics from Plotinus to Kepler," in *Osiris,* 2nd series, ii (1986), 5-42.

Antoni Malet, "Keplerian Illusions: Geometrical Pictures *vs* Optical Images in Kepler's Visual Theory," *Studies in History and Philosophy of Science,* xxi (1990), 1-40.

Of Related Interest

Brian S. Baigrie, "Kepler's Laws of Planetary Motion, before and after Newton's *Principia:* an Essay on the Transformation of Scientific Problems," in *Studies in History and Philosophy of Science,* xviii (1987), 177-208.

Brian S. Baigrie, "The Justification of Kepler's Ellipse," in *Studies in History and Philosophy of Science,* xxi (1990), 633-64.

Stillman Drake, *Galileo Studies* (Ann Arbor, 1970).

Owen Gingerich, "Ptolemy, Copernicus, Kepler," in *Great Ideas Today 1983,* edited by Mortimer Adler and John Van Doren (Chicago, 1983), pp. 137-80.

Owen Gingerich and Robert S. Westman, *The Wittich Connection: Conflict and Priority in Late Sixteenth-Century Cosmology, Transactions of the American Philosophical Society,* lxxviii, pt. 7 (Philadelphia, 1988).

Richard A. Jarrell, "Astronomy at the University of Tübingen: The Work of Michael Mästlin," in *Wissenschaftsgeschichte um Wilhelm Schickard,* edited by Friedrich Seck (*Contubernium,* xxvi, Tübingen, 1981), pp. 9-19.

Friedrich Seck, "Johannes Kepler und der Buchdruck," in *Archiv für Geschichte des Buchwesens,* xi (1970), 610-726.

Victor Thoren, *The Lord of Uraniborg: A Biography of Tycho Brahe* (Cambridge, 1991).

Albert Van Helden, *Measuring the Universe: Cosmic Dimensions from Aristarchus to Halley* (Chicago, 1985).

BIBLIOGRAPHICAL CITATIONS

THROUGHOUT Max Caspar's text there are hundreds of direct quotations as well as numerous paraphrases, very few of which he cited to the original sources. These have now been added in this section according to page and line number of the foregoing text. The lines are counted beginning with the first line under the running head. Empty spaces and larger subheads are counted as if ordinary lines were present, so that a standard scale can be used against any page.

Most citations are from *Johannes Kepler Gesammelte Werke,* abbreviated *GW.* Numbers following the colons are line numbers on the page, except for letters where the line number within the entire letter follows the number of the letter. (This corresponds to the numeration within the *GW* itself.) Some material, especially biographical details, is still found only in the earlier Frisch edition, *Joannis Kepleri astronomi opera omnia* (Frankfurt and Erlangen, 1858-71), abbreviated *Fr.* The other abbreviations are:

BK for *Bibliographia Kepleriana;*
NK for *Nova Kepleriana;*
OGG for *Le Opere di Galileo Galilei* (A. Favaro, ed., the so-called National Edition);
OS for "Old Style," i.e., dates in the Julian calendar.

29:4	*GW* xix, 320, nr. 7.12, the famous "Selbstcharakteristik."
30:19	*GW* xix, 313, nr. 7.1.
30:21	K to Bianchi, 17 February 1619, *GW* xvii, nr. 827:9-17.
30:24	*GW* xix, 313-14, nr. 7.2.
33:26	*GW* xix, 380-1, nr. 7.122.
34:9f	*Fr* viii, 670-71.
35:3	*Fr* viii, 672.
35:16	*Fr* viii, 828-29.
35:31	*Fr* viii, 935-36.
35:39	*Fr* viii, 671.
37:34	*Fr* viii, 672.
37:41	K to Fabricius, 4 July 1603, *GW* xiv, nr. 262:287-89.
38:1	*Astronomiae pars optica, GW* ii, 239:32-34.
39:6	K to Fabricius, 4 July 1603, *GW* xiv, nr. 262.
39:11	*Fr* viii, 673.
39:32	*GW* xix, 328-37, nr. 7.30.
40:8	*GW* xix, 337.
40:13	*GW* xix, 338.
40:27	*GW* xix, 337.
40:34	*GW* xix, 336.

41:16   *GW* xix, 329.
41:20   *GW* xix, 327.
41:30   *GW* xix, 315, nr. 7.5.
41:16   *GW* xix, 316, nr. 7.7.

43:14   *GW* xix, 316, nr. 7.8.
43:18   *GW* xix, 319, nr. 7.10; Mayor and Council of Weil der Stadt to Tübingen University, 22 May 1590, *GW* xiii, nr. 1.
43:29   *Fr* viii, 676; K to Fabricius, 1 October 1602, *GW* xix, 316, nr. 226:479.
43:37   *Fr* viii, 676.
43:38   *GW* xix, 320-21, nr. 7.12, from Röslin, 17 October 1592.
43:41   *GW* xix, 319, nr. 7.11.

44:.5   *GW* xiii, 4/14 November 1591, nr. 3:1055.
44:23   *GW* xix, 325, nr. 7.30.
44:36   *Mysterium cosmographicum, GW* i, 23; *GW* viii, 44.
44:40   *Mysterium cosmographicum, GW* i, 23:13; *GW* viii, 44:12.

45:2   *Mysterium cosmographicum, GW* viii, 15:6-7; *Fr* viii, 673.
45:10   *GW* xix, 329, nr. 7.30.
45:16   See Friedrich Seck, "Marginalien zum Thema Kepler und Tübingen," pp. 3-19 in *Attempto, Nachrichten für die Freunde der Universität Tübingen*, Heft 41/42, 1971, esp. pp. 11-13.
45:21   Maestlin to K, 12 April 1599 OS, *GW* xiii, nr. 119:209f.
45:25   *GW* xix, 329, nr. 7.30; see Edward Rosen, "Kepler's Rake Was Not a Hoe," *Classical Outlook*, xliv (1966), 6-7.
45:41   See R. A. Jarrell, "The Life and Work of the Tübingen Astronomer M. Maestlin (1550-1631)," *Thesis*, University of Toronto, 1972, and also "Maestlin's Place in Astronomy," *Physis*, xvii (1975), 5-20.

46:19   *GW* xix, 329, nr. 7.30.
46:30   *Mysterium cosmographicum, GW* i, 16:39-40; *GW* viii, 33:20-21. See also Robert Westman, "The Comet and the Cosmos: Kepler, Mästlin and the Copernican Hypothesis," in Jerzy Dobrzycki (ed.), *The Reception of Copernicus' Heliocentric Theory* (Dordrecht, 1972) = *Studia Copernicana*, v (1972), 7-30.
47:1f   *Mysterium cosmographicum, GW* i, 9:11-21; *GW* viii, 23:11-21.
47:17f *Mysterium cosmographicum, GW* i, 9:22-23; *GW* viii, 23:22-23.
47:35   See Edward Rosen, *Kepler's Somnium* (Madison, 1967), Appendix C, pp. 207-8.
47:37   A facsimile reprint of Kepler's annotated copy of Copernicus' *De revolutionibus* in the Leipzig University Library has been issued by Johnson Reprints (New York, 1965).   The poem is Kepler's translation of a Greek poem by Joachimus Camerarius—see *GW* xii, 257 and 417-19.

48:2   K to Maestlin, *GW* xiii, nr. 23:241-44.
48:31   *NK* 6, 13:26-27.

49:8   *GW* xix, 329, nr. 7.30.
49:15f *NK* vi, 13:33-14:7.

50:5   K to Maestlin, 1/11 June 1598, *GW* xiii, nr. 99:499.
50:6   *GW* xix, 3, nr. 1.1; *Fr* viii, 677.

50:25f   The description of Kepler's move from Tübingen to Graz is directly paraphrased from *Astronomia nova*, ch. 7, *GW* iii, 108-9.

50:33   K to Tübingen Theological Faculty, 28 February 1594 OS, *GW* xiii, nr. 8:20.

51:6   *Astronomia nova*, *GW* iii, 108:17-18.

51:37   *Astronomia nova*, *GW* iii, 108:21.

51:38   *Mysterium cosmographicum*, *GW* i, 9:24; *GW* viii, 23:24.

52:23   *GW* xix, 3, nr. 1.2.

52:26   K to Gerlach, 19/29 October 1594, *GW* xiii, nr. 13.

53:16   *Fr* viii, 677; see Owen Gingerich, "The Civil Reception of the Gregorian Calendar Reform," in G. V. Coyne, M. A. Hoskin and O. Pedersen (eds.), *Gregorian Reform of the Calendar* (Vatican City, 1983), pp. 265-79.

54:41   See B. Sutter, *Johannes Kepler und Graz* (Graz, 1975), 53-74.

56:6   *GW* xix, 5, nr. 1.6; 11-12, nr. 1.21.

56:7   *GW* xix, 4, nr. 1.5; *Fr* viii, 679.

56:10f   *GW* xix, 4, nr. 1.4.

56:25   Zehentmair to K, 24 December 1598, *GW* xiii, nr.108:42f.

56:31   K to Maestlin, 10 February 1597, *GW* xiii, nr. 60:78f.

56:40   *GW* xix, 8, nr. 1.16.

57:5   *GW* xix, 30, nr. 1.66.

57:8   Papius to K, *GW* xiii, nr. 41, 45, 127, etc.

57:10f   *GW* xix, 335, nr. 7.30.

57:12   Papius to K, 7 June 1596, *GW* xiii, nr. 45.

57:16f   *GW* xix, 8, nr. 1.16.

57:29   *GW* xix, 328, nr. 7.30; the rest of this paragraph and onto p. 58 is also taken from this "Selbstcharakteristik."

58:25   *GW* xix, 5, nr. 1.8.

58:26   Since Caspar's book appeared, a copy of the 1597 *Practica* has been found; its text was published by B. Sutter in *Johannes Kepler 1571-1971 Gedenkschrift der Universität Graz* (Graz, 1975), pp. 343-52.

59:2f   *GW* iv, 9.16-10.1.

59:32f   *Fr* i, 400. The calendars will be published in *GW* xi,2.

59:37f   *De fundamentis astrologiae certioribus*, *GW* iv, 10:11-16.

60:11f   K to Maestlin, 8 December 1, *GW* xiii, nr. 106:59f.

60:23   *GW* xix, 5, nr. 1.8.

60:25   K to Maestlin, 8/18 January 1595, *GW* xiii, nr. 16:16f.

60:39   *cupiditas speculandi:* see above, p. 57. The 1596 edition of *Mysterium cosmographicum* appears in *GW* i, 3-80; the 1618 edition, augmented by Kepler's notes, appears in *GW* viii, 7-128. In the following section, when a reference is given only for *GW* viii, the citation is to these added notes. For a complete English translation, see *The Secret of the Universe*, A. M. Duncan (trans.) with notes by E. J. Aiton (New York, 1981).

61:18   *Mysterium cosmographicum*, *GW* i, 16:29-31; *GW* viii, 33:10-12.

61:33f *Mysterium cosmographicum*. *GW* i, 5, 9, 26 (Pythagoras); 9, 23, 26 (Plato); 23 (Cusanus) there is no mention of Augustine, but his influence is powerful; see, for example, J. Hübner, *Die Theologie Johannes Keplers* (Tübingen, 1975), *passim*.

61:40 *Mysterium cosmographicum*, *GW* i, 10; GW viii, 24.

62:11f *Mysterium cosmographicum*, *GW* i, 6:7-10; GW viii, 17:10-14.

62:24f K to Maestlin, 9 April 1597, *GW* xiii, nr. 64:10f.

62:36f *Mysterium cosmographicum*, *GW* i, 11:34-35; *GW* viii, 25:31-32.

62:41f *Mysterium cosmographicum*, *GW* i, 13:1-4; *GW* viii, 26:10-13.

63:11f *Mysterium cosmographicum*, *GW* i, 13:18-23; *GW* viii, 27:13-18.

63:22 *Mysterium cosmographicum*, *GW* viii, 9:20.

63:33f *Mysterium cosmographicum*, *GW* i, 13:25-37; *GW* viii, 27:20-32.

64:15 K to Maestlin, 2 August 1595, *GW* xiii, nr. 21 and continuing with nrs. 22, 23, 24, 32, 35, 36, 38, 39, and 47.

64:20 *Fr* viii, 688.

64:28 K to Maestlin, March 1596, *GW* xiii, nr. 32:25-26.

64:34 *Fr* viii, 683; K to Fabricius, 1 October 1602, *GW* xiv, nr. 226:517f—this long, technical letter contains in one section, for astrological purposes, many small biographical details.

64:38 K to Duke Friedrich, 17 February 1596, *GW* xiii, nr. 28.

65:11f *GW* xix, 323, nr. 7.17.

65:22 *GW* xix, 11, nr. 1.11.

65:33f Maestlin to University of Tübingen Prorector, end of May 1596, *GW* xiii, nr. 43.

66:5 Maestlin to K, 10 January 1597 OS, *GW* xiii, nr. 58:28-30, etc.

66:10 K to Maestlin, 10 February 1597, *GW* xiii, nr. 60:24f.

66:19 Maestlin to K, 9 March 1597 OS, *GW* xiii, nr. 63:5f.

66:28 Maestlin to K, 9 March 1597 OS, *GW* xiii, nr. 63:153f; the price of the *Mysterium cosmographicum* could hardly be as small as a nickel, since 200 copies cost Kepler 33 gulden, and his annual salary was only 150 gulden.

67:16 *Mysterium cosmographicum*, *GW* i, 30:11-12; *GW* viii, 52:11-12.

67:17 *Mysterium cosmographicum*, *GW* i, 30:8-9; *GW* viii, 52:8-9.

67:19 *Mysterium cosmographicum*, *GW* viii, 62:30-33.

67:23 *Mysterium cosmographicum*, *GW* viii, 72:16-17.

67:29 *Mysterium cosmographicum*, *GW* i, 80; *GW* viii, 127-28.

67:42 *Mysterium cosmographicum*, *GW* i, 71:2; *GW* viii, 111:28.

68:1 K to Maestlin, 14 September 1595, *GW* xiii, nr. 22:28-33.

68:11f Hafenreffer to K, 12 April 1598 OS, *GW* xiii, nr. 93:27f.

68:41f K to Maestlin, 1/11 June 1598. *GW* xiii, nr. 99:502f.

69:7 *Astronomia nova*, *GW* iii, 29-34.

69:12 *Mysterium cosmographicum*, *GW* viii, 20:19-20.

69:20 Prätorius to Herwart von Hohenburg, 23 April 1598, *GW* xiii, nr. 95.

69:27 Limnäus to K, 24 April 1598, *GW* xiii, nr. 96.

69:31 Galilei to K, 4 August 1597, *GW* xiii, nr. 73.

69:40   K to Galilei, 13 October 1597, *GW* xiii, nr. 76.

70:2   Bruce to K, 15 August 1502, *GW* xiv, nr. 222.
70:22   K to Tycho Brahe, 13 December 1597, *GW* xiii, nr. 82.
70:27   Tycho Brahe to K, 1 April 1598 OS, *GW* xiii, nr. 92.
70:37   Tycho Brahe to Maestlin, 21 April 1598 OS, *GW* xiii, nr. 94:18-22.

71:15   *Mysterium cosmographicum*, *GW* viii, 21:29-30.
71:16f   *Mysterium cosmographicum*, *GW* viii, 9:25-35.
71:28f   *Mysterium cosmographicum*, *GW* viii, 9:12-16.
71:40   K to Fabricius, 1 October 1602, *GW* xiv, nr. 226:517f; see above, p. 64.

72:4   These medallions (one of Barbara Kepler, one of Kepler himself) are reproduced by W. Gerlach and M. List, *Dokumente zu Leben und Werk* (Munich, 1971), p. 77, and in color but in rather small scale as the frontispiece to *Vistas in Astronomy*, xviii (1975).
72:37   Papius to K, 17/27 May 1597, *GW* xiii, nr. 41.

73:12   *GW* xiii, 11 September 1596, nr. 51.

74:10   Papius to K, 7/17 June 1596, *GW* xiii, nr. 45.
74:21   *GW* xix, 335, nr. 7.30.
74:38   K to the church office in Graz, 17/27 January 1597, *GW* xiii, nr. 57.
74:40   *GW* xix, 15, nr. 1.29; *GW* xiii, 12 April 1597, nr. 65.

75:6   *GW* xix, 15, nr. 1.29.
75:8   K to the School Inspector of Graz, 30 June 1597, *GW* xiii, nr. 70.
75:11f   K to Maestlin, 9 April 1597, *GW* xiii, nr. 64:239.
75:34   K to Fabricius, 1 January 1602, *GW* xiv, nr. 226:522-23.
75:36   *Fr* viii, 689.

76:2f   K to Herwart von Hohenburg, 9/10 April 1599, *GW* xiii, nr. 117:240f; a partial English translation is found in *Sidereal Messenger*, vi (1887), 109f.
76:33   K to Maestlin, 15 March 1598, *GW* xiii, nr. 89:180f.

77:2   K to Maestlin, 1/11 June 1598, *GW* xiii, nr. 99:360f.
77:6f   K to Maestlin, 19/29 August 1599, *GW* xiv, nr. 132.
77:21   K to Maestlin, 1/11 June 1598, *GW* xiii, nr. 99:369f.

78:16f   K to Maestlin, 1/11 June 1598, *GW* xiii, nr. 99:379-82. This passage is quoted in J. Loserth, *Akten und Korrespondenzen zur Geschichte der Gegenreformation in Innerösterreich unter Ferdinand II* (*Fontes Rerum Austriacarum*, Bd. 58, Vienna, 1906), p. XLVI; Loserth's work is a key source for this period in Graz.
78:23f   K to Maestlin, 8 December 1598, *GW* xiii, nr. 106:517f; the rest of the section comes from this same letter.

79:2   *GW* xix, 21, nr. 1.44.
79:3   K to Maestlin, 8 December 1598, *GW* xiii, nr. 106:535f.
79:7   *GW* xix, 23, nr. 1.46.
79:13   *GW* xix, 23, nr. 1.47.
79:26   K to Maestlin, 8 December 1598, *GW* xiii, nr. 106:553.
79:28   *GW* xix, 24, nr. 1.49.

79:31   Zehentmair to K, 15 November 1598, *GW* xiii, nr. 105.

79:35f   *GW* xix, 31, nr. 1.66.

80:3f   *GW* xix, 24, nr. 1.49; 8/18 December 1598, *GW* xiii, nr. 106:558f.

80:20   K to Maestlin, 8 December 1598, *GW* xiii, nr. 106:563.

80:24   K to Fickler, 24 October 1597, *GW* xiii, nr. 78.

80:25   K to Herwart von Hohenburg, 12 September 1597, *GW* xiii, nr. 74:6f.

80:40   K to Fickler, 24 October 1597, *GW* xiii, nr. 78.

80:41   Fickler to K, 4 November 1597, *GW* xiii, nr. 79.

81:2   K to Fickler, 24 October 1597, *GW* xiii, nr. 78:110f.

81:25   K to Maestlin, 1/11 June 1598, *GW* xiii, nr. 99:387-88.

81:29   K to Hafenreffer, 11 April 1619, *GW* xvii, nr. 835:112-14.

81:32   K to Hafenreffer, 11 April 1619, *GW* xvii, nr. 835:115-19.

81:39f   K to Georg Friedrich von Baden, 10 October 1607, *GW* xvi, nr. 451.

82:10   Maestlin to K, 11/12 January 1599, *GW* xiii, nr. 110:12f.

82:16   *GW* xix, 337, nr. 7.30.

83:21   *Ad epistolam Hafenrefferi*, *GW* xii, 50:16-17.

83:26   K to Herwart von Hohenburg, 16 December 1598, *GW* xiii, nr. 107:195f.

83:38   *Fr* viii, 713.

84:3f   Zehentmair to K, 20 April 1599, *GW* xiii, nr. 118.

84:8   Zehentmair to K, 23 June 1599, *GW* xiii, nr. 125:10.

84:33   K to Herwart von Hohenburg, 16 December 1598, *GW* xiii, nr. 107:200f.

84:42f   K to Herwart von Hohenburg, 16 December 1598, *GW* xiii, nr. 107:202f.

85:47   K to Maestlin, 8 December 1598, *GW* xiii, nr. 106:560f.

85:28   K to Maestlin, 19/29 August 1599, *GW* xiv, nr. 132:601f.

85:37   K to Johann Georg Brengger, 17 January 1605, *GW* xv, nr. 317:20-21.

86:9f   K to Herwart von Hohenburg, 26 March 1598, *GW* xiii, nr. 91:61f.

86:26   K to Duke Friedrich von Württemberg, 17 February 1596 OS, *GW* xiii, nr. 28.

86:33   K to Ursus, 15 November 1595, *GW* xiii, nr. 26.

87:4   Caspar rather underestimates Ursus' achievement: see the monograph by Owen Gingerich and Robert Westman cited in our notes to p. 108.

87:16   Brahe to K, 1/10 April 1598, *GW* xiii, nr. 92.

87:26f   K to Herwart von Hohenburg, 16 December 1598, *GW* xiii, nr. 107:213f.

87:31f   K to Maestlin, 16/26 February 1599, *GW* xiii, nr. 113.

88:5   K to Herwart von Hohenburg, 29 January 1599, *GW* xiii, nr. 111.

88:13   K to Herwart von Hohenburg, 9/10 April 1599, *GW* xiii, nr. 117:749f.

88:32   See Christine Schofield, "The Tychonic and Semi-Tychonic World Systems," pp. 33-44 in *The General History of Astronomy*, vol. 2A, edited by R. Taton and C. Wilson (Cambridge, 1989).

88:34f   K to Herwart von Hohenburg, 26 March 1598, *GW* xiii, nr. 91:182f.

89:7f   K to Herwart von Hohenburg, 26 March 1598, *GW* xiii, nr. 91:182f.

89:17   Tycho Brahe to K, 9 December 1599, *GW* xiv, nr. 145:228f.

89:22   K to Galilei, 13 October 1597, *GW* xiii, nr. 76:57f; K to Maestlin, early October 1597, *GW* xiii, nr. 75:90f; K to Maestlin, 6 January 1598, *GW* xiii, nr. 85:136f.

89:23   K to Maestlin, 6 January 1598, *GW* xiii, nr. 85:136f.

89:27   K to Herwart von Hohenburg, 16 December 1598, *GW* xiii, nr. 107:101f.

89:36   K to Maestlin, 8 December 1598, *GW* xiii, nr. 106.

90:2   Herwart von Hohenburg to K, 12 September 1597, *GW* xiii, nr. 74:35f; K to Herwart von Hohenburg, 24 December 1597, *GW* xiii, nr. 83.

90:8   K to Herwart von Hohenburg, 24 December 1597, *GW* xiii, nr. 83:182.

90:10   K to Herwart von Hohenburg, 15 October 1598, *GW* xiii, nr. 104; K to Herwart von Hohenburg, 16 December 1598, *GW* xiii, nr. 107.

90:14   K to Herwart von Hohenburg, 2 January 1599, *GW* xiii, nr. 109; K to Herwart von Hohenburg, 29 January 1599, *GW* xiii, nr. 111; K to Herwart von Hohenburg, 10 March 1599, *GW* xiii, nr. 114:122f; K to Herwart von Hohenburg, 9/10 April 1599, *GW* xiii, nr. 117:433f.

90:42   K to Herwart von Hohenburg, 26 March 1598, *GW* xiii, nr. 91; K to Herwart von Hohenburg, 30 May 1599, *GW* xiii, nr. 123:434f.

91:3   K to Herwart von Hohenburg, 9/10 April 1599, *GW* xiii, nr. 117:117f. See above 76:2.

91:4   Zehentmair to K, 23 March 1599, *GW* xiii, nr. 115:28f; Zehentmair to K, 24 May 1599, *GW* xiii, nr. 122; etc.

91:17f   K to Maestlin, 19/29 August 1599, *GW* xiv, nr. 132.

91:29   Copernicus, *De revolutionibus,* I,10. The word translated as symmetry might better be "commensurability" or "common measure."

92:41   This text, which comes from Proclus (see G. Friedlein (ed.), *In Euclidem* (Leipzig, 1873), pp. 22-23) is quoted by Kepler on the subtitle for Book I of the *Harmonice mundi, GW* vi, 207.

93:13   K to Herwart von Hohenburg, 14 September 1599, *GW* xiv, nr. 134:508-509.

93:21f   K to Herwart von Hohenburg, 9/10 April 1599, *GW* xiii, nr. 117:174-79.

93:28   *Dissertatio cum Nuncio sidereo, GW* iv, 308:9-10.

93:33   Kepler addresses this point in ch. 12 of *Mysterium cosmographicum,* but he notes in the second edition, *GW* viii, 76, that the concept of "knowable" polygons is fully developed only in Book III of the *Harmonice mundi.*

94:8f   K to Maestlin, 19/29 August 1599, *GW* xiv, nr. 132:239-40.

94:30   *Calendarium in annum 1599, Fr* i, 402.

94:41   *Calendarium in annum 1599, Fr* i, 402.

95:2f   *Calendarium in annum 1599, Fr* i, 402-3.

95:14f   A similar quotation is in K to Herwart von Hohenburg, 9/10 April 1599, *GW* xiii, nr. 117:393f.

95:32   K to Herwart von Hohenburg, 6 August 1599, *GW* xiv, nr. 130:226f.

95:35   K to Herwart von Hohenburg, 6 August 1599, *GW* xiv, nr. 130:240f.

95:38   K to Herwart von Hohenburg, 6 August 1599, *GW* xiv, nr. 130:220.

95:40f   *Mysterium cosmographicum, GW* i, 68f; *GW* viii, 109.

96:9   K to Maestlin, 19/29 August 1599, *GW* xiv, nr. 132:466-68.

96:11   K to Herwart von Hohenburg, 14 September 1599, *GW* xiv, nr. 134:438f.

96:24f K to Herwart von Hohenburg, 6 August 1599, *GW* xiv, nr. 130:309-12.

96:28f K to Herwart von Hohenburg, 14 December 1599, *GW* xiv, nr. 148.

97:25 K to Maestlin, 19/29 August 1599, *GW* xiv, nr. 132:572; the entire paragraph is based on this letter.

98:10 K to Maestlin, 19/29 August 1599, *GW* xiv, nr. 132:560-63.

98:14 K to Maestlin, 19/29 August 1599, *GW* xiv, nr. 132:527-28.

98:18 K to Maestlin, 19/29 August 1599, *GW* xiv, nr. 132:601f.

98:28 K to Maestlin, 19/29 August 1599, *GW* xiv, nr. 132:532-35.

98:31f K to Maestlin, 19/29 August 1599, *GW* xiv, nr. 132:535f.

98:38 K to Maestlin, 19/29 August 1599, *GW* xiv, nr. 132:635f.

99:3f Maestlin to K, 15 January 1600 OS, *GW* xiv, nr. 153.

99:11 K to Maestlin, 8 December 1598, *GW* xiii, nr. 106:572-74.

99:22 Herwart von Hohenburg to K, 29 August 1599, *GW* xiv, nr. 133:5f.

99:27 Brahe to Johann Friedrich von Hoffmann, 26 January 1600, *GW* xiv, nr. 155.

99:37f Brahe to K, 9 December 1599, *GW* xiv, nr. 145:242f.

100:5 Brahe to K, 26 January 1600, *GW* xiv, nr. 154.

100:10 Brahe to Johann Friedrich von Hoffmann, 3 February 1600, *GW* xiv, nr. 156.

101:42f K to Longomontanus, 1605, *GW* xiv, nr. 323.

102:18 K to Herwart von Hohenburg, 12 July 1600, *GW* xiv, nr. 168:105f.

102:27f *GW* xix, 37, nr. 2.1 (April 1600); Brahe himself mentions this opinion in 8 April 1600, *GW* xiv, nr. 161:54-57.

102:34f K to Herwart von Hohenburg, 12 July 1600, *GW* xiv, nr. 168:109-11.

103:19f *GW* xix, 37-41, nr. 2.1.

104:41f *GW* xix, 41-42, nr. 2.2; the entire page derives from this source.

105:11 Brahe to Baron Hoffman, 6 March 1600, *GW* xiv, nr. 157.

105:38 *GW* xix, 44-47, nr. 2.4.

106:14f Brahe to Jessenius, 8 April 1600, *GW* xiv, nr. 161.

106:19 *GW* xix, 336, nr. 7.30.

107:3f K to Brahe, April 1600, *GW* xiv, nr. 162.

107:21 K to an anonymous woman, 1612, *GW* xvii, nr. 643:8.

107:28 Brahe to Rosenkrantz, 3 June 1600, *GW* xiv, nr. 163.

107:33 K to Herwart von Hohenburg, 12 July 1600, *GW* xiv, nr. 168:21f.

108:4 Brahe to K, 28 August 1600, *GW* xiv, nr. 173:34f.

108:18 Brahe to K, 28 August 1600, *GW* xiv, nr. 173:129. Recently several works have shed new and unexpected light on the Ursus-Brahe quarrel: see N. Jardine, *The Birth of History and Philosophy of Science: Kepler's* A Defence of Tycho against Ursus (Cambridge, 1988); E. Rosen, *Three Imperial Mathematicians: Kepler Trapped between Tycho Brahe and Ursus* (New York, 1986); O. Gingerich and R. S. Westman, *The Wittich Connection: Conflict and Priority in Late Sixteenth-Century Cosmology, Transactions of the American Philosophical Society*, lxxviii, pt. 7 (Philadelphia, 1988).

108:34   *GW* xix, 48-49, nr. 2.6.
108:41f K to Herwart von Hohenburg, 12 July 1600, *GW* xiv, nr. 168:28f.

109:13   K to Herwart von Hohenburg, 12 July 1600, *GW* xiv, nr. 168:54f.
109:36f K to Herwart von Hohenburg, 14 September 1599, *GW* xiv, nr. 134:19-23.

110:5   Herwart von Hohenburg to K, 25 July 1600, *GW* xiv, nr. 169:14-15.
110:11f K to Ferdinand, July 1600, *GW* xiv, nr. 166.
110:30   K to Ferdinand, July 1600, *GW* xiv, nr. 166:56-57.
110:33   K to Ferdinand, July 1600, *GW* xiv, nr. 166:154-55.

111:5   K to Ferdinand, July 1600, *GW* xiv, nr. 166:22f.
111:14   K to Maestlin, 9 September 1600, *GW* xiv, nr. 175.
111:29   J. Loserth, *Akten und Korrespondenzen zur Geschichte der Gegenreformation in Innerösterreich unter Ferdinand II* (*Fontes Rerum Austriacarum,* Bd. 60, Vienna, 1907), pp. 11-12.
111:39   K to Maestlin, 9 September 1600, *GW* xiv, nr. 175:30f.

112:3   *GW* xix, 28, nr. 1.61.
112:7   *GW* xix, 29, nr. 1.64.
112:13   *GW* xix, 30, nr. 1.65.
112:14   *GW* xix, 30-31, nr. 1.66.
112:19   *GW* xix, 28, nr. 1.62.

114:14   K to Brahe, 17 October 1600, *GW* xiv, nr. 177:17f.
114:16   Brahe to K, 28 August 1600, *GW* xiv, nr. 173.
114:21   Brahe to K, 28 August 1600, *GW* xiv, nr. 173:78-79.
114:32   K to Maestlin, 9 September 1600, *GW* xiv, nr. 175:41.

115:3f   K to Maestlin, 9 September 1600, *GW* xiv, nr. 175:52-56.

117:5   K to Brahe, 17 October 1600, *GW* xiv, nr. 177.
117:7   Brahe to K, 28 August 1600, *GW* xiv, nr. 173:51.
117:17   K to Brahe, 17 October 1600, *GW* xiv, nr. 177:48.
117:26   K to Maestlin, 6/16 December 1600, *GW* xiv, nr. 180:1-2.
117:33   K to Maestlin, 6/16 December 1600, *GW* xiv, nr. 180:9f.

118:4   Maestlin to K, 9 October 1600 OS, *GW* xiv, nr. 178:17-18.
118:5   K to Maestlin, 6/16 December 1600, *GW* xiv, nr. 180:6f.
118:10   K to Maestlin, 8 February 1601, *GW* xiv, nr. 183:5f.
118:15f Hafenreffer to K, 28 March 1601 OS, *GW* xiv, nr. 184:22f.
118:22   Lyser to K, 19 January 1601 OS, *GW* xiv, nr. 181.

119:14   K to Maestlin, 8 February 1601, *GW* xiv, nr. 183:161f.
119:16   *Fr* i, 236-76; two modern editions have recently appeared: N. Jardine (see note 108:18), the other by V. Bialas, *GW* xx,1, 17-62 and some previously unpublished text, 65-82.
119:28f K to Frau Barbara, 30 May 1601, *GW* xiv, nr. 187.
119:35   K to Fabricius, 2 December 1602, *GW* xiv, nr. 239:512-41.
119:36   Barbara Kepler to K, 31 May 1601, *GW* xiv, nr. 188.

120:3f Eriksen to K, 13 June 1601, *GW* xiv, nr. 191.

120:19 K to Maestlin, 10/20 December 1601, *GW* xiv, nr. 203:24-26.

120:23 K to Maestlin, 8 February 1601, *GW* xiv, nr. 183:170-72.

120:27 K to Maestlin, 10/20 December 1601, *GW* xiv, nr. 203:22-23.

120:32 K to Magini, 1 June 1601, *GW* xiv, nr. 190:29-31.

120:36 K to Maestlin, 8 February 1601, *GW* xiv, nr. 183:174-76; K to Magini, 1 June 1601, *GW* xiv, nr. 190:44.

120:39 *GW* xix, 48, nr. 2.5.

120:42f K to Magini, 1 June 1601, *GW* xiv, nr. 190:450-52.

121:7 K to Magini, 1 June 1601, *GW* xiv, nr. 190.

121:29 *Astronomia nova, GW* iii, 89:7-11; see also the text published by J. L. E. Dreyer in *Tychonis Brahe Opera Omnia,* xiii, 283.

121:37 *Fr* viii, 138-42.

122:6 K to Maestlin, 10/20 December 1601, *GW* xiv, nr. 203:7f.

122:22 Hafenreffer to K, 14 November 1601 OS, *GW* xiv, nr. 198.

122:23 Rollenhagen to K, 22 February 1602 OS, *GW* xiv, nr. 209.

122:29f Herwart von Hohenburg to Barwitz, 23 February 1602, *GW* xiv, nr. 207:8-14.

122:39 Herwart von Hohenburg to Barwitz, 23 February 1602, *GW* xiv, nr. 207:19-21.

123:2f Herwart von Hohenburg to K, 2 December 1601, *GW* xiv, nr. 200:102-7.

123:11 *GW* xix, 49, nr. 2.7.

123:12 *GW* xix, 49, nr. 2.8.

123:32 K to Maestlin, 10/20 December 1601, *GW* xiv, nr. 203:18-21.

124:36 A flight of rhetorical fancy on Caspar's part: Kepler mentions neither Columbus nor Magellan.

126:41 K to Longomontanus, 1605, , *GW* xv, nr. 323:188-89.

127:5f K to Herwart von Hohenburg, 12 July 1600, *GW* xiv, nr. 168:102-4.

127:7f K to Herwart von Hohenburg, 12 July 1600, *GW* xiv, nr. 168:108-9.

127:17f *Astronomia nova, GW* iii, 109:18-20.

128:1 *Astronomia nova, GW* iii, 156:4-5.

128:25f *Astronomia nova, GW* iii, 178:1f.

129:14f *Astronomia nova, GW* iii, 191f (ch. 22).

130:14 *Astronomia nova, GW* iii; the phrase is part of the subtitle for part III.

131:8 *Astronomia nova, GW* iii, 236f (ch. 33).

132:4f *Astronomia nova, GW* iii, 264:3f.

132:15 Kepler's so-called second law or area law is never stated so clearly in the *Astronomia nova,* although it is introduced as a working technique in ch. 40; see, for example, Alexandre Koyré, *The Astronomical Revolution* (Paris, 1973), pp. 235-36, or Bruce Stephenson, *Kepler's Physical Astronomy* (New York, 1987), pp. 80f.

133:30 K to Fabricius, 4 July 1603, *GW* xiv, nr. 262:48-50.

133:40 *Astronomia nova, GW* iii, 288:13.

134:12 *Astronomia nova, GW* iii, 346:3.

134:17   Kepler announces at the very end of chapter 58 (*Astronomia nova, GW* iii, 366:37) that for Mars, "there is no figure left for the orbit except a perfect ellipse," but only by implication is the so-called first law established for planets in general.

135:30f K to Longomontanus, 1605, *GW* xv, nr. 323:105-8.

135:36   K to Longomontanus, 1605, *GW* xv, nr. 323:108-9.

136:9   K to Herwart von Hohenburg, 10 February 1605, *GW* xv, nr. 325:57-62.

138:10f K to Fabricius, 11 October 1605, *GW* xv, nr. 358:67-74.

138:30   *Astronomia nova, GW* iii, 246:19f.

139:12f K to the Tübingen University Senate, 12 December 1604, *GW* xv, nr. 304:42-43.

139:29f For nearly ten months after Tycho's death in December, 1601, no letters from Kepler survive except one to Maestlin informing him of the event and its aftermath. Tycho's *Progymnasmata* was published in August, and finally Kepler pours out the story of his difficulties with the heirs in K to Herwart von Hohenburg, 7 October 1602, *GW* xiv, nr. 228. He refers to the heirs as "the Tychonians" in K to Fabricius, February 1604, *GW* xv, nr. 281:225.

140:21f K to Fabricius, February 1604, *GW* xv, nr. 281:226.

141:23   *GW* xix, 53, nr. 2.16.2.

141:29   K to the Imperial Treasurer, August 1608, *GW* xix, 56, nr. 2.24.

141:31   The printer was Gotthard Vögelin, not Erhart; see Friedrich Seck, *Archiv für Geschichte des Buchwesens*, xi (1970), 644.

141:38   K to Brengger, 4 October 1607, *GW* xvi, nr. 448:7-8.

143:21f "Hipparchus," *GW* xx,1, 181-268.

144:31   *Astronomiae pars optica, GW* ii.

144:32   K to Fabricius, 4 July 1603, *GW* xiv, nr. 262:177-82; K to Herwart von Hohenburg, 5 July 1603, *GW* xiv, nr. 263:87f.

144:33   K to Maestlin, 10/20 January 1604, *GW* xv, nr. 278:32-33.

146:3   *Astronomiae pars optica, GW* ii, 16:27-31.

148:36   Cristini to Magini, 26 February 1605, *GW* xv, nr. 333.

149:14   See R. J. W. Evans, *Rudolf II and His World: A Study in Intellectual History 1576-1612,* (Oxford, 1973); see also *Prag um 1600: Kunst und Kultur am Hofe Rudolfs II* (Freren, 1988).

150:35   K to Herwart von Hohenburg 1/11 June 1598, *GW* xiii, nr. 199:471f.

151:3f   K to Herwart von Hohenburg 1/11 June 1598, *GW* xiii, nr. 99:473-82.

152:4f   *Prognosticum* for 1604, *NK* i, 12.

152:12f "De directionibus" (1601), *Fr* viii, 295.

152:18f *NK* i, 18-19.

152:37   See above p. 135.

153:3   *Fr* viii, 331-43.

153:4   *Fr* viii, 300-20.

153:11   See Frank D. Prager, "Kepler als Erfinder," *Internationales Kepler-Symposium: Weil der Stadt 1971* (Hildesheim, 1973), pp. 385-405.

153:15 "Judicium de trigono igneo 1603," *Fr* i, 439-50.

153:36 *De stella nova, GW* i, 332:21-24.

154:7 *De stella nova, GW* i, 158:10-159:6.

154:13 *De stella nova, GW* i, 159:26-27.

154:27 *Bericht vom neuen Stern, GW* i, 391-99. For an English translation, see J. V. Field and A. Postl, *Vistas in Astronomy,* xx (1977), 333-39.

155:14 *De stella nova, GW* i, 322:10.

155:15 *De stella nova, GW* i, 347f.

155:21f *De stella nova, GW* i, 151:31-36.

155:30 Herwart von Hohenburg to K, 6 March 1607, *GW* xv, nr. 412:45-46.

155:32f *De stella nova, GW* i, 354:9-19.

155:37f *De stella nova, GW* i, 355:38f.

156:3f K to anonymous, 4 December 1604, *GW* xv, nr. 306:49-50.

156:19 *De stella nova, GW* i, 359:6f.

156:34 Hafenreffer to K, 1 January 1607 OS, *GW* xv, nr. 406:41f; K to Herwart von Hohenburg, January 1607, *GW* xv, nr. 409:8f.

157:2f *GW* xix, 61-62, nr. 2.32.

157:22 *GW* xix, 86, nr. 2.68 (July, 1616).

157:29 K to Herwart von Hohenburg, 24 November 1607, *GW* xvi, nr. 461:187-88.

157:31 K to Duke Johann Friedrich, 9/19 March 1611, *GW* xvi, nr. 609:7.

157:34 *GW* xix, 61-67, nr. 2:32-40 etc.

158:4 K to Herwart von Hohenburg, 12 January 1603, *GW* xiv, nr. 242:30f.

158:7 K to Edmund Bruce, *GW* xiv, nr. 268:20f.

158:14 K to Duke Maximilian, end of 1604, *GW* xv, nr. 312:36f.

158:25 K to Herwart von Hohenburg, 22 January 1605, *GW* xv, nr. 318.

158:36 Maestlin to K, 28 January 1605 OS, *GW* xv, nr. 322.

159:1f K to Maestlin, 5 March 1605, *GW* xv, nr. 335:5-12.

159:30 K to Herwart von Hohenburg, 13 January 1606, *GW* xv, nr. 368:132.

159:32 *Fr* viii, 616-25.

159:34f K to Herwart von Hohenburg, 24 November 1607, *GW* xvi, nr. 461:164f.

160:13 *De stella Cygni, GW* i, nr. 294:18-28

161:8 *De fundamentis astrologiae certioribus, GW* iv, 5-35. English translations have been given by J. B. Brackenridge and M. A. Rossi, *Proceedings of the American Philosophical Society,* cxxiii (1979), 85-116, and J. V. Field, *Archive for History of the Exact Sciences,* xxxi (1984), 229-68.

161:35 *Strena seu de nive sexangula, GW* iv, 259-80. There is an English translation by C. Hardie, *The Six-Cornered Snowflake* (Oxford, 1966), and a French one by R. Halleux, *L'étrenne ou la neige sexangulaire* (Paris, 1975).

162:2 K to Besold, 8/18 June 1607, *GW* xv, nr. 432.

162:17 *Prognosticum* for 1605, *Fr* i, 453.

162:20 *Prognosticum* for 1605, *Fr* i, 454.

163:9 Pistorius to K, 14 March 1607, *GW* xv, nr. 413:5-7.

163:17 K to Pistorius, 15 June 1607, *GW* xv, nr. 431:24.

163:29 K to Pistorius, 15 June 1607, *GW* xv, nr. 431:37f.

164:12f Pistorius to K, 12 July 1607, *GW* xv, nr. 433.

165:13 K to Duke Friedrich, end of 1604, *GW* xv, nr. 314:21f; August von Anhalt to K, 9 July 1607, *GW* xvi, nr. 434; August von Anhalt to K, 10/20 July 1607, *GW* xvi, nr. 435; K to August von Anhalt, undated, *GW* xvi, nr. 436; etc.

165:17 K to Duke Friedrich, end of 1604, *GW* xv, nr. 314:34; Burgi's role is deduced from Ludwig von Dietrichstein's two letters to K, Kepler's own reply being lost: *GW* xv, 10 July 1604, nr. 291 and 20 October 1604, nr. 295.

165:31f *De stella nova*, *GW* i, 307:16-19.

166:10 *Astronomiae pars optica*, *GW* ii, 144f.

166:12 Jessenius to K, 30 November 1617, *GW* xvii, nr. 776.

166:28 Bachacek to Hofman von Saaz, 2 March 1605, *GW* xv, nr. 334.

166:31 *Phaenomenon singulare*, *GW* iv, 77-98.

166:41 *Phaenomenon singulare*, *GW* iv, 92-93.

167:7f *Phaenomenon singulare*, *GW* iv, 92-93.

167:20f *Phaenomenon singulare*, *GW* iv, 93.

167:29 *Ephemerides* for 1618; *GW* xi.1, 28:10-11.

167:32 *Ephemerides* for 1618; *GW* xi.1, 28:19-21.

167:36 *De solis deliquio*, *GW* iv, 37-53.

168:5 *Astronomiae pars optica*, *GW* ii, 221:33f, 300:12f, 305:31f, 307.25f.

168:9 *De solis deliquio*, *GW* iv, 43-44.

168:37 *GW* iv, 424.

168:33 Martini to K, 1 January 1606, *GW* xv, nr. 364; Schele to K, 29 December 1605, *GW* xv, nr. 366; Ziegler to K, 12 January 1606, *GW* xv, nr. 367; Crzistanowic to K, 16 January 1606, *GW* xv, nr. 369.

169:41 Fabricius to K, 23 June 1601, *GW* xiv, nr. 193.

170:4 Fabricius to K, 11 March 1608 OS, *GW* xvi, nr. 485.

170:5 K to Fabricius, November 1608, *GW* xvi, nr. 509.

170:21 K to Fabricius, 11 October 1605, *GW* xv, nr. 358:312, 390f.

170:26f K to Fabricius, August/October 1608, *GW* xvi, nr. 508:280f.

171:3f K to Fabricius, August/October 1608, *GW* xvi, nr. 508:419-25, 449-60.

171:28f K to Brengger, 30 November 1607, *GW* xvi, nr. 463:22-24.

171:33 K to Brengger, 30 November 1607, *GW* xvi, nr. 463:5f.

172:2 Brengger to K, 23 December 1604, *GW* xv, nr. 310:20f.

172:18f *Antwort auf Röslin*, *GW* iv, 128:7-14. At the end of the quotation, the text should read, instead of "with such super-astrologers": "with such people on astrological matters."

172:28f *Tertius interveniens*, *GW* iv, 218:1-9.

172:40 *De stella nova*, i, 325-35, esp. 330-32; see N. Jardine, *The Birth of History and Philosophy of Science* (Cambridge, 1984), pp. 276-80.

173:12 *De stella nova*, i, 332:4.

173:33 *GW* xix, 344, nr. 7.51.

173:37 *GW* xix, 455, nr. 8.35.

174:2 K to Herwart von Hohenburg, 24 November 1607, *GW* xvi, nr. 461:204.

174:9 *Fr* viii, 745-46.

174:10 K to Fabricius, 18 December 1604, *GW* xv, nr. 308:85; *Fr* viii, 758.

174:11 *Fr* viii, 775.

174:27f K to Herwart von Hohenburg, 10 December 1604, *GW* xv, nr. 302:32-34.

174:32 *GW* xix, 412, nr. 8.7, section 12.

174:34 *Fr* viii, 502.

174:35 K to Maestlin, 31 March 1606, *GW* xv, nr. 376.

174:36 K to Emperor Matthias, 1 October 1615, *GW* xvii, nr. 663.

174:38 Heinrich von Strahlendorf to K, 28 January 1615, *GW* xvii, nr. 708:7-15; *GW* xix, 356, nr. 7.78; *GW* xix, 370, nr. 7.106, section 4.

174:42 K to Fabricius, November 1608, *GW* xvi, nr. 509:2-3.

175:2 K to Rudolf II, April 1608, *GW* xvi, nr. 490; K to Fabricius, *GW* xvi, nr. 508:266-67; *GW* xix, 428-435, nr. 8.22

175:10 K to Wacker von Wackenfels, early 1618, *GW* xvii, nr. 783:15-21.

175:17 K to an anonymous woman, 1612, *GW* xvii, nr. 643.

175:19 *GW* xix, 452-57, nr. 8.35.

176:2 K to an anonymous woman, 1612, *GW* xvii, nr. 643:7f.

176:7 *GW* xix, 455, nr. 8:35.

176:11 K to an anonymous woman, 1612, *GW* xvii, nr. 643:115f.

176:16 K to an anonymous woman, 1612, *GW* xvii, nr. 643:133f.

176:19 K to an anonymous woman, 1612, *GW* xvii, nr. 643:125-6.

176:22 *GW* xix, 455, nr. 8:35.

176:27 K to an anonymous woman, 1612, *GW* xvii, nr. 643:180f.

176:37f *GW* xix, 455, nr. 8:35.

176:42 *GW* xix, 454, nr. 8:35.

177:10 *GW* xix, 412, nr. 8:7, section 11.

177:11 K to Bruce, 4 September 1603, *GW* xiv, nr. 268:281.

177:12 Herwart von Hohenburg to K, 17 June 1603, *GW* xiv, nr. 259.

177:13 K to Herwart von Hohenburg, 5 July 1603, *GW* xiv, nr. 263:10, where Kepler envisions a journey; see later, K to Herwart von Hohenburg, September 1603, *GW* xiv, nr. 270:8f.

177:21 *GW* xix, 413, nr. 8.8, section 15.

177:24 K to Fabricius, 1 August 1607, *GW* xvi, nr. 438:663.

177:28 *Astronomia nova, GW* iii, 10.

177:39 K to Duke Johann Friedrich, April/May 1609, *GW* xvi, nr. 527.

179:6 K to Magini, 22 March 1610, *GW* xvi, nr. 560:41, where he says 1583 instead of 1582, but see K to Marius, 10 November 1612, *GW* xvii, nr. 640:124-28.

179:25 K to Magini, 22 March 1610, *GW* xvi, nr. 560:60f.

179:27 K to Magini, 1 June 1601, *GW* xiv, nr. 190:481f.

179:29 Magini to K, 20 June 1610, *GW* xvi, nr. 569:15f.

180:2 *GW* xx,1, 183-268.

180:3 *Ausführlicher Bericht, GW* iv, 59-76.

180:8 *De vero anno, GW* i, 357-90.

180:26 K to Heydon, October 1605, *GW* xv, nr. 357:91-92.

181:1f *GW* iv, 105:24-28.

181:5 The first one is called *Antwort auff... D. Helisaei Röslini... Discurs von heutiger Zeit Beschaffenheit und wie es ins Künfftig ergehen werde* (Prague, 1609) *GW* iv, 101-44; the other is *Tertius interveniens* (Frankfurt am Main, 1610)—see p. 182.

181:15 *Antwort auf Röslin, GW* iv, 115 and i, 270-72.

181:21 *De stella nova, GW* i, 313-56.

181:28 *Antwort auf Röslin, GW* iv, 116:6-7.

181:29f *Tertius interveniens, GW* iv, 149:14-21.

181:32 *Antwort auf Röslin, GW* iv, 106-108.

181:37 *Antwort auf Röslin, GW* iv, 138:35-37.

181:40 *Antwort auf Röslin, GW* iv, 127:15-16; 131:41; 134:8.

182:1 *Antwort auf Röslin, GW* iv, 127:19-23.

182:3 *Antwort auf Röslin, GW* iv, 129:9-10.

182:24 *Tertius interveniens, GW* iv, 147-258.

183:30f *Tertius interveniens, GW* iv, 161:32f.

184:12 The notion that stars impel but do not compel is developed at length in *Harmonice mundi,* IV,7, *GW* vi, 279f—see especially 283:19.

184:16f *Tertius interveniens, GW* iv, 209:30-210:2.

185:9f *Tertius interveniens, GW* iv, 215.

185:26f *Tertius interveniens, GW* iv, 245:40-246:1.

185:35f *Tertius interveniens, GW* iv, 246:23-24.

187:23 K to Gerlach, 8/18 July 1609, *GW* xvi, nr. 532:14-15.

187:30 K to Heydon, October 1605, *GW* xv, nr. 357:106.

188:11 K to Gerlach, 8/18 July 1609, *GW* xvi, nr. 532:45f.

188:17f K to Duke Johann Friedrich, April/May 1609, *GW* xvi, nr. 527:19-22.

188:23 K to Duke Johann Friedrich, April/May 1609, *GW* xvi, nr. 527:48-49.

188:31f K to Duke Johann Friedrich, May 1609, *GW* xvi, nr. 528.

189:4 K to Duke Johann Friedrich, May 1609, *GW* xvi, nr. 528:118-121.

189:36 *Dissertatio cum Nuncio sidereo, GW* iv, 288-89.

190:4f *Dissertatio cum Nuncio sidereo, GW* iv, 289.

190:18 Note the new translation by A. Van Helden, *Sidereus Nuncius or the Sidereal Messenger* (Chicago, 1989); for the original Latin text see *OGG* iii, 53-96.

192:1 *Dissertatio cum Nuncio sidereo, GW* iv, 281-311. See E. Rosen's translation: *Kepler's Conversation with Galileo's Sidereal Messenger,* (New York, 1955).

192:4 *Dissertatio cum Nuncio sidereo, GW* iv, 285; as Rosen pointed out (note 25), Kepler received the book on April 8, but the request for a personal opinion took place on April 13.

192:10f *Dissertatio cum Nuncio sidereo, GW* iv, 289:34-37.

193:5 *OGG* iii, 62:14.

193:20 See A. Van Helden, "The Invention of the Telescope," in *Transactions of the American Philosophical Society,* lxvii (1977), pt. 4.

193:27 Horky to K, 27 April 1610, *GW* xvi, nr. 570:86f; Galilei to Vinta, 26 May 1610, *GW* xvi, nr. 572:5-9.

193:38 *Peregrinatio contra Nuncium sidereum, OGG* iii, 129-45.

194:4 *Dissertatio cum Nuncio sidereo, GW* iv, 289:26-29, paraphrasing Galileo.

194:9f *Dissertatio cum Nuncio sidereo, GW* iv, 290:9-12.

194:14 *Dissertatio cum Nuncio sidereo, GW* iv, 287:5-7.

194:20 *Dissertatio cum Nuncio sidereo, GW* iv, 288:8-16.

195:6 *Dissertatio cum Nuncio sidereo, GW* iv, 295:10f.

195:11 *Dissertatio cum Nuncio sidereo, GW* iv, 291:10-12.

195:16 *Dissertatio cum Nuncio sidereo, GW* iv, 305:32-34.

195:30f *Dissertatio cum Nuncio sidereo, GW* iv, 308:9-10.

196:6f *Dissertatio cum Nuncio sidereo, GW* iv, 286:39-287:4.

196:13 Galilei to K, 19 August 1610, *GW* xvi, nr. 587.

196:16 Galilei to Vinta, 7 May 1610, *GW* xvi, nr. 572.

196:19 Magini to K, 26 May 1610, *GW* xvi, nr. 576.

196:22 Fugger to K, 28 May 1610, *GW* xvi, nr. 578.

196:25 Maestlin to K, 7/17 September 1610, *GW* xvi, nr. 592:11-13.

196:28 *OGG* iii, 129-45.

196:32 Hasdale to Galilei, 19 December 1610, *GW* xvi, nr. 601.

196:33 K to Horky, 9 August 1610, *GW* xvi, nr. 585:32.

196:42 Giuliano de' Medici to Galilei, 19 April 1610, *GW* xvi, nr. 568.

197:7   K to Horky, 9 August 1610, *GW* xvi, nr. 585:115f.

197:11   Galilei to K, 19 August 1610, *GW* xvi, nr. 587:30f.

197:21   *Narratio de Jovis satellitibus, GW* iv, 318:37f.

197:30   *Narratio de Jovis satellitibus, GW* iv, 319:3f.

197:36   *Narratio de Jovis satellitibus, GW* iv, 315-25; there is an Italian translation: E. Pasoli and G. Tabarroni, *Discussione col Nunzio Sidereo e Relazione sui Quattro Satelliti di Giove* (Turin, 1972).

198:6   *Mundus Jovialis:* see the English translation by A. O. Prichard published in *Observatory*, xxxix (1916), 367-81, 403-12, 443-52 and 498-503. On Marius' alleged priority, see K to Marius, 10 October 1612, *GW* vii, nr. 640:61-80.

198:15   *Dioptrice, GW* iv, 327-414.

198:26   *Dioptrice, GW* iv, 331:32.

199:6f   *Dioptrice, GW* iv, 387:6-7.

199:20f   *Dioptrice, GW* iv, 334:5-8.

199:30   K to Galilei, December 1610, *GW* xvi, nr. 603:63-65.

200:4   K to Galilei, December 1610, *GW* xvi, nr. 603:60f; K to Galilei, 9 January 1611, *GW* xvi, nr. 604:48-51.

200:26   Kepler reproduced (and translated) Galileo's letters in his *Dioptrice, GW* iv, 345-54.

200:37   *Dioptrice, GW* iv, 344:36.

200:38   *Dioptrice, GW* iv, 347:4.

201:2   *Dioptrice, GW* iv, 319:27, 344:39.

201:7   *Dioptrice, GW* iv, 345:10.

201:10f   K to Galilei, 9 January 1611, *GW* xvi, nr. 604:6-8

201:14   *Dioptrice, GW* iv, 348:26-27.

201:21f   *Dioptrice, GW* iv, 344:15-16.

201:37   K to Galilei, 13 October 1597, *GW* xiii, nr. 6. Kepler did not get an answer from Galileo.

201:40   Galilei to K, 28 August 1627, *GW* xviii, nr. 1054.

202:1   Galilei to K, 19 August 1610, *GW* xvi, nr. 587.

202:3   Giuliano de' Medici to K, 7 February 1611, *GW* xvi, nr. 606:1.

202:15   *Tychonis Brahei Dani Hyperaspistes adversus Scipionis Claramontii Anti-Tychonem* (Frankfurt am Main, 1625); *GW* viii, 263-437, esp. *Appendix Hyperaspistis,* 417-18, section 7. The appendix has been translated in S. Drake and C.D. O'Malley, *The Controversy on the Comets of 1618* (Philadelphia, 1960), pp. 337-55.

202:27   K to Crüger, 1 March 1615, *GW* xvii, nr. 710:13—"Annus enim 1611 luctuosus undiquaque fuit et funestus."

202:35   K to Crüger, 1 March 1615, *GW* xvii, nr. 710:13f.

202:42f   *Eclogae chronicae, GW* v, 224:14-17.

204:9f   K to an anonymous nobleman, 3 April 1611, *GW* xvi, nr. 612:24-26.

204:35   K to Duke Johann Friedrich, 9/19 March 1611, *GW* xvi, nr. 609.

205:14   K to Duke Johann Friedrich, 9/19 March 1611, *GW* xvi, nr. 609:73f.

205:19f   Printed in *GW* xvi, 464-65.

206:3   Galilei to Giuliano de' Medici, 1 October 1610, *GW* xvi, nr. 593:25-32.

206:7   Jörger to K, 20 December 1610, *GW* xvi, nr. 602.

206:10   K to the Upper Austrian Estates, 10 June 1610, *GW* xvi, nr. 617; *GW* xix, 123-24, nr. 3.3.

206:14   *GW* xix, 349, nr. 7:63.

206:17   *GW* xix, 349, nr. 7:64.

206:20 *Eclogae chronicae, GW* v, 225:1-3.
206:27 K to Crüger, 1 March 1615, *GW* xvii, nr. 710:25f.
206:30f *Eclogae chronicae, GW* v, 224:24-30.
207:2 *Funera domestica duo luctuosissima, GW* xii, 217:6.
207:6 *Funera domestica duo luctuosissima, GW* xii, 218:32-33.
207:7 *Funera domestica duo luctuosissima, GW* xii, 217:27.
207:11 *Funera domestica duo luctuosissima, GW* xii, 217:39.
207:15 *Funera domestica duo luctuosissima,* a leaflet published by Kepler in 1616, of which only two copies have been discovered (see *Bibliographia Kepleriana,* nr. 50).
207:20f *Funera domestica duo luctuosissima, GW* xii, 214-15. This text was contributed by Kepler to an autograph book of one of his friends, Jakob Zoller, when he left Tübingen University for Graz in March 1594—see *GW* xix, 322, nr. 7:14, and see F. Seck, "J. Kepler als Dichter," *Internationales Kepler Symposium: Weil der Stadt 1971* (Hildesheim, 1973), pp. 427-50 (esp. pp. 444-45).
208:11 K to Vicke, August 1619, *GW* xvi, nr. 619:10-11.
208:13 K to Crüger, 1 March 1615, *GW* xvii, nr. 710:27-28, etc.
208:19 K to Crüger, 1 March 1615, *GW* xvii, nr. 710:28-29.
208:27 *Eclogae chronicae, GW* v, 221-370.
208:37 *GW* xix, 296, nr. 6.57.
208:42 These texts are now printed in *GW* xix, part 8, 403-73.

209:40 K to Matthias Bernegger, 7 February 1617, *GW* xvii, nr. 754:17.

210:9f K to anonymous nobleman, 23 October 1613, *GW* xvii, nr. 669:84-88.
210:18f K to Upper Austrian Estates, 10 June 1611, *GW* xvi, nr. 617:32, 40-44.

211:9f K to Upper Austrian Estates, 10 June 1611, *GW* xvi, nr. 617:21f.
211:18 K to Crüger, 1 March 1615, *GW* xvii, nr. 710:31f.
211:19 *GW* xix 74-79, nr. 2.49-2.53.
211:27 K to Upper Austrian Estates, 10 June 1611, *GW* xvi, nr. 617:27-28

212:19 K to Ursus, 20/30 October 1595, *GW* xiii, nr. 26:7-16.
212:35 *GW* xix, 123-24, nr. 3.2.
212:40f *GW* xix, 123-24, nr. 3.2.

213:23 K to Bernegger, 7 February 1617, *GW* xvii, nr. 754:15.
213:40 *Notae ad epistolam Hafenrefferi, GW* xii, 50.

214:14 K to Hoffmann, 26 April 1612, *GW* xvii, nr. 715:8-9; Papius to K, 17 August 1615, nr. 718; 18 August 1612, nr. 719.
214:23 Stuttgart Consistory to K, 25 September 1612 OS, *GW* xvii, nr. 638.

215:30 Stuttgart Consistory to K, 25 September 1612 OS, *GW* xvii, nr. 638:140-41.
215:41 Stuttgart Consistory to K, 25 September 1612 OS, *GW* xvii, nr. 638:184-85.

216:26 Stuttgart Consistory to K, 25 September 1612 OS, *GW* xvii, nr. 638:163.
216:31 K to Hafenreffer, 11 April 1619, *GW* xvii, nr. 835:162-76.
216:34 K to Hafenreffer, 11 April 1619, *GW* xvii, nr. 835:291.

217:8f *Glaubenbekenntnis, GW* xii, 23:33-38.
217:28 Christoph von Schallenberg to K, 2 January 1617, *GW* xvii, nr. 752:17f.
217:31f *Glaubenbekenntnis, GW* xii, 27:12-26.

218:9f *Glaubenbekenntnis, GW* xii, 28:44-47.
218:18 K to Hafenreffer, 28 November 1618, *GW* xvii, nr. 808:55-56.
218:19f *Glaubenbekenntnis, GW* xii, 28:3-5.

218:21f   *Glaubenbekenntnis, GW* xii, 28:17-21.
218:28f   K to an anonymous woman, 1612, *GW* xvii, 643:183-88.
218:38   Maestlin to K, 21 September 1616 OS, *GW* xvii, nr. 744:53f.

219:4f   K to Maestlin, 12/22 December 1616, *GW* xvii, nr. 750:234-38.
219:13f   K to Maestlin, 12/22 December 1616, *GW* xvii, nr. 750:260-66.
219:32   *Notae ad epistolam Hafenrefferi, GW* xii, 50:30f.
219:38f   K to Bernegger, 7 February 1617, *GW* xvii, nr. 754:11-13; "in the farmhouse"
          should probably read "in the house of the Estates."
219:42f   *Epitome astronomiae Copernicanae, GW* vii, 9:18-20.

220:6   *GW* xix, 132, nr. 3.17.
220:10f   K to Bernegger, 7 February 1617, *GW* xvii, nr. 754:33-40.
220:22   Roffenus to K, 1 March 1617, *GW* xvii, nr. 757.
220:27f   K to Roffenus 17 April 1617, *GW* xvii, nr. 761:27-32.
220:42   K to anonymous nobleman, 23 October 1613, *GW* xvii, nr. 669:31f.

221:19   K to anonymous nobleman, 23 October 1613, *GW* xvii, nr. 669:6-7.
221:22f   K to anonymous nobleman, 23 October 1613, *GW* xvii, nr. 669:15-19, 22-24.
221:31f   K to anonymous nobleman, 23 October 1613, *GW* xvii, nr. 669:19-22.

222:13   K to anonymous nobleman, 23 October 1613, *GW* xvii, nr. 669:64-65.
222:17f   K to anonymous nobleman, 23 October 1613, *GW* xvii, nr. 669:72-77.
222:25f   K to anonymous nobleman, 23 October 1613, *GW* xvii, nr. 669:273-76.
222:36f   K to anonymous nobleman, 23 October 1613, *GW* xvii, nr. 669:282-85.
222:38   K to Emperor Matthias, 17 October 1613, *GW* xvii, nr. 666.
222:40   *GW* xix, 353-54, nr. 7.72.
222:41f   Pisani to K, 23 October 1613, *GW* xvii, nr. 670:276-82.

223:4f   Pisani to K, 23 October 1613, *GW* xvii, nr. 670:28-29.
223:6f   Regina to K, 3 September 1612, *GW* xvii, nr. 635:16.
223:20   *GW* xix, 296, nr. 6.57.
223:25   Margareta Regina: K to Matthaus Wacker von Wackenfels, beginning of 1618, *GW*
          xvii, nr. 783:44-45; K to Crüger, 11 March 1619, *GW* xvii, nr. 831:3-5.
223:26   Katharina: Gringalletus to K, 20 November 1617, *GW* xvii, nr. 773:4-8; K to
          Wacker von Wackenfels, beginning of 1618, *GW* xvii, nr. 783:44-46.
223:28   Sebald: K to Crüger, 11 March 1619, *GW* xvii, nr. 831:5-6; see *GW* xix, 481.
223:29   Cordula: *GW* xix, 360, nr. 7.92.
223:30   Fridmar: Besold to K, 2/12 April 1623, *GW* xviii, nr. 945:17-20; see also *GW* xix,
          251, nr. 6.20.
223:31f   Hildebert: K to Crüger, 1 May 1626, *GW* xviii, nr. 1026:4-11.
223:35   *GW* xix, 354, nr. 7.74; 357, nr. 7.79.
223:38   *GW* xix, 374, nr. 7.111.
223:40   K to Oberndorffer, 17/27 May 1622, *GW* xviii, nr. 931:1-2.

224:3   *GW* xix, 374, nr. 7.111.
224:6   *GW* xix, 360, nr. 7.92.
224:12   *Fr* viii, 205-10.
224:13   Ludwig Kepler, *Des fürtrefflichen Weltweisen Römers Cornelii Taciti Historischer
          Beschreibung. Das Erste Buch ...* (Linz, 1625), *GW* xii, 103-75.
224:15f   *GW* xii, 105.
224:26   Eichler to K, 23 May 1607, *GW* xv, nr. 427:8-9.
224:30   *Unterricht Vom H. Sacrament des Leibs u. Blut Jesu Christi unsers Erlösers* (Linz?,
          1618), *GW* xii, 11-18.
224:33f   *GW* xii, 13:27-30.

225:25 K to Crüger, 1 March 1615, *GW* xvii, nr. 710:32-33.
225:28 *GW* xix, 130, nr. 3.12.
225:35 Ursinus to K, 13 August 1612, *GW* xvii, nr. 633; 644 etc.
226:3 K to Bernegger, 7 February 1617, *GW* xvii, nr. 754.
226:14f K to Pisani, 18 April 1618, *GW* xvii, nr. 788:7f.
226:19f K to Pisani, 16 December 1613, *GW* xvii, nr. 675:35f.
226:24 *GW* xix, 351, nr. 7.69.
227:11 K to Bernegger, end 1612/beginning 1613, *GW* xvii, nr. 645; 26 March 1613, *GW* xvii, nr. 646.
227:17 K to Crüger, 1 March 1615, *GW* xvii, nr. 710:33-35; Preface to his *Eclogae chronicae, GW* v, 225:8-13.
227:32 *Bericht vom Geburtsjahr Christi, GW* v, 127-201.
227:36 Röslin had published at least three books on the question; see Franz Hammer, *GW* v, 444, nrs. 8, 9, 10. The one dedicated to Matthias I is *Prodromus dissertationum chronologicarum* ... (Frankfurt, 1612).
228:2f *Bericht vom Geburtsjahr Christi* (1613), *GW* v, 149:40-150:3; the passage is in Latin in *De vero anno* (1614), *GW* v, 18:18-29.
228:22 *De vero anno, GW* v, 7-126.
228:31 *GW* xix, 124-27, nr. 3.5.
228:32 K to Upper Austrian Estates, 25 July 1613, *GW* xvii, nr. 659 = *GW* xix, 127-28, nr. 3.7.
228:36 *GW* xix, 127-28, nr. 3.7(2).
231:3 *De calendario Gregoriano, Fr* iv, 11-64; a Latin translation by M. G. Hansch was published in Frankfurt and Leipzig in 1726; see Max Caspar, *BK*, nr. 105; it will be printed in *GW* xi,2.
231:5f *De calendario Gregoriano*, from the title page as it was rendered into Latin by Hansch, *Fr* iv, 10; see also *BK*, nr. 105.
231:12 K to Maestlin, 9 April 1597, *GW* xiii, nr. 64.
231:19f *Judicium de calendario Gregoriano, Fr* iv, 58.
231:28f K to Maestlin, 9 April 1597, *GW* xiii, nr. 64:50-53.
231:39 *Judicium de calendario Gregoriano, Fr* iv, 26.
231:42f *De calendario Gregoriano, Fr* iv, 13-14.
232:30 *Judicium de calendario Gregoriano, Fr* iv, 59:1-5.
232:41 *Admonitio ad astronomos* (1629), *Fr* vii, 594; to be printed in *GW* xi,2.
233:4 K to Wacker von Wackenfels, 1612, *GW* xvii, nr. 627:19-21.
233:8 K to Malcotius, 18 July 1613, *GW* xvii, nr. 658.
233:10f K to Wacker von Wackenfels, 1612, *GW* xvii, nr. 627:200-210; for the *Astronomia nova,* see *GW* iii, 244f; 18 July 1613, *GW* xvii, nr. 658:26-30.
233:29 *GW* ix, 9:17-22.
233:40 *GW* ix, 9:22-31.
234:6 *Nova stereometria, GW* ix, 7-133.
234:35 *Nova stereometria, GW* ix, 13:8-9.
235:4 Paul Guldin, *De centro gravitatis* (Vienna, 1635-41); cf. *GW* ix, 484-86.
235:25 Welser to K, 11 Feb 1614, *GW* xvii, nr. 680:1-10.
235:29 *Nova stereometria, GW* ix, 36:12-25.
235:36 *GW* ix, 433-34.
235:41 Crüger to K, 31 Mar 1616 OS, *GW* xix, nr. 727:9-20.

236:7   *Messekunst Archimedis, GW* ix, 137-274.

236:15   *Messekunst Archimedis, GW* ix, 139:9-11.

236:19   *Messekunst Archimedis, GW* ix, 140:1-2, 20-21.

236:23   *Messekunst Archimedis, GW* ix, 140:22.

236:28   *Messekunst Archimedis, GW* ix, 270-71.

236:40   *GW* xvii, K to Upper Austrian Estates, 9 May 1616, nr. 734:5-12.

237:6   *GW* xix, 131-32, nr. 3.16 (20 May 1616).

237:9   *GW* xix, 133, nr. 3.19 (16 December 1616).

237:11   *GW* xvii, K to Upper Austrian Estates, 9 May 1616, nr. 734:107-61.

237:20   *GW* xvii, K to Upper Austrian Estates, 9 May 1616, nr. 734:138-42.

237:26   *GW* xvii, K to Upper Austrian Estates, 9 May 1616, nr. 734:121-22.

237:33   *GW* xvii, K to Upper Austrian Estates, 9 May 1616, nr. 734:160-61.

238:3   *GW* xvii, K to Upper Austrian Estates, 9 May 1616, nr. 734:39-45.

238:13   *GW* xvii, K to Upper Austrian Estates, 9 May 1616, nr. 734:76-82.

239:17   K to Wacker von Wackenfels, early 1618, *GW* xvii, nr. 783:4f.

239:24   See *GW* xix, 87, nr. 2.71; 195, nr. 5.4.

239:32   *Epitome astronomiae Copernicanae, GW* vii, 7:8.

239:33   *Epitome astronomiae Copernicanae, GW* vii, 7:9-11.

239:40   *Epitome astronomiae Copernicanae, GW* vii, 7:17-30; K to Schickard, 11 March 1618, *GW* xvii, nr. 785:6-11.

240:7   The pages that follow give the most detailed account in English of Katharina Kepler's witchcraft trial, based primarily on the "Judicium matris Kepleri," *Fr* viii, 361-562; Kepler's letters relating to the trial are the main part now found in *GW*, and we give those citations rather than the counterparts in *Fr*. See also Berthold Sutter, *Der Hexenprozesz gegen Katharina Kepler* (Weil der Stadt, 1979).

240:14   K to Leonberg Senate, 2 January 1616, *GW* xvii, nr. 725:4.

241:31   K to Vice-chancellor Faber, early 1617, *GW* xvii, nr. 756:27f, esp. 35.

242:2   K to Vice-chancellor Faber, early 1617, *GW* xvii, nr. 756:37f; *Fr* viii, 391-93.

242:14   *Fr* viii, 405.

242:22   *Fr* viii, 428-30.

242:26   *Fr* viii, 413-16.

242:29   *Fr* viii, 407-8.

242:33   *Fr* viii, 411-12.

243:1   *Fr* viii, 416-18.

243:15   K to Leonberg Senate, early 1617, *GW* xvii, nr. 756:42f.

243:39   K to Leonberg Senate, early 1617, *GW* xvii, nr. 756:50.

244:8   K to Leonberg Senate, early 1617, *GW* xvii, nr. 756:54f.

244:11   K to Leonberg Senate, 2 January 1616, *GW* xvii, nr. 725:26.

244:12   K to Leonberg Senate, 2 January 1616, *GW* xvii, nr. 725:20.

244:19   K to Leonberg Senate, early 1617, *GW* xvii, nr. 756:58.

244:41   K to Leonberg Senate, 2 January 1616, *GW* xvii, nr. 725:4-5.

245:5   K to Leonberg Senate, 2 January 1616, *GW* xvii, nr. 725:14-16.

245:9   K to Leonberg Senate, 2 January 1616, *GW* xvii, nr. 725:68-77.

245:16 K to Leonberg Senate, 2 January 1616, *GW* xvii, nr. 725:86f.

245:32 *Somnium, Fr* viii, 27-123; see note to 351.9.

246:12 K to Leonberg Senate, early 1617, *GW* xvii, nr. 756:69f; *Fr* viii, 365f.

246:18 K to Leonberg Senate, early 1617, *GW* xvii, nr. 756:74--for some reason Kepler refers to an eight-year-old daughter, whereas the report to the chancery gives the age as 12, *Fr* viii, 365.

246:24 K to Leonberg Senate, early 1617, *GW* xvii, nr. 756:94f.

246:32 K to Leonberg Senate, early 1617, *GW* xvii, nr. 756:104.

246:38 K to Leonberg Senate, early 1617, *GW* xvii, nr. 756:112f.

246:42 *Fr* viii, 365-6.

247:3   K to Leonberg Senate, early 1617, *GW* xvii, nr. 756:128f.

247:9   K to Leonberg Senate, early 1617, *GW* xvii, nr. 756:119f.

247:18 *Fr* viii, 366.

247:32 K to Duke Johann Friedrich, 1 September 1617, *GW* xvii, nr. 768.

248:8   *Fr* viii, 367-71.

248:16 K to Duke Johann Friedrich, 1 September 1617, *GW* xvii, nr. 768:18.

248:32 K to Wacker von Wackenfels, early 1618, *GW* xvii, nr. 783:15f.

248:37 K to Wacker von Wackenfels, early 1618, *GW* xvii, nr. 783:21f.

248:41 K to Wacker von Wackenfels, early 1618, *GW* xvii, nr. 783:29f.

249:16 K to Duke Johann Friedrich, spring of 1620, *GW* xvii, nr. 880:71f.

249:22 K to Duke Johann Friedrich, spring of 1620, *GW* xvii, nr. 880:80f.

249:38 K to Duke Johann Friedrich, spring of 1620, *GW* xvii, nr. 880:92f.

250:15 K to Duke Johann Friedrich, spring of 1620, *GW* xvii, nr. 880.

250:21 K to Duke Johann Friedrich, spring of 1620, *GW* xvii, nr. 880:128.

250:28 K to Duke Johann Friedrich, August 1620, *GW* xvii, nr. 889:17-20.

251:1   K to Maestlin, 19 June 1620, *GW* xviii, nr. 884:720f.

252:5   For example, Kepler had dedicated his *Harmonice mundi* to James I, *GW* vi, 9-12.

252:7   Wotton to Francis Bacon, 1620, *GW* xviii, nr. 892.

252:9   K to Bernegger, 29 August 1620, *GW* xviii, nr. 891.

252:11 K to Bernegger, 29 August 1620, *GW* xviii, nr. 891:11-13.

252:20 K to Duke Johann Friedrich, August 1620, *GW* xviii, nr. 889:27-38.

252:30 K to Bernegger, 29 August 1620, *GW* xviii, nr. 891:14-15;
        K to Bernegger, 25 September 1620, *GW* xviii, nr. 893.

252:35 K to Duke Johann Friedrich, October 1620, *GW* xviii, nr. 898:7f.

252:38 *Fr* viii, 435-36.

253:14 K to Duke Johann Friedrich, October 1620, *GW* xviii, nr. 898:14-15.

253:21 K to Duke Johann Friedrich, November 1620, *GW* xviii, nr. 902:41-48.

253:27 K to Duke Johann Friedrich, November 1620, *GW* xviii, nr. 902:9.

253:29 K to Duke Johann Friedrich, later in 1620, *GW* xviii, nr. 905:24f.

253:37 K to Duke Johann Friedrich, later in 1620, *GW* xviii, nr. 905:41f.

254:10 *Fr* viii, 448-51.

254:21 *Fr* viii, 462-87.
254:24 *Fr* viii, 493-509; see also K to Duke Johann Friedrich, May 1621, *GW* xviii, nr. 916.
254:28 *Fr* viii, 510-18.
254:36 *Fr* viii, 519-47.
254:42 *Fr* viii, 491.

255:18 *Fr* viii, 547.
255:24 *Fr* viii, 548-49.
255:32 *Fr* viii, 549-50.

256:10 *Fr* viii, 550-51.
256:16 *GW* xix, 361-62, nr. 7.96 (4 October 1621).
256:20 K to Oberndorffer, 17/27 May 1622, *GW* xviii, nr. 931.
256:23 K to Bernegger, 5/15 February 1622, *GW* xviii, nr. 909:8-9.
256:26 K to Bernegger, 5/15 February 1622, *GW* xviii, nr. 909:23-28.
256:31 K to Bernegger, 5/15 February 1622, *GW* xviii, nr. 909:18-19.
256:38 K to Bernegger, 25 September 1620, *GW* xviii, nr. 893.

257:2   Florian Crusius to K, 5/15 May 1621, *GW* xviii, nr. 914:5-12; compare *GW* xix, 195, nr. 5.4.
257:8   Besold to K, 29 July 1622, *GW* xviii, nr. 935:79-82.
257:14 *Fr* i, 486; see also pp. 263-64.
257:17 K to Guldin, summer 1620, *GW* xviii, nr. 890:3-8 and p. 474.
257:37 K to Crüger, 28 February 1624, *GW* xviii, nr. 974:160-64.

258:3   Hebenstreit to K, 23 March 1622, *GW* xviii, nr. 927:12 and p. 483.
258:6   *GW* xix, 145-47, nr. 3.45 (10 October 1625).
258:35 *Notae ad epistolam Hafenrefferi, GW* xii, 51:5-7.

259:1   The actual letter is lost, but is cited in K to Hafenreffer, 11 April 1619, *GW* xvii, nr. 835:230-31.
259:8   K to Hafenreffer, 11 April 1619, *GW* xvii, nr. 835:16f, 268f.
259:12 K to Hafenreffer, 28 November 1618, *GW* xvii, nr. 808:3-7.
259:21 K to Hafenreffer, 28 November 1618, *GW* xvii, nr. 808:7-25.
259:42 K to Hafenreffer, 28 November 1618, *GW* xvii, nr. 808:41-44.

260:6   Hafenreffer to K, 17 February 1619 OS, *GW* xvii, nr. 829.
260:11 Hafenreffer to K, 17 February 1619 OS, *GW* xvii, nr. 829:19.
260:12 Hafenreffer to K, 17 February 1619 OS, *GW* xvii, nr. 829:34-35.
260:14 Hafenreffer to K, 17 February 1619 OS, *GW* xvii, nr. 829:37-39.
260:17 K to Hafenreffer, 11 April 1619, *GW* xvii, nr. 835.
260:29 K to Hafenreffer, 11 April 1619, *GW* xvii, nr. 835:14-16.
260:32 K to Hafenreffer, 11 April 1619, *GW* xvii, nr. 835:9-11.
260:38 K to Hafenreffer, 11 April 1619, *GW* xvii, nr. 835:47-49.
260:40 K to Hafenreffer, 11 April 1619, *GW* xvii, nr. 835:71-76.

261:4   K to Hafenreffer, 11 April 1619, *GW* xvii, nr. 835:131f.
261:7   K to Hafenreffer, 11 April 1619, *GW* xvii, nr. 835:271-79.

261:19 Hafenreffer to K, 31 July 1619 OS, *GW* xvii, nr. 847:157-59.

261:26 Hafenreffer to K, 31 July 1619 OS, *GW* xvii, nr. 847:70.

262:34 Grüninger to Osiander, 1 July 1619 OS, *GW* xvii, nr. 843:10-19.

263:21 K to Hafenreffer, 28 November 1618, *GW* xvii, nr. 808:59-60.

263:24 K to Hafenreffer, 28 November 1618, *GW* xvii, nr. 808:65-67.

263:32 *Prognosticon auff 1619, Fr* i, 486.

263:39 *Prognosticon auff 1619, Fr* i, 487.

264:3 *Fr* i, 659-60.

264:8 K to Remus Quietanus, end of October, 1619, *GW* xvii, nr. 859:38-40.

264:21 *Glaubenbekandtnus* (*Glaubenbekenntnis*), known in only two copies (Vienna and Wittenberg), *GW* xii, 21-38.

264:23 K to Bernegger, 21 August 1623, *GW* xviii, nr. 958.

264:27 Hafenreffer to K, published in *Acta Mentzeriana* (Tübingen, 1625), 62-68; *GW* xii, 41-45.

264:29 *Notae ad epistolam Hafenrefferi, GW* xii, 46-62.

265:26 K to Wacker von Wackenfels, *GW* xvii, nr. 783:46-48.

265:41 *Harmonice mundi, GW* vi, 104:37-105:3.

266:7 Herwart von Hohenburg to K, 25 July 1600, *GW* xiv, nr. 169:35; Herwart to K, 6 March 1607, about sending the book, *GW* xv, nr. 412; K to Herwart, April 1607, on receipt of book, *GW* xv, nr. 424. See also Ulrich Klein, "Johannes Keplers Bemühungen um die Harmonieschriften des Ptolemaios und Porphyrios," pp. 51-60 in *Johannes Kepler Werk und Leistung* (Linz, 1971).

266:9 *Harmonice mundi, GW* vi, 289:20-22.

266:15 *Harmonice mundi, GW* vi, 289:29-31.

266:19 *Harmonice mundi, GW* vi, 289:35-39.

266:34f K to Fabricius, 1 October 1602, *GW* xiv, nr. 239; K to Heydon, May 1605, *GW* xv, nr. 357; K to Brengger, 5 April 1608, *GW* xvi, nr. 488; K to Tanckius, 12 May 1608, *GW* xvi, nr. 493.

267:23 Quoted without citation by Caspar in *GW* vi, 480.

267:27f *Harmonice mundi, GW* vi, 290:1-9.

269:15 *Harmonice mundi, GW* vi, 212:29-32, "They do not have the relation they are said to have unless the presence of some mind is assumed to relate one to another."

269:21 *Harmonice mundi, GW* vi, 215:30-33.

270:2 *Harmonice mundi, GW* vi, 226:22f.

270:6 *Harmonice mundi, GW* vi, 277:11-12.

270:13 *Harmonice mundi, GW* vi, 216:1f.

271:3f *Harmonice mundi, GW* vi, 223:26-35.

271:25 *Harmonice mundi, GW* vi, 55:14f.

272:12 *Harmonice mundi, GW* vi, 224:12f.

273:10 *Harmonice mundi, GW* vi, 16:35-38.

273:22 *Harmonice mundi, GW* vi, 16-19.

273:34 *Harmonice mundi, GW* vi, 47f.

274:7   *Harmonice mundi, GW* vi, 63:2.

274:36   *Harmonice mundi, GW* vi, 88, and see plate on p. 79.

275:31   *Harmonice mundi,* II.2, *GW* vi, 114-18.

276:11   *Harmonice mundi, GW* vi, 118.

277:6   *Harmonice mundi, GW* vi, 120:1f.

277:23f *Harmonice mundi, GW* vi, 114:10-15.

278:27f *Harmonice mundi,* especially IV.7, *GW* vi, 264f.

278:34f *Harmonice mundi, GW* vi, 241:5-9.

279:1f   *Harmonice mundi, GW* vi, 277:26-32.

279:22f *Harmonice mundi, GW* vi, 280:8-32.

280:11f *Harmonice mundi, GW* vi, 271:21-24.

280:31   *Harmonice mundi, GW* vi, 239:22-23.

281:32   *Harmonice mundi, GW* vi, 261:25-26.

282:16f *Harmonice mundi, GW* vi, 330:13-23.

283:16   *Harmonice mundi, GW* vi, 314:2-3.

283:25f *Harmonice mundi, GW* vi, 316:29-34.

283:30f *Harmonice mundi,* V.5, *GW* vi, 317f.

284:9f   *Harmonice mundi, GW* vi, 328:20-25.

284:21f *Harmonice mundi, GW* vi, 320:16-19.

284:26f *Harmonice mundi, GW* vi, 328:25-31.

285:18f *Harmonice mundi, GW* vi, 354:4-9.

285:42   *Harmonice mundi, GW* vi, 296.

286:3f   *Harmonice mundi, GW* vi, 302:9-22.

288:1   *Harmonice mundi, GW* vi, 359:30-31.

288:8   *Harmonice mundi, GW* vi, 368.

288:24   K to Wacker von Wackenfels, early 1618, *GW* xvii, 783:36; Schickard to K, 22 March 1618 OS, *GW* vii, nr. 787:8; Schickard to K, 15/25 October 1608, *GW* xvii, nr. 803:5f.

289:35   *Tertius interveniens, GW* iv, 245:40-246:1

289:38   K to Herwart von Hohenburg, 12 July 1600, *GW* xiv, nr. 168:108-109.

290:19f *Appendix* to the *Harmonice, GW* vi, 369-77.

290:26   *Appendix* to the *Harmonice, GW* vi, 374:36.

290:30   Robert Fludd, *Veritatis proscenium* (Frankfurt, 1621).

291:30   *Bibliographia Kepleriana,* 79.

292:3f   *Pro suo Opere harmonices mundi apologia, GW* vi, 379-457.

292:11   *Pro suo Opere harmonices mundi apologia, GW* vi, 446:24-25.

292:16f *Pro suo Opere harmonices mundi apologia, GW* vi, 445:32f.

292:21f *Discurs von der grossen Conjunction 1623, Fr* vii, 710.

292:29 K to Philipp Müller, after 13 September 1622, *GW* xviii, nr. 938:78-79.

292:31 K to Crüger, 28 February 1624, *GW* xviii, nr. 974:285-88.

292:38f *Appendix* to the *Harmonice, GW* vi, 374:18-22.

293:5 K to Seussius, 15 July 1622, *GW* xviii, nr. 934:30.

293:9f *Epitome astronomiae Copernicanae, GW* vii.

293:26 K to Bernegger, 11/21 August 1621, *GW* xviii, nr. 919:3-7; K to Bernegger, 5/15 February 1621, *GW* xviii, nr. 909:33-34.

293:42 *Fr* viii, 878.

294:1 *GW* xix, nr. 7.92.

294:5 *Fr* viii, 878f.

294:7 K to Remus Quietanus, 31 August 1619, *GW* xvii, 850:364-65.

294:33f See *Epitome astronomiae Copernicanae*, Book I, Part V, "On the Earth's Daily Motion," especially near the end.

295:6f *Epitome astronomiae Copernicanae*, Books IV and V; note the English translation of most of these sections in *Great Books of the Western World*, xvi, pp. 843-1004. See Bruce Stephenson, *Kepler's Physical Astronomy*, (New York, 1987), ch. 4.

296:26f *Epitome astronomiae Copernicanae, GW* vii, 267f.

297:31 Junius' very rare work is cited in *BK*, 126.

298:10 *Epitome astronomiae Copernicanae, GW* vii, 8:26-28.

298:17 *Epitome astronomiae Copernicanae, GW* vii, 9:10-12.

298:25f Remus Quietanus to K, 23 July 1619, *GW* xvii, nr. 845:10-12.

298:31 K to Remus Quietanus, 4 August 1619, *GW* xvii, nr. 846:22-30.

298:36 Remus Quietanus to K, 13 August 1619, *GW* xvii, nr. 848:14-15.

298:41 Bianchi to K, 20 January 1619, *GW* xvii, nr. 825:109-14.

299:6f *Harmonice mundi, GW* vi, 543-44.

299:20 *Mysterium cosmographicum, GW* viii, 40:5-6.

299:38f *Astronomia nova, GW* iii, 29f, and *Epitome, GW* vii, 99-100.

300:7 K to Herwart von Hohenburg, 28 March 1605, *GW* xv, nr. 340:77-81.

300:15 *Ephemerides, GW* xi,1.

300:26 Until the texts of all six Linz calendars are printed in *GW* xi,2, four of them may be at least partially found as follows: 1619, *Fr* i, 479-94; 1620, *NK* vii; 1621, *Fr* viii, 3-20; 1623, partly in *Fr* vii, 685-713; 1624, *NK* i.

301:1 *Fr* i, 483.

301:3 *Fr* i, 482.

301:17 *Bericht vom Kometen, GW* iv, 57-76.

301:20 *Bericht vom Kometen, GW* iv, 75:5-8.

301:25 *De cometis libelli tres, GW* viii, 131-262.

301:41 *De cometis libelli tres, GW* viii, 225:1f.

302:3 *De cometis libelli tres, GW* viii, 225:18-20.

302:7 *De cometis libelli tres, GW* viii, 226:13f.

302:19f *De cometis libelli tres, GW* viii, 230:6-19.

302:28 *De cometis libelli tres, GW* viii, 230:25-30.

302:37 *De cometis libelli tres*, *GW* viii, 239:11.
302:41f *De cometis libelli tres*, *GW* viii, 239:8-10.

303:9   *De cometis libelli tres*, *GW* viii, 233:33-36.
303:12f *Bericht vom Kometen*, *GW* iv, 64:2-7.
303:29 *De cometis libelli tres*, *GW* viii, 260:37-40.
303:33 *De cometis libelli tres*, *GW* viii, 261:33.
303:37 *De cometis libelli tres*, *GW* viii, 240:1-6.

304:8   *Hyperaspistes*, *GW* viii, 265-412.
304:12 *Appendix*, *GW* xviii, 413-425.

305:23 *Ephemerides novae*, *GW* xi,1, 11:20.
305:26f *Mysterium cosmographicum*, *GW* viii, 11:23-26.
305:32f *Mysterium cosmographicum*, *GW* viii, 11:33-35.
305:36 *Mysterium cosmographicum*, *GW* viii, 12:3.
305:40f *Epitome astronomiae Copernicanae*, *GW* vii, 9:10-12.

306:6f *Mysterium cosmographicum*, *GW* viii, 12:14-18.
306:16f K to Crüger, 15 July 1623, *GW* xviii, 955:13-17.

307:5f *Glaubenbekenntnis*, *GW* xii, 26:45-27:4.
307:29 K to Schickard, 9/19 April 1627, *GW* xviii, nr. 1042:74-75.
307:35 Schickard to K, 25 February 1624 OS, *GW* xviii, nr. 975:362f.
307:38 Crüger to K, 15 July 1624 OS, *GW* xviii, nr. 990:362f.
307:40 *Prognosticum* for 1624, *NK* i, 54.
307:41 K to Crüger, 9 September 1624, *GW* xviii, nr. 993:576-80.

308:10 K to Bernegger, 4 December 1623, *GW* xviii, nr. 963:33-34.
308:15 Notes to a horoscope for Wallenstein, *Fr* viii, 358.
308:18f K to Upper Austrian Estates, 9 May 1616, *GW* xvii, nr. 734:66.
308:22f K to Bianchi, 17 February 1619, *GW* xvii, nr. 827:249-51.
308:28 K to Schickard, 25 April 1626, *GW* xviii, nr. 1025:52-53.

309:10 *Supplementum Chiliadis logarithmorum*, *GW* ix, 355.
309:30f Schickard to K, 20 September 1623 OS, *GW* xviii, nr. 962.

310:1   K to Landgrave Philipp, December 1623, xviii, nr. 968:148.
310:9   K to Gunther, 4 December 1623, *GW* xviii, nr. 964:59.
310:12f K to Bernegger, 20 May 1624, *GW* xviii, nr. 983:24-26.
310:38 *GW* xix, 190-93.

311:8   *GW* xix, 193-97.
311:11 *GW* xix, 189-90; 198-99.
311:18 *GW* xix, 202-3.
311:24 *GW* xix, 200.
311:31f K to Bernegger, 20 May 1624, *GW* xviii, nr. 983
311:40 K to Bernegger, 20 May 1624, *GW* xviii, nr. 983:36.

312:3   K to Georg Brahe, 25 January 1626, *GW* xviii, nr. 1023:32-33.
312:6   Hebenstreit to K, 11 May 1624 OS, *GW* xviii, nr. 984.

312:18f *Fr* viii, 889-90.
312:29 *GW* xix, 144, nr. 3.42.
312:33 *GW* xix, 199-202, 204, 211-12.

313:3 *GW* xix, 197-214, esp. 203.
313:7f *Fr* viii, 890.
313:17 *Fr* viii, 890; *GW* xix, 375, nr. 7.113.
313:27f K to Guldin, 1 October 1626, *GW* xviii, nr. 1031:22f.

314 The details in these two paragraphs are paraphrased from K to Guldin, August 1625, *GW* xviii, nr. 1014.

315:18f K to Georg Brahe, 25 January 1626, *GW* xviii, nr. 1023:80.
315:24 K to Commissioners, middle of 1627, *GW* xix, 221 (first lines of that page).
315:27 K to Guldin, 1 October 1626, *GW* xviii, nr. 1031:36-37.

316:4f *GW* xix, 145-47, nr. 3.45.
316:29f K to Guldin, 7 February 1626, *GW* xviii, nr. 1024, 100-101.

317:1 K to Guldin, 7 February 1626, *GW* xviii, nr. 1024.
317:5 K to Guldin, 7 February 1626, *GW* xviii, nr. 1024:51-53.
317:10 Rector of Tübingen Senate to Duke of Württemberg, 6 October 1626 OS, *Fr* viii, 900.
317:20 K to Guldin, 7 February 1626, *GW* xviii, nr. 1024:28-29.
317:24 K to Guldin, 7 February 1626, *GW* xviii, nr. 1024:31; the rest of the page is from this letter.

318:4 Crüger to K, 15 July 1624 OS, *GW* xviii, nr. 990:211-14.
318:7 K to Crüger, 1 May 1626, *GW* xviii, nr. 1026:56-60.

319:18f K to Bernegger, 8 February 1627, *GW* xviii, nr. 1036:7-8.
319:20f *Responsio ad epistolam Bartschii, GW* xi,1, 470:4-5.
319:28f K to Guldin, 1 October 1626, *GW* xviii, nr. 1031:38f; the rest of the page is from this letter.

321:34f K to Brenegger, 8 February 1627, *GW* xviii, nr. 1036:13-14.

322:7f K to Landgrave Georg II, November 1627, *GW* xviii, nr. 1066:30-31.
322:18 Hebenstreit to K, 1 October 1625 OS, *GW* xviii, nr. 1019:46.
322:36 *GW* xix, 382-83, nr. 7.126.

323:2f K to Schickard, 10 February 1627, *GW* xviii, nr. 1037:29f; the rest of the paragraph is from this letter.
323:28 K to Schickard, 10/20 March 1627, *GW* xviii, nr. 1039:5-7.
323:35 K to Schickard, 19 April 1627, *GW* xviii, nr. 1042:42-48.
323:41f K to Georg Brahe, 25 January 1626, *GW* xviii, nr. 1023.

324:3 *GW* xix, 216-17, nr. 5.15, 16 May 1627.
324:5f *GW* xix, 220-22, nr. 5.19.
324:16f See K to Schleicher, 14 November 1627, *GW* xviii, nr. 1061; *GW* xix, 382, nr. 7.125.

325:8f K to Bernegger, 2 October 1627, *GW* xviii, nr. 1056.

325:17 K to Bernegger, 6 April 1627, *GW* xviii, nr. 1040:38-40.

326:2 Response from Tampach, 7 April 1628, *GW* xix, 226, nr. 5.23.

326:9 *Tabulae Rudolphinae, GW* x.

326:33 K to Upper Austrian Estates, 9 May 1616, *GW* xvii, 734:44.

327:7 See Ernst Zinner (Ezra Brown, transl.), *Regiomontanus: His Life and Works,* (Amsterdam, 1990), pp. 119-25.

327:17 Fragments of the unfinished "Hipparchus" are found in *GW* xx,1, 181-268.

327:41 While only ten columns are visible, the introductory poem describes twelve (for the zodiacal signs); see Owen Gingerich, "Johannes Kepler and the *Rudolphine Tables,"* *Sky and Telescope,* xlii (1971), 328-33.

328:29 K to Bernegger, 8 February 1627, *GW* xviii, nr. 1036:21-24.

328:37 K to Bernegger, 6 April 1627, *GW* xviii, nr. 1040:22-27.

329:6f *Tabulae Rudolphinae, GW* x, 39-40; see the translation of this introduction by Owen Gingerich and William Walderman, *Quarterly Journal of the Royal Astronomical Society,* xiii (1972), 360-73.

329:27 Bernegger to K, 23 February 1627 OS, *GW* xviii, nr. 1038:32-34.

329:35 Bernegger to K, 7/17 October 1627, *GW* xviii, nr. 1057:29-32.

330:1f K to Schickard, 10 February 1627, *GW* xviii, nr. 1037:60-62.

330:8 See *Fr* viii, 909; *Chilias logarithmorum, GW* ix, 275-426.

330:9f *Responsio ad epistolam... Bartschii, GW* xi.1, 469f.

330:20 K to Landgrave Georg II, November 1627, *GW* xviii, nr. 1066:18-19.

331:5 K to Landgrave Georg II, November 1627, *GW* xviii, nr. 1066:4f

331:8f K to Landgrave Georg II, November 1627, *GW* xviii, nr. 1066:4f

331:22 See K to Bernegger, 5/15 April 1628, *GW* xviii, nr. 1082.

331:31 Slightly different dates are given in *Fr* viii, 909.

331:41 *R.P. Joannis Terrentii Epistolium cum Commentatiuncula Joannis Keppleri, Fr* vii, 667-83 (to be printed in *GW* xi,2)

332:2 See *GW* xix, 225.

334:1f K to Elector Johann Georg of Saxony, 19/29 February 1628, *GW* xviii, nr. 1073.

334:13 Upper Austrian Estates to K, 20 October 1627, *GW* xviii, nr. 1058.

334:35 Curtius (Kurz) to K, 10 June 1627, *GW* xviii, nr. 1048; 25 June 1627, *GW* xviii, nr. 1050; 26 August 1627, *GW* xviii, nr. 1053.

334:40 Guldin to K, no longer extant.

334:41 K to Guldin, 24 February 1628, *GW* xviii, nr. 1072.

335:5f K to Guldin, 24 February 1628, *GW* xviii, nr. 1072:39-44.

335:14f K to Guldin, 24 February 1628, *GW* xviii, nr. 1072:45-49.

335:21 K to Guldin, Spring 1628, *GW* xviii, nr. 1083:85-86.

335:23 K to Guldin, Spring 1628, *GW* xviii, nr. 1083:87-90.

335:36 K to Guldin, 24 February 1628, *GW* xviii, nr. 1072:61-62.

336:3f K to Guldin, 24 February 1628, *GW* xviii, nr. 1072:65f.

336:22f K to Guldin, 24 February 1628, *GW* xviii, nr. 1072:104-110.

337:1   Upper Austrian Estates to K, 20 October 1627, *GW* xviii, nr. 1058.

337:8f   K to Guldin, 24 February 1628, *GW* xviii, nr. 1072:114-133.

337:29   Guldin to K, 29 March 28, *GW* xviii, nr. 1080.

337:34   K to Guldin, Spring 1628, *GW* xviii, nr. 1083.

338:9   K to Guldin, Spring 1628, *GW* xviii, nr. 1083:37-38.

338:20f   See Francis Watson, *Wallenstein: Soldier under Saturn* (New York, 1938) and also Angelika Geiger, *Wallensteins Astrologie* (Graz, 1983), pp. 88-110 and 395-421.

339:14   *Fr* viii, 348.

339:26   *Fr* i, 387, near the bottom of the page.

339:37f   *Fr* i, 388-89.

340:37   Gerhard von Taxis to K, 14 December 1614, *GW* xvii, nr. 704; 1 May 1615, *GW* xvii, nr. 717.

341:13   Gerhard von Taxis to K, 20 November 1624, *GW* xviii, nr. 999; 16 December 1624, *GW* xviii, nr. 1000.

341:23f   *Fr* viii, 351-52.

342:3f   *Fr* viii, 352.

342:16f   *Fr* viii, 357.

342:25   *Fr* viii, 355.

342:35   Gerhard von Taxis to K, 25 September 1625, *GW* xviii, nr. 1016:18-20.

343:6f   *Responsio ad epistolam Bartschii, GW* xi,1, 470:23-31.

343:26f   K to Elector Johann Georg of Saxony, 19/29 February 1628, *GW* xviii, nr. 1073:46-57.

344:6f   *GW* xix, 165, nr. 4.1.

344:11   *GW* xix, 165, nr. 4.2.

344:17   See *Fr* viii, 909.

345:12   K to Bernegger, 5/15 April 1628, *GW* xviii, nr. 1082:20-22.

345:27f   K to Upper Austrian Estates, 11 February 1628, *GW* xviii, nr. 1069.

345:36   *GW* xix, 149-50, nr. 3.52, 3 July 1628.

345:40   *Responsio ad epistolam Bartschii, GW* xi,1, 470:31-38.

346:2   K to Georg Brahe, 17 August 1628, *GW* xviii, nr. 1088:2-3 (in fact, they arrived on the 26th).

346:10f   K to Bernegger, 12/22 July 1629, *GW* xviii, nr. 1111:116-17.

346:22f   K to Bernegger, 2 March 1629, *GW* xviii, nr. 1102:3-6.

346:29   Müller to K, 3 August 1622 OS, *GW* xviii, nr. 936:1-3.

347:8   Wallenstein to Becker, 7 April 1630, *GW* xviii, nr. 1131.

347:15   K to Bernegger, 12/22 July 1629, *GW* xviii, nr. 1111:42-48, 59-79.

347:30f   K to Bernegger, 12/22 July 1629, *GW* xviii, nr. 1111:76-79.

348:17   K to Gerhard von Taxis, 10 October 1629, *GW* xviii, nr. 1115:71-84.

348:34f   K to Müller, 17/27 October 1629, *GW* xviii, nr. 1116:2-9.

349:1   *Responsio ad epistolam Bartschii, GW* xi,1, 473:43-46.

349:10   K to Gerhard von Taxis, 10 October 1629, *GW* xviii, nr. 1115:31-42.

349:13 K to Remus Quietanus, 2 March 1629, *GW* xviii, nr. 1103:28-29.

349:15 K to Wallenstein, 24 February 1629, *GW* xviii, nr. 1101:44-52.

349:21 K to Müller, 17/27 October 1629, *GW* xviii, nr. 1116:10-24.

349:26 K to Müller, 16/26 January 1630, *GW* xviii, nr. 1122:7-11; see also *GW* xix, 173, nr. 4.23.

349:35 *Sportula genethliacis missa de Tabularum Rudolphi usu* included in the *Tabulae Rudolphinae*, *GW* x, 244-54.

349:37f "elderly mother and thankless daughter": images frequently used by Kepler—see for example in *Tertius interveniens*, *GW* iv, 161:10-17 or *De stella nova*, *GW* i, 211:6-13.

349:40 *GW* xix, 168, nr. 4.8, 25 May 1629, 168; nr. 4.9, etc.; see also K to Wallenstein, April 1630, *GW* xviii, nr. 1133.

349:42 *GW* xix, 177, nr. 4.34, 10 March 1630; 178, nr. 4.37, 27 March 1630.

350:1 K to Müller, 4 January 1630, *GW* xviii, nr. 1120:39-43.

350:31 *R.P. Joannis Terrentii Epistolium cum Commentatiuncula Joannis Keppleri*, *Fr* vii, 667-83 (to be printed in *GW* xi,2); see Kurz to K, 3 September 27, *GW* xviii, nr. 1055:10-12 and Kepler to Müller, 16/26 January 1630, nr. 1122:18-21.

350:35 K to Müller, 16/26 January 1630, *GW* xviii, nr. 1122:21-18.

350:36 K to Müller, 26 August 1630, *GW* xviii, nr. 1137:21-33.

350:37 K to Wallenstein, 18 October 1630, *GW* xviii, nr. 1144.

350:40 *Ephemeridum pars secunda*, *GW* xi,1, 137-38.

350:42 *Ephemeridum pars tertia*, *GW* xi,1, 303.

351:1 *Ephemeridum pars secunda*, *GW* xi,1, 135: *Sagani Silesiorum, In Typographeio Ducali, Sumptibus Authoris.*

351:2 On Tampach's role, see *BK*, 94-95 and K to Müller, 26 August 1630, *GW* xviii, nr. 1137:21-24.

351:9 The Latin text of the *Somnium* will be included in *GW* xi,2; meanwhile, there is a facsimile reprint by M. List and W. Gerlach (Osnabrück, 1969) and the text in *Fr* viii, 27-123. There is a good English translation by E. Rosen (Madison, 1967).

351:11 *Somnium*, note 2, p. 29 (Rosen, translation, p. 32 and Appendix, pp. 107-208); *Dissertatio cum Nuncio sidereo*, *GW* iv, 299:5f.

351:17 *Dissertatio cum Nuncio sidereo*, *GW* iv, 297:37f.

351:33 K to Bernegger, 4 December 1623, *GW* xviii, nr. 963:28.

352:1-2 K to Bernegger, 2 March 1629, *GW* xviii, nr. 1102:45-48.

352:8 *Somnium*, p. 1 (Rosen, translation, p. 11).

352:9 *Somnium*, note 3, p. 30 (Rosen, p. 35).

352:14 *Somnium*, p. 5 (Rosen, pp. 12-13).

352:17 *Somnium*, p. 5 (Rosen, p. 15).

352:26 *Somnium*, pp. 8-16 (Rosen, pp. 17-22).

352:36 *Somnium*, note 96, p. 46 (Rosen, p. 82).

353:3 *Somnium*, pp. 97-184 (not translated by Rosen).

353:8 On Jacob Bartsch, see E. Rosen, *Somnium*, Appendix A, 177-93.

353:15 *Responsio ad epistolam Bartschii*, *GW* xi,1, 470:26-29.

353:20 Bartsch to K, end of April 1625, *GW* xviii, nr. 1004.

353:26 Bartsch to Müller, 30 April 1629, *GW* xviii, nr. 1107:1-14.

353:35 K to Bernegger, 5/15 April 1628, *GW* xviii, nr. 1082:15-16; *GW* xix, 386 nr. 7. 136, 8 June 1628.

353:41 K to Bernegger, 10 April 1629, *GW* xviii, nr. 1105.

354:1 K to Bernegger, 10 April 1629, *GW* xviii, nr. 1105:22-24.

354:5 K to Bernegger, 10 April 1629, *GW* xviii, nr. 1105:20-21.

354:22 Bernegger to K, 12/22 March 1630, *GW* xviii, nr. 1127:12-46.

354:36 K to Müller, 22 April 1630, *GW* xviii, nr. 1134:4.

356:20 K to Wallenstein, 18 October 1630, *GW* xviii, nr. 1144:14-15.

356:31f *GW* xix, 398, nr. 7:155; 129, nr. 3:10; 130, nr. 3:12; 150, nr. 3:53; etc.

357:28 Bartsch to Müller, 3 January 1631, *GW* xix, 231, nr. 6.1.

357:40 *Fr* viii, 912.

358:4 K to Bernegger, 21 October 1630, *GW* xviii, nr. 1145, in reply to Bernegger to K, 8 September 1630 OS, *GW* xviii, nr. 1142.

358:10f K to Bernegger, 21 October 1630, *GW* xviii, nr. 1145:34-35.

358:20 Lansius to anonymous (an unknown friend in Tübingen), 24 January 1631 OS, *GW* xviii, nr. 1146:8f. (= *GW* xix, 236-37, nr. 6.6).

358:24 Lansius to anonymous, 24 January 1631 OS, *GW* xviii, nr. 1146:9 (= *GW* xix, 236-37, nr. 6.6:8-10).

358:31 Lansius to anonymous, 24 January 1631 OS, *GW* xviii, nr. 1146:32-35 (= *GW* xix, 237, nr. 6.6:8-10).

358:36 *GW* xix, 7:147, 7/17 November 1630,.

358:37 Bartsch to Wallenstein, 9 December 1630, *GW* xix, 182, nr. 4.48.

358:38 See Schickard to Bernegger, 15 November 1630, *GW* xix, 394, nr. 7.152; *Bartschii Appendix...*, *GW* x, 257:35-36.

359:2 Lansius to anonymous, 24 January 1631 OS, *GW* xviii, nr. 1146:22 (= *GW* xix, 237, nr. 6.6).

359:11 *GW* xix, 393, nr. 7.149; *Tabulae Rudolphinae*, *GW* x, 258:6-8.

359:23 Fischer's letter is mentioned in Lansius' letter (see our note 358:20).

359:26 *GW* xix, 360, nr. 7.90, 2 January 1620.

359:33 *GW* xix, 394, nr. 7.151.

359:41 *GW* xix, 394, nr. 7.150.

360:9f Lansius to anonymous, 24 January 1631 OS, *GW* xviii, nr. 1146:27-32.

360:15 Lansius to anonymous, 24 January 1631 OS, *GW* xviii, nr. 1146:21-22.

361:31f Bartsch to Müller, 3 January 1631, *GW* xix, 231, nr. 6.1.

362:1 Susanna Kepler on 10 January 1631, *GW* xix, 233, nr. 6.2.

362:4 Bartsch to Wallenstein, 9 December 1630, *GW* xix, 183-85, nr. 4.49.

362:15 Susanna Kepler on 10 January 1631, *GW* xix, 233, nr. 6.2.

362:23 *GW* xix, 238, nr. 6:7; 240, nr. 6:9; 242, nr. 6.10.

362:27 Bernegger to Bartsch, 16 January 1631 OS, *GW* xix, 233-34, nr. 6.5.

362:35 Bartsch to Müller, 5 May 1631, *GW* xix, 237, nr. 6.7.

362:38 Bartsch to Müller, 19/29 June 1631, *GW* xix, 240-41, nr. 6.9.

363:7   Bartsch to Müller, 3 September 1631, *GW* xix, 242:3-7, nr. 6.10.

363:12   Wallenstein to Stosch, 21 September 1631, *GW* xix, 242, nr. 6.11.

363:20   Inventory made on 13 December 1630 OS: *GW* xix, 395-99, 7.155.

363:23   *GW* xix, 153-56, nr. 3.62-69.

363:28   *GW* xix, 244-45, nr. 6.14, between 8-20 June 1632.

363:32   *GW* xix, 253-55, nr. 6.24, 27 April 1633.

363:34   *GW* xix, 303, nr. 6.62, 8 April 1717.

363:38   *GW* xix, 308-9, nr. 6.68, 9 July 1717.

363:40   *GW* xix, 309-10, nr. 6.69, 31 July 1717.

364:1   *GW* xix, 265, nr. 6.35, precise date unknown.

364:6   Crüger to Müller, 10 April 1635, *GW* xix, 259, nr. 6.30.

364:9   *Somnium*, preface of L. Kepler, unnumbered folio (= Rosen, p. 6-7).

364.11   *GW* xix, 297, nr. 6.57.

364:20   *GW* xix, 261, nr. 6.32.

364:22   *GW* xix, 251, nr. 6.20; Fridmar died 19 February 1633 at Lauban, at age 11.

364:25   *GW* xix, 263, nr. 6.34; Frau Kepler died 30 August 1636 at Regensburg; she was 40, not 47.

364:27   *GW* xix, 288, nr. 6.49, 6 August 1638.

364:31   *GW* xix, 293-94, nr. 6.52-53; Cordula died 29 June 1654 in Vienna at age 30. Her children: *GW* xix, 481.

364:35   *GW* xix, 541—see in the index *Kepler, Susanna,* Tochter; and see p. 481.

364:36   *GW* xix, 296-97, nr. 6.57.

365:1   Crüger to Müller, 10 April 1635, *GW* xix, 259, nr. 6.30

365:6   Scheiner to Tengnagel, 4 November 1621, *GW* xix, 362-63, nr. 7.98.

365:14   L. Kepler to Galilei, 6 February 1638, *GW* xix, 283-85, nr. 6.46.

365:33   *GW* xix, 291-92, nr. 6.51, June 1644.

365:38   On the history of Kepler's literary legacy, see *BK,* 104-106.

366:23   On Gottlieb von Murr's efforts, see *BK,* 106.

368:9   Persius, *Sat.* I 1, quoted many times by Kepler—see, for example, K to Herwart von Hohenburg, 9/10 April 1599, *GW* xiii, nr. 117:31; it was a favorite quotation for inscribing in friends' autograph albums, such as *GW* xix, 324, nr. 7.21, 22 July 1596; 339, nr. 7.38, 5 February 1602; 345, nr. 7.55, 19 May 1609 OS; 346, nr. 7.56, 20 June 1609 OS; 349, 7.62, 13 June 1611; 351, 7.68, 3 May 1612; 351, nr. 7.69, 17 July 1612; 355, nr. 7.75, 20 September 1615; 357, nr. 7.80, 31 Oct 1617; 360 7.90, 2 January 1621; 360, nr. 7.91, 2 January 1621; 361, nr. 7.94, 13 April 1621; 375, nr. 7.112, 30 April 1625; 377, nr. 7.116, 1 July 1625; 384, nr. 7.131, 21 November 1627; 392, nr. 7.145, 14-21 October 1630; etc. Variations are found in *GW* xix, 376, nr. 7.114, 5 May 1625, against alchemy; 378, nr. 7.119, 14 August 1625, against astrology, 382, nr. 7.124, 5 April 1627.

369:9   Valesius to K, 5 August 1614, *GW* xvii, 693:7-9.

369:19   *GW* xix, 336, 7.30; most of the details of this page are taken from the famous "Selbstcharakteristik," Graz, 1597.

369:27   *Paralipomena in Vitellionem, GW* ii, 182:13-14; *De stella nova, GW* i, 221:4-5.

371:8   *Epitome astronomiae Copernicanae, GW* vii, 8:42-45.

371:16 K to an anonymous physician, 7 May 1607, *GW* xv, nr. 425:61-62.

371:22 K to Herwart von Hohenburg, 16 December 1598, *GW* xiii, nr. 107:198.

372:7   *Epitome astronomiae Copernicanae, GW* vii, 255:36-37.

372:14f *Dissertatio cum Nuncio sidereo, GW* iv, 287:9-13.

372:21f K to V. Bianchi, 13 April 1616, *GW* xvii, nr. 729:17-23.

372:31f Same thought in a letter to Bruce, 4 September 1603, *GW* xiv, nr. 268:7-12.

372:37 K to Magini, 1 June 1601, *GW* xiv, nr. 190:450-52.

372:40 Bruce to K, 21 August 1603, *GW* xiv, nr. 265:19-21.

372:41 K to Bruce, 4 September 1603, *GW* xiv, nr. 268:12-19.

373:2f   K to Bruce, 4 September 1603, *GW* xiv, nr. 268:12-19.

373:7f   *Sylva chronologica* (part IV of *De stella nova*), *GW* i, 361:4-7.

373:22f *Astronomia nova, GW* iii, 33:17-26.

374:23 *Mysterium cosmographicum, GW* i, 6:25-26.

374:37 K to an anonymous nobleman, 23 October 1613, *GW* xvii, nr. 669:19f.

374:40 K to Maestlin, 3 October 1595, *GW* xiii, nr. 23:258.

375:5f   *Harmonice mundi, GW* vi, 362:40-363:11.

375:19f *Harmonice mundi, GW* vi, 368:18-22.

375:36f K to Maestlin, 3 October 1595, *GW* xiii, nr. 23:256-58.

375:40 K to Herwart von Hohenburg, 26 March 1598, *GW* xiii, nr. 91:182f.

375:41 *Epitome astronomiae Copernicanae, GW* vii, 9:10-16.

376:18 *Responsio ad epistolam... Bartschii, GW* xi,1, 471:28-29.

376:20f *Prognosticum auff das Jahr 1604, NK* i,15.

376:25 *Ephemerides novae, GW* xi,1, 11:20.

376:28f K to anonymous, 18 December 1610, *GW* xvi, nr. 600:266-72.

377:12 Copernicus, *De revolutionibus*, author's preface to Pope Paul III.

377:15 Rheticus, *Narratio prima, GW* i, 105:120f.

378:4f   K to Herwart von Hohenburg, 9/10 April 1599, *GW* xiii, nr. 117:295-96.

378:23 *Wisdom* 11:21, quoted for instance in *Harmonice mundi, GW* vi, 81:22.

378:29f *Tertius interveniens, GW* iv, 246:23-24.

378:32 *Mysterium cosmographicum, GW* i, 23:20f.

379:1   *Epitome astronomiae Copernicanae, GW* vii, 47:30f.

379:2   *Mysterium cosmographicum, GW* i, 23:14-23.

379:4   *Mysterium cosmographicum, GW* i, 23:20f, 25:1f; *Paralipomena in Vitellionem, GW* ii, 19:8f; *Harmonice mundi, GW* vi, 224:12f; *Epitome astronomiae Copernicanae, GW* vii, 51:1f, etc.

379:26 *Paralipomena in Vitellionem, GW* ii, 19:11-12.

380:7f   K to Herwart von Hohenburg, 9/10 April 1599, *GW* xiii, nr. 117:174-79.

380:15 *Dissertatio cun Nuncio sidereo, GW* iv, 308:9-10.

380:22 *De stella nova, GW* i, 285:34-35.

380:38 *Harmonice mundi, GW* vi, 218:20-21.

380:39 *Harmonice mundi, GW* vi, 364:38f (and passim).

380:40f *Harmonice mundi, GW* vi, 223:32-35.

381:5   *Prognosticum auff 1604,* p. A4v, *NK* i, 15; *GW* xii,2 (in press).

381:16   *Epitome astronomiae Copernicanae, GW* vii, 208:19-29.

383:1   *Mysterium cosmographicum, GW* viii, 62:29-63:9.

383:17   *Mysterium cosmographicum, GW* i, 29:28f.

383:36   *Harmonice mundi, GW* vi, 264-86; 4:25-26; *GW* xv, K to Wilhelm von Neuburg; 21 February 1605, nr. 332:111-63, etc.

384:2   *Astronomia nova, GW* iii, 67:13-68:25.

384:28   *Epitome atronomiae Copernicanae, GW* vii, 8:26-29.

385:18   *Astronomiae pars optica, GW* ii, 253:8-27; *Dissertatio cun Nuncio sidereo, GW* iv, 289:14-23.

385:22   *De stella nova, GW* i, 253:15.

387:36   *Astronomia nova, GW* iii, 141:3-4; Leibniz' text is quoted by I. B. Cohen in *Vistas in Astronomy,* xviii (1975), 12-13.

INDEX OF NAMES

Index of Names

434

INDEX OF SUBJECTS AND PLACES

Bold face designates principal entries or the beginning of a major section.

A CATALOG OF SELECTED DOVER
BOOKS IN ALL FIELDS OF INTEREST

CONCERNING THE SPIRITUAL IN ART, Wassily Kandinsky. Pioneering work by father of abstract art. Thoughts on color theory, nature of art. Analysis of earlier masters. 12 illustrations. 80pp. of text. 5⅜ × 8½. 23411-8 Pa. $3.95

ANIMALS: 1,419 Copyright-Free Illustrations of Mammals, Birds, Fish, Insects, etc., Jim Harter (ed.). Clear wood engravings present, in extremely lifelike poses, over 1,000 species of animals. One of the most extensive pictorial sourcebooks of its kind. Captions. Index. 284pp. 9 × 12. 23766-4 Pa. $11.95

CELTIC ART: The Methods of Construction, George Bain. Simple geometric techniques for making Celtic interlacements, spirals, Kells-type initials, animals, humans, etc. Over 500 illustrations. 160pp. 9 × 12. (USO) 22923-8 Pa. $9.95

AN ATLAS OF ANATOMY FOR ARTISTS, Fritz Schider. Most thorough reference work on art anatomy in the world. Hundreds of illustrations, including selections from works by Vesalius, Leonardo, Goya, Ingres, Michelangelo, others. 593 illustrations. 192pp. 7⅛ × 10¼. 20241-0 Pa. $8.95

CELTIC HAND STROKE-BY-STROKE (Irish Half-Uncial from "The Book of Kells"): An Arthur Baker Calligraphy Manual, Arthur Baker. Complete guide to creating each letter of the alphabet in distinctive Celtic manner. Covers hand position, strokes, pens, inks, paper, more. Illustrated. 48pp. 8¼ × 11.
24336-2 Pa. $3.95

EASY ORIGAMI, John Montroll. Charming collection of 32 projects (hat, cup, pelican, piano, swan, many more) specially designed for the novice origami hobbyist. Clearly illustrated easy-to-follow instructions insure that even beginning papercrafters will achieve successful results. 48pp. 8¼ × 11. 27298-2 Pa. $2.95

THE COMPLETE BOOK OF BIRDHOUSE CONSTRUCTION FOR WOOD-WORKERS, Scott D. Campbell. Detailed instructions, illustrations, tables. Also data on bird habitat and instinct patterns. Bibliography. 3 tables. 63 illustrations in 15 figures. 48pp. 5¼ × 8½. 24407-5 Pa. $1.95

BLOOMINGDALE'S ILLUSTRATED 1886 CATALOG: Fashions, Dry Goods and Housewares, Bloomingdale Brothers. Famed merchants' extremely rare catalog depicting about 1,700 products: clothing, housewares, firearms, dry goods, jewelry, more. Invaluable for dating, identifying vintage items. Also, copyright-free graphics for artists, designers. Co-published with Henry Ford Museum & Greenfield Village. 160pp. 8¼ × 11. 25780-0 Pa. $9.95

HISTORIC COSTUME IN PICTURES, Braun & Schneider. Over 1,450 costumed figures in clearly detailed engravings—from dawn of civilization to end of 19th century. Captions. Many folk costumes. 256pp. 8⅜ × 11¾. 23150-X Pa. $11.95

STICKLEY CRAFTSMAN FURNITURE CATALOGS, Gustav Stickley and L. & J. G. Stickley. Beautiful, functional furniture in two authentic catalogs from 1910. 594 illustrations, including 277 photos, show settles, rockers, armchairs, reclining chairs, bookcases, desks, tables. 183pp. 6½ × 9¼. 23838-5 Pa. $8.95

AMERICAN LOCOMOTIVES IN HISTORIC PHOTOGRAPHS: 1858 to 1949, Ron Ziel (ed.). A rare collection of 126 meticulously detailed official photographs, called "builder portraits," of American locomotives that majestically chronicle the rise of steam locomotive power in America. Introduction. Detailed captions. xi + 129pp. 9 × 12. 27393-8 Pa. $12.95

AMERICA'S LIGHTHOUSES: An Illustrated History, Francis Ross Holland, Jr. Delightfully written, profusely illustrated fact-filled survey of over 200 American lighthouses since 1716. History, anecdotes, technological advances, more. 240pp. 8 × 10¾. 25576-X Pa. $11.95

TOWARDS A NEW ARCHITECTURE, Le Corbusier. Pioneering manifesto by founder of "International School." Technical and aesthetic theories, views of industry, economics, relation of form to function, "mass-production split" and much more. Profusely illustrated. 320pp. 6⅛ × 9¼. (USO) 25023-7 Pa. $8.95

HOW THE OTHER HALF LIVES, Jacob Riis. Famous journalistic record, exposing poverty and degradation of New York slums around 1900, by major social reformer. 100 striking and influential photographs. 233pp. 10 × 7⅞.
22012-5 Pa $10.95

FRUIT KEY AND TWIG KEY TO TREES AND SHRUBS, William M. Harlow. One of the handiest and most widely used identification aids. Fruit key covers 120 deciduous and evergreen species; twig key 160 deciduous species. Easily used. Over 300 photographs. 126pp. 5⅜ × 8½. 20511-8 Pa. $3.95

COMMON BIRD SONGS, Dr. Donald J. Borror. Songs of 60 most common U.S. birds: robins, sparrows, cardinals, bluejays, finches, more—arranged in order of increasing complexity. Up to 9 variations of songs of each species.
Cassette and manual 99911-4 $8.95

ORCHIDS AS HOUSE PLANTS, Rebecca Tyson Northen. Grow cattleyas and many other kinds of orchids—in a window, in a case, or under artificial light. 63 illustrations. 148pp. 5⅜ × 8½. 23261-1 Pa. $3.95

MONSTER MAZES, Dave Phillips. Masterful mazes at four levels of difficulty. Avoid deadly perils and evil creatures to find magical treasures. Solutions for all 32 exciting illustrated puzzles. 48pp. 8¼ × 11. 26005-4 Pa. $2.95

MOZART'S DON GIOVANNI (DOVER OPERA LIBRETTO SERIES), Wolfgang Amadeus Mozart. Introduced and translated by Ellen H. Bleiler. Standard Italian libretto, with complete English translation. Convenient and thoroughly portable—an ideal companion for reading along with a recording or the performance itself. Introduction. List of characters. Plot summary. 121pp. 5¼ × 8½.
24944-1 Pa. $2.95

TECHNICAL MANUAL AND DICTIONARY OF CLASSICAL BALLET, Gail Grant. Defines, explains, comments on steps, movements, poses and concepts. 15-page pictorial section. Basic book for student, viewer. 127pp. 5⅜ × 8½.
21843-0 Pa. $3.95

BRASS INSTRUMENTS: Their History and Development, Anthony Baines. Authoritative, updated survey of the evolution of trumpets, trombones, bugles, cornets, French horns, tubas and other brass wind instruments. Over 140 illustrations and 48 music examples. Corrected and updated by author. New preface. Bibliography. 320pp. 5⅜ × 8½.   27574-4 Pa. $9.95

HOLLYWOOD GLAMOR PORTRAITS, John Kobal (ed.). 145 photos from 1926–49. Harlow, Gable, Bogart, Bacall; 94 stars in all. Full background on photographers, technical aspects. 160pp. 8⅜ × 11¼.   23352-9 Pa. $11.95

MAX AND MORITZ, Wilhelm Busch. Great humor classic in both German and English. Also 10 other works: "Cat and Mouse," "Plisch and Plumm," etc. 216pp. 5⅜ × 8½.   20181-3 Pa. $5.95

THE RAVEN AND OTHER FAVORITE POEMS, Edgar Allan Poe. Over 40 of the author's most memorable poems: "The Bells," "Ulalume," "Israfel," "To Helen," "The Conqueror Worm," "Eldorado," "Annabel Lee," many more. Alphabetic lists of titles and first lines. 64pp. 5³⁄₁₆ × 8¼.   26685-0 Pa. $1.00

SEVEN SCIENCE FICTION NOVELS, H. G. Wells. The standard collection of the great novels. Complete, unabridged. First Men in the Moon, Island of Dr. Moreau, War of the Worlds, Food of the Gods, Invisible Man, Time Machine, In the Days of the Comet. Total of 1,015pp. 5⅜ × 8½. (USO)   20264-X Clothbd. $29.95

AMULETS AND SUPERSTITIONS, E. A. Wallis Budge. Comprehensive discourse on origin, powers of amulets in many ancient cultures: Arab, Persian, Babylonian, Assyrian, Egyptian, Gnostic, Hebrew, Phoenician, Syriac, etc. Covers cross, swastika, crucifix, seals, rings, stones, etc. 584pp. 5⅜ × 8½. 23573-4 Pa. $12.95

RUSSIAN STORIES/PYCCKNE PACCKA3bl: A Dual-Language Book, edited by Gleb Struve. Twelve tales by such masters as Chekhov, Tolstoy, Dostoevsky, Pushkin, others. Excellent word-for-word English translations on facing pages, plus teaching and study aids, Russian/English vocabulary, biographical/critical introductions, more. 416pp. 5⅜ × 8½.   26244-8 Pa. $8.95

PHILADELPHIA THEN AND NOW: 60 Sites Photographed in the Past and Present, Kenneth Finkel and Susan Oyama. Rare photographs of City Hall, Logan Square, Independence Hall, Betsy Ross House, other landmarks juxtaposed with contemporary views. Captures changing face of historic city. Introduction. Captions. 128pp. 8¼ × 11.   25790-8 Pa. $9.95

AIA ARCHITECTURAL GUIDE TO NASSAU AND SUFFOLK COUNTIES, LONG ISLAND, The American Institute of Architects, Long Island Chapter, and the Society for the Preservation of Long Island Antiquities. Comprehensive, well-researched and generously illustrated volume brings to life over three centuries of Long Island's great architectural heritage. More than 240 photographs with authoritative, extensively detailed captions. 176pp. 8¼ × 11.   26946-9 Pa. $14.95

NORTH AMERICAN INDIAN LIFE: Customs and Traditions of 23 Tribes, Elsie Clews Parsons (ed.). 27 fictionalized essays by noted anthropologists examine religion, customs, government, additional facets of life among the Winnebago, Crow, Zuni, Eskimo, other tribes. 480pp. 6⅛ × 9¼.   27377-6 Pa. $10.95

FRANK LLOYD WRIGHT'S HOLLYHOCK HOUSE, Donald Hoffmann. Lavishly illustrated, carefully documented study of one of Wright's most controversial residential designs. Over 120 photographs, floor plans, elevations, etc. Detailed perceptive text by noted Wright scholar. Index. 128pp. 9¼ × 10¾.
27133-1 Pa. $11.95

THE MALE AND FEMALE FIGURE IN MOTION: 60 Classic Photographic Sequences, Eadweard Muybridge. 60 true-action photographs of men and women walking, running, climbing, bending, turning, etc., reproduced from rare 19th-century masterpiece. vi + 121pp. 9 × 12. 24745-7 Pa. $10.95

1001 QUESTIONS ANSWERED ABOUT THE SEASHORE, N. J. Berrill and Jacquelyn Berrill. Queries answered about dolphins, sea snails, sponges, starfish, fishes, shore birds, many others. Covers appearance, breeding, growth, feeding, much more. 305pp. 5¼ × 8¼. 23366-9 Pa. $7.95

GUIDE TO OWL WATCHING IN NORTH AMERICA, Donald S. Heintzelman. Superb guide offers complete data and descriptions of 19 species: barn owl, screech owl, snowy owl, many more. Expert coverage of owl-watching equipment, conservation, migrations and invasions, etc. Guide to observing sites. 84 illustrations. xiii + 193pp. 5⅜ × 8½. 27344-X Pa. $7.95

MEDICINAL AND OTHER USES OF NORTH AMERICAN PLANTS: A Historical Survey with Special Reference to the Eastern Indian Tribes, Charlotte Erichsen-Brown. Chronological historical citations document 500 years of usage of plants, trees, shrubs native to eastern Canada, northeastern U.S. Also complete identifying information. 343 illustrations. 544pp. 6½ × 9¼. 25951-X Pa. $12.95

STORYBOOK MAZES, Dave Phillips. 23 stories and mazes on two-page spreads: Wizard of Oz, Treasure Island, Robin Hood, etc. Solutions. 64pp. 8¼ × 11.
23628-5 Pa. $2.95

NEGRO FOLK MUSIC, U.S.A., Harold Courlander. Noted folklorist's scholarly yet readable analysis of rich and varied musical tradition. Includes authentic versions of over 40 folk songs. Valuable bibliography and discography. xi + 324pp. 5⅜ × 8½. 27350-4 Pa. $7.95

MOVIE-STAR PORTRAITS OF THE FORTIES, John Kobal (ed.). 163 glamor, studio photos of 106 stars of the 1940s: Rita Hayworth, Ava Gardner, Marlon Brando, Clark Gable, many more. 176pp. 8⅜ × 11¼. 23546-7 Pa. $10.95

BENCHLEY LOST AND FOUND, Robert Benchley. Finest humor from early 30s, about pet peeves, child psychologists, post office and others. Mostly unavailable elsewhere. 73 illustrations by Peter Arno and others. 183pp. 5⅜ × 8½.
22410-4 Pa. $5.95

YEKL and THE IMPORTED BRIDEGROOM AND OTHER STORIES OF YIDDISH NEW YORK, Abraham Cahan. Film Hester Street based on Yekl (1896). Novel, other stories among first about Jewish immigrants on N.Y.'s East Side. 240pp. 5⅜ × 8½. 22427-9 Pa. $6.95

SELECTED POEMS, Walt Whitman. Generous sampling from *Leaves of Grass*. Twenty-four poems include "I Hear America Singing," "Song of the Open Road," "I Sing the Body Electric," "When Lilacs Last in the Dooryard Bloom'd," "O Captain! My Captain!"—all reprinted from an authoritative edition. Lists of titles and first lines. 128pp. 5³⁄₁₆ × 8¼. 26878-0 Pa. $1.00

THE BEST TALES OF HOFFMANN, E. T. A. Hoffmann. 10 of Hoffmann's most important stories: "Nutcracker and the King of Mice," "The Golden Flowerpot," etc. 458pp. 5⅜ × 8½. 21793-0 Pa. $8.95

FROM FETISH TO GOD IN ANCIENT EGYPT, E. A. Wallis Budge. Rich detailed survey of Egyptian conception of "God" and gods, magic, cult of animals, Osiris, more. Also, superb English translations of hymns and legends. 240 illustrations. 545pp. 5⅜ × 8½. 25803-3 Pa. $11.95

FRENCH STORIES/CONTES FRANÇAIS: A Dual-Language Book, Wallace Fowlie. Ten stories by French masters, Voltaire to Camus: "Micromegas" by Voltaire; "The Atheist's Mass" by Balzac; "Minuet" by de Maupassant; "The Guest" by Camus, six more. Excellent English translations on facing pages. Also French-English vocabulary list, exercises, more. 352pp. 5⅜ × 8½. 26443-2 Pa. $8.95

CHICAGO AT THE TURN OF THE CENTURY IN PHOTOGRAPHS: 122 Historic Views from the Collections of the Chicago Historical Society, Larry A. Viskochil. Rare large-format prints offer detailed views of City Hall, State Street, the Loop, Hull House, Union Station, many other landmarks, circa 1904–1913. Introduction. Captions. Maps. 144pp. 9⅜ × 12¼. 24656-6 Pa. $12.95

OLD BROOKLYN IN EARLY PHOTOGRAPHS, 1865–1929, William Lee Younger. Luna Park, Gravesend race track, construction of Grand Army Plaza, moving of Hotel Brighton, etc. 157 previously unpublished photographs. 165pp. 8⅞ × 11¾. 23587-4 Pa. $13.95

THE MYTHS OF THE NORTH AMERICAN INDIANS, Lewis Spence. Rich anthology of the myths and legends of the Algonquins, Iroquois, Pawnees and Sioux, prefaced by an extensive historical and ethnological commentary. 36 illustrations. 480pp. 5⅜ × 8½. 25967-6 Pa. $8.95

AN ENCYCLOPEDIA OF BATTLES: Accounts of Over 1,560 Battles from 1479 B.C. to the Present, David Eggenberger. Essential details of every major battle in recorded history from the first battle of Megiddo in 1479 B.C. to Grenada in 1984. List of Battle Maps. New Appendix covering the years 1967–1984. Index. 99 illustrations. 544pp. 6½ × 9¼. 24913-1 Pa. $14.95

SAILING ALONE AROUND THE WORLD, Captain Joshua Slocum. First man to sail around the world, alone, in small boat. One of great feats of seamanship told in delightful manner. 67 illustrations. 294pp. 5⅜ × 8½. 20326-3 Pa. $5.95

ANARCHISM AND OTHER ESSAYS, Emma Goldman. Powerful, penetrating, prophetic essays on direct action, role of minorities, prison reform, puritan hypocrisy, violence, etc. 271pp. 5⅜ × 8½. 22484-8 Pa. $5.95

MYTHS OF THE HINDUS AND BUDDHISTS, Ananda K. Coomaraswamy and Sister Nivedita. Great stories of the epics; deeds of Krishna, Shiva, taken from puranas, Vedas, folk tales; etc. 32 illustrations. 400pp. 5⅜ × 8½. 21759-0 Pa. $9.95

BEYOND PSYCHOLOGY, Otto Rank. Fear of death, desire of immortality, nature of sexuality, social organization, creativity, according to Rankian system. 291pp. 5⅜ × 8½. 20485-5 Pa. $7.95

A THEOLOGICO-POLITICAL TREATISE, Benedict Spinoza. Also contains unfinished Political Treatise. Great classic on religious liberty, theory of government on common consent. R. Elwes translation. Total of 421pp. 5⅜ × 8½. 20249-6 Pa. $8.95

MY BONDAGE AND MY FREEDOM, Frederick Douglass. Born a slave, Douglass became outspoken force in antislavery movement. The best of Douglass' autobiographies. Graphic description of slave life. 464pp. 5⅜ × 8½.   22457-0 Pa. $8.95

FOLLOWING THE EQUATOR: A Journey Around the World, Mark Twain. Fascinating humorous account of 1897 voyage to Hawaii, Australia, India, New Zealand, etc. Ironic, bemused reports on peoples, customs, climate, flora and fauna, politics, much more. 197 illustrations. 720pp. 5⅜ × 8½.   26113-1 Pa. $15.95

THE PEOPLE CALLED SHAKERS, Edward D. Andrews. Definitive study of Shakers: origins, beliefs, practices, dances, social organization, furniture and crafts, etc. 33 illustrations. 351pp. 5⅜ × 8½.   21081-2 Pa. $8.95

THE MYTHS OF GREECE AND ROME, H. A. Guerber. A classic of mythology, generously illustrated, long prized for its simple, graphic, accurate retelling of the principal myths of Greece and Rome, and for its commentary on their origins and significance. With 64 illustrations by Michelangelo, Raphael, Titian, Rubens, Canova, Bernini and others. 480pp. 5⅜ × 8½.   27584-1 Pa. $9.95

PSYCHOLOGY OF MUSIC, Carl E. Seashore. Classic work discusses music as a medium from psychological viewpoint. Clear treatment of physical acoustics, auditory apparatus, sound perception, development of musical skills, nature of musical feeling, host of other topics. 88 figures. 408pp. 5⅜ × 8½.   21851-1 Pa. $9.95

THE PHILOSOPHY OF HISTORY, Georg W. Hegel. Great classic of Western thought develops concept that history is not chance but rational process, the evolution of freedom. 457pp. 5⅜ × 8½.   20112-0 Pa. $9.95

THE BOOK OF TEA, Kakuzo Okakura. Minor classic of the Orient: entertaining, charming explanation, interpretation of traditional Japanese culture in terms of tea ceremony. 94pp. 5⅜ × 8½.   20070-1 Pa. $2.95

LIFE IN ANCIENT EGYPT, Adolf Erman. Fullest, most thorough, detailed older account with much not in more recent books, domestic life, religion, magic, medicine, commerce, much more. Many illustrations reproduce tomb paintings, carvings, hieroglyphs, etc. 597pp. 5⅜ × 8½.   22632-8 Pa. $10.95

SUNDIALS, Their Theory and Construction, Albert Waugh. Far and away the best, most thorough coverage of ideas, mathematics concerned, types, construction, adjusting anywhere. Simple, nontechnical treatment allows even children to build several of these dials. Over 100 illustrations. 230pp. 5⅜ × 8½.   22947-5 Pa. $7.95

DYNAMICS OF FLUIDS IN POROUS MEDIA, Jacob Bear. For advanced students of ground water hydrology, soil mechanics and physics, drainage and irrigation engineering, and more. 335 illustrations. Exercises, with answers. 784pp. 6⅛ × 9¼.   65675-6 Pa. $19.95

SONGS OF EXPERIENCE: Facsimile Reproduction with 26 Plates in Full Color, William Blake. 26 full-color plates from a rare 1826 edition. Includes "The Tyger," "London," "Holy Thursday," and other poems. Printed text of poems. 48pp. 5¼ × 7.   24636-1 Pa. $4.95

OLD-TIME VIGNETTES IN FULL COLOR, Carol Belanger Grafton (ed.). Over 390 charming, often sentimental illustrations, selected from archives of Victorian graphics—pretty women posing, children playing, food, flowers, kittens and puppies, smiling cherubs, birds and butterflies, much more. All copyright-free. 48pp. 9¼ × 12¼.   27269-9 Pa. $5.95

PERSPECTIVE FOR ARTISTS, Rex Vicat Cole. Depth, perspective of sky and sea, shadows, much more, not usually covered. 391 diagrams, 81 reproductions of drawings and paintings. 279pp. 5⅜ × 8½. 22487-2 Pa. $6.95

DRAWING THE LIVING FIGURE, Joseph Sheppard. Innovative approach to artistic anatomy focuses on specifics of surface anatomy, rather than muscles and bones. Over 170 drawings of live models in front, back and side views, and in widely varying poses. Accompanying diagrams. 177 illustrations. Introduction. Index. 144pp. 8⅜ × 11¼. 26723-7 Pa. $7.95

GOTHIC AND OLD ENGLISH ALPHABETS: 100 Complete Fonts, Dan X. Solo. Add power, elegance to posters, signs, other graphics with 100 stunning copyright-free alphabets: Blackstone, Dolbey, Germania, 97 more—including many lower-case, numerals, punctuation marks. 104pp. 8⅛ × 11. 24695-7 Pa. $7.95

HOW TO DO BEADWORK, Mary White. Fundamental book on craft from simple projects to five-bead chains and woven works. 106 illustrations. 142pp. 5⅜ × 8. 20697-1 Pa. $4.95

THE BOOK OF WOOD CARVING, Charles Marshall Sayers. Finest book for beginners discusses fundamentals and offers 34 designs. "Absolutely first rate . . . well thought out and well executed."—E. J. Tangerman. 118pp. 7¾ × 10⅝. 23654-4 Pa. $5.95

ILLUSTRATED CATALOG OF CIVIL WAR MILITARY GOODS: Union Army Weapons, Insignia, Uniform Accessories, and Other Equipment, Schuyler, Hartley, and Graham. Rare, profusely illustrated 1846 catalog includes Union Army uniform and dress regulations, arms and ammunition, coats, insignia, flags, swords, rifles, etc. 226 illustrations. 160pp. 9 × 12. 24939-5 Pa. $10.95

WOMEN'S FASHIONS OF THE EARLY 1900s: An Unabridged Republication of "New York Fashions, 1909," National Cloak & Suit Co. Rare catalog of mail-order fashions documents women's and children's clothing styles shortly after the turn of the century. Captions offer full descriptions, prices. Invaluable resource for fashion, costume historians. Approximately 725 illustrations. 128pp. 8⅜ × 11¼. 27276-1 Pa. $11.95

THE 1912 AND 1915 GUSTAV STICKLEY FURNITURE CATALOGS, Gustav Stickley. With over 200 detailed illustrations and descriptions, these two catalogs are essential reading and reference materials and identification guides for Stickley furniture. Captions cite materials, dimensions and prices. 112pp. 6½ × 9¼. 26676-1 Pa. $9.95

EARLY AMERICAN LOCOMOTIVES, John H. White, Jr. Finest locomotive engravings from early 19th century: historical (1804–74), main-line (after 1870), special, foreign, etc. 147 plates. 142pp. 11⅜ × 8¼. 22772-3 Pa. $8.95

THE TALL SHIPS OF TODAY IN PHOTOGRAPHS, Frank O. Braynard. Lavishly illustrated tribute to nearly 100 majestic contemporary sailing vessels: Amerigo Vespucci, Clearwater, Constitution, Eagle, Mayflower, Sea Cloud, Victory, many more. Authoritative captions provide statistics, background on each ship. 190 black-and-white photographs and illustrations. Introduction. 128pp. 8⅞ × 11¾. 27163-3 Pa. $13.95

EARLY NINETEENTH-CENTURY CRAFTS AND TRADES, Peter Stockham (ed.). Extremely rare 1807 volume describes to youngsters the crafts and trades of the day: brickmaker, weaver, dressmaker, bookbinder, ropemaker, saddler, many more. Quaint prose, charming illustrations for each craft. 20 black-and-white line illustrations. 192pp. 4⅝ × 6. 27293-1 Pa. $4.95

VICTORIAN FASHIONS AND COSTUMES FROM HARPER'S BAZAR, 1867–1898, Stella Blum (ed.). Day costumes, evening wear, sports clothes, shoes, hats, other accessories in over 1,000 detailed engravings. 320pp. 9⅜ × 12¼.
22990-4 Pa. $13.95

GUSTAV STICKLEY, THE CRAFTSMAN, Mary Ann Smith. Superb study surveys broad scope of Stickley's achievement, especially in architecture. Design philosophy, rise and fall of the Craftsman empire, descriptions and floor plans for many Craftsman houses, more. 86 black-and-white halftones. 31 line illustrations. Introduction. 208pp. 6½ × 9¼. 27210-9 Pa. $9.95

THE LONG ISLAND RAIL ROAD IN EARLY PHOTOGRAPHS, Ron Ziel. Over 220 rare photos, informative text document origin (1844) and development of rail service on Long Island. Vintage views of early trains, locomotives, stations, passengers, crews, much more. Captions. 8⅞ × 11¾. 26301-0 Pa. $13.95

THE BOOK OF OLD SHIPS: From Egyptian Galleys to Clipper Ships, Henry B. Culver. Superb, authoritative history of sailing vessels, with 80 magnificent line illustrations. Galley, bark, caravel, longship, whaler, many more. Detailed, informative text on each vessel by noted naval historian. Introduction. 256pp. 5⅜ × 8½. 27332-6 Pa. $6.95

TEN BOOKS ON ARCHITECTURE, Vitruvius. The most important book ever written on architecture. Early Roman aesthetics, technology, classical orders, site selection, all other aspects. Morgan translation. 331pp. 5⅜ × 8½. 20645-9 Pa. $8.95

THE HUMAN FIGURE IN MOTION, Eadweard Muybridge. More than 4,500 stopped-action photos, in action series, showing undraped men, women, children jumping, lying down, throwing, sitting, wrestling, carrying, etc. 390pp. 7⅞ × 10⅝. 20204-6 Clothbd. $24.95

TREES OF THE EASTERN AND CENTRAL UNITED STATES AND CANADA, William M. Harlow. Best one-volume guide to 140 trees. Full descriptions, woodlore, range, etc. Over 600 illustrations. Handy size. 288pp. 4½ × 6⅜.
20395-6 Pa. $5.95

SONGS OF WESTERN BIRDS, Dr. Donald J. Borror. Complete song and call repertoire of 60 western species, including flycatchers, juncoes, cactus wrens, many more—includes fully illustrated booklet. Cassette and manual 99913-0 $8.95

GROWING AND USING HERBS AND SPICES, Milo Miloradovich. Versatile handbook provides all the information needed for cultivation and use of all the herbs and spices available in North America. 4 illustrations. Index. Glossary. 236pp. 5⅜ × 8½. 25058-X Pa. $5.95

BIG BOOK OF MAZES AND LABYRINTHS, Walter Shepherd. 50 mazes and labyrinths in all—classical, solid, ripple, and more—in one great volume. Perfect inexpensive puzzler for clever youngsters. Full solutions. 112pp. 8⅛ × 11.
22951-3 Pa. $3.95

PIANO TUNING, J. Cree Fischer. Clearest, best book for beginner, amateur. Simple repairs, raising dropped notes, tuning by easy method of flattened fifths. No previous skills needed. 4 illustrations. 201pp. 5⅜ × 8½.   23267-0 Pa. $5.95

A SOURCE BOOK IN THEATRICAL HISTORY, A. M. Nagler. Contemporary observers on acting, directing, make-up, costuming, stage props, machinery, scene design, from Ancient Greece to Chekhov. 611pp. 5⅜ × 8½.   20515-0 Pa. $11.95

THE COMPLETE NONSENSE OF EDWARD LEAR, Edward Lear. All nonsense limericks, zany alphabets, Owl and Pussycat, songs, nonsense botany, etc., illustrated by Lear. Total of 320pp. 5⅜ × 8½. (USO)   20167-8 Pa. $6.95

VICTORIAN PARLOUR POETRY: An Annotated Anthology, Michael R. Turner. 117 gems by Longfellow, Tennyson, Browning, many lesser-known poets. "The Village Blacksmith," "Curfew Must Not Ring Tonight," "Only a Baby Small," dozens more, often difficult to find elsewhere. Index of poets, titles, first lines. xxiii + 325pp. 5⅜ × 8¼.   27044-0 Pa. $8.95

DUBLINERS, James Joyce. Fifteen stories offer vivid, tightly focused observations of the lives of Dublin's poorer classes. At least one, "The Dead," is considered a masterpiece. Reprinted complete and unabridged from standard edition. 160pp. 5³⁄₁₆ × 8¼.   26870-5 Pa. $1.00

THE HAUNTED MONASTERY and THE CHINESE MAZE MURDERS, Robert van Gulik. Two full novels by van Gulik, set in 7th-century China, continue adventures of Judge Dee and his companions. An evil Taoist monastery, seemingly supernatural events; overgrown topiary maze hides strange crimes. 27 illustrations. 328pp. 5⅜ × 8½.   23502-5 Pa. $7.95

THE BOOK OF THE SACRED MAGIC OF ABRAMELIN THE MAGE, translated by S. MacGregor Mathers. Medieval manuscript of ceremonial magic. Basic document in Aleister Crowley, Golden Dawn groups. 268pp. 5⅜ × 8½.
                   23211-5 Pa. $8.95

NEW RUSSIAN-ENGLISH AND ENGLISH-RUSSIAN DICTIONARY, M. A. O'Brien. This is a remarkably handy Russian dictionary, containing a surprising amount of information, including over 70,000 entries. 366pp. 4½ × 6⅛.
                   20208-9 Pa. $9.95

HISTORIC HOMES OF THE AMERICAN PRESIDENTS, Second, Revised Edition, Irvin Haas. A traveler's guide to American Presidential homes, most open to the public, depicting and describing homes occupied by every American President from George Washington to George Bush. With visiting hours, admission charges, travel routes. 175 photographs. Index. 160pp. 8¼ × 11. 26751-2 Pa. $10.95

NEW YORK IN THE FORTIES, Andreas Feininger. 162 brilliant photographs by the well-known photographer, formerly with *Life* magazine. Commuters, shoppers, Times Square at night, much else from city at its peak. Captions by John von Hartz. 181pp. 9¼ × 10¾.   23585-8 Pa. $12.95

INDIAN SIGN LANGUAGE, William Tomkins. Over 525 signs developed by Sioux and other tribes. Written instructions and diagrams. Also 290 pictographs. 111pp. 6⅛ × 9¼.   22029-X Pa. $3.50

ANATOMY: A Complete Guide for Artists, Joseph Sheppard. A master of figure drawing shows artists how to render human anatomy convincingly. Over 460 illustrations. 224pp. 8⅜ × 11¼. 27279-6 Pa. $9.95

MEDIEVAL CALLIGRAPHY: Its History and Technique, Marc Drogin. Spirited history, comprehensive instruction manual covers 13 styles (ca. 4th century thru 15th). Excellent photographs; directions for duplicating medieval techniques with modern tools. 224pp. 8⅜ × 11¼. 26142-5 Pa. $11.95

DRIED FLOWERS: How to Prepare Them, Sarah Whitlock and Martha Rankin. Complete instructions on how to use silica gel, meal and borax, perlite aggregate, sand and borax, glycerine and water to create attractive permanent flower arrangements. 12 illustrations. 32pp. 5⅜ × 8½. 21802-3 Pa. $1.00

EASY-TO-MAKE BIRD FEEDERS FOR WOODWORKERS, Scott D. Campbell. Detailed, simple-to-use guide for designing, constructing, caring for and using feeders. Text, illustrations for 12 classic and contemporary designs. 96pp. 5⅜ × 8½. 25847-5 Pa. $2.95

OLD-TIME CRAFTS AND TRADES, Peter Stockham. An 1807 book created to teach children about crafts and trades open to them as future careers. It describes in detailed, nontechnical terms 24 different occupations, among them coachmaker, gardener, hairdresser, lacemaker, shoemaker, wheelwright, copper-plate printer, milliner, trunkmaker, merchant and brewer. Finely detailed engravings illustrate each occupation. 192pp. 4⅝ × 6. 27398-9 Pa. $4.95

THE HISTORY OF UNDERCLOTHES, C. Willett Cunnington and Phyllis Cunnington. Fascinating, well-documented survey covering six centuries of English undergarments, enhanced with over 100 illustrations: 12th-century laced-up bodice, footed long drawers (1795), 19th-century bustles, 19th-century corsets for men, Victorian "bust improvers," much more. 272pp. 5⅜ × 8¼. 27124-2 Pa. $9.95

ARTS AND CRAFTS FURNITURE: The Complete Brooks Catalog of 1912, Brooks Manufacturing Co. Photos and detailed descriptions of more than 150 now very collectible furniture designs from the Arts and Crafts movement depict davenports, settees, buffets, desks, tables, chairs, bedsteads, dressers and more, all built of solid, quarter-sawed oak. Invaluable for students and enthusiasts of antiques, Americana and the decorative arts. 80pp. 6½ × 9¼. 27471-3 Pa. $7.95

HOW WE INVENTED THE AIRPLANE: An Illustrated History, Orville Wright. Fascinating firsthand account covers early experiments, construction of planes and motors, first flights, much more. Introduction and commentary by Fred C. Kelly. 76 photographs. 96pp. 8¼ × 11. 25662-6 Pa. $8.95

THE ARTS OF THE SAILOR: Knotting, Splicing and Ropework, Hervey Garrett Smith. Indispensable shipboard reference covers tools, basic knots and useful hitches; handsewing and canvas work, more. Over 100 illustrations. Delightful reading for sea lovers. 256pp. 5⅜ × 8½. 26440-8 Pa. $7.95

FRANK LLOYD WRIGHT'S FALLINGWATER: The House and Its History, Second, Revised Edition, Donald Hoffmann. A total revision—both in text and illustrations—of the standard document on Fallingwater, the boldest, most personal architectural statement of Wright's mature years, updated with valuable new material from the recently opened Frank Lloyd Wright Archives. "Fascinating"—*The New York Times*. 116 illustrations. 128pp. 9¼ × 10¾. 27430-6 Pa. $10.95

PHOTOGRAPHIC SKETCHBOOK OF THE CIVIL WAR, Alexander Gardner. 100 photos taken on field during the Civil War. Famous shots of Manassas, Harper's Ferry, Lincoln, Richmond, slave pens, etc. 244pp. 10⅝ × 8¼.
22731-6 Pa. $9.95

FIVE ACRES AND INDEPENDENCE, Maurice G. Kains. Great back-to-the-land classic explains basics of self-sufficient farming. The one book to get. 95 illustrations. 397pp. 5⅜ × 8½. 20974-1 Pa. $7.95

SONGS OF EASTERN BIRDS, Dr. Donald J. Borror. Songs and calls of 60 species most common to eastern U.S.: warblers, woodpeckers, flycatchers, thrushes, larks, many more in high-quality recording. Cassette and manual 99912-2 $8.95

A MODERN HERBAL, Margaret Grieve. Much the fullest, most exact, most useful compilation of herbal material. Gigantic alphabetical encyclopedia, from aconite to zedoary, gives botanical information, medical properties, folklore, economic uses, much else. Indispensable to serious reader. 161 illustrations. 888pp. 6½ × 9¼. 2-vol. set. (USO) Vol. I: 22798-7 Pa. $9.95
Vol. II: 22799-5 Pa. $9.95

HIDDEN TREASURE MAZE BOOK, Dave Phillips. Solve 34 challenging mazes accompanied by heroic tales of adventure. Evil dragons, people-eating plants, bloodthirsty giants, many more dangerous adversaries lurk at every twist and turn. 34 mazes, stories, solutions. 48pp. 8¼ × 11. 24566-7 Pa. $2.95

LETTERS OF W. A. MOZART, Wolfgang A. Mozart. Remarkable letters show bawdy wit, humor, imagination, musical insights, contemporary musical world; includes some letters from Leopold Mozart. 276pp. 5⅜ × 8½. 22859-2 Pa. $6.95

BASIC PRINCIPLES OF CLASSICAL BALLET, Agrippina Vaganova. Great Russian theoretician, teacher explains methods for teaching classical ballet. 118 illustrations. 175pp. 5⅜ × 8½. 22036-2 Pa. $4.95

THE JUMPING FROG, Mark Twain. Revenge edition. The original story of The Celebrated Jumping Frog of Calaveras County, a hapless French translation, and Twain's hilarious "retranslation" from the French. 12 illustrations. 66pp. 5⅜ × 8½. 22686-7 Pa. $3.95

BEST REMEMBERED POEMS, Martin Gardner (ed.). The 126 poems in this superb collection of 19th- and 20th-century British and American verse range from Shelley's "To a Skylark" to the impassioned "Renascence" of Edna St. Vincent Millay and to Edward Lear's whimsical "The Owl and the Pussycat." 224pp. 5⅜ × 8½. 27165-X Pa. $4.95

COMPLETE SONNETS, William Shakespeare. Over 150 exquisite poems deal with love, friendship, the tyranny of time, beauty's evanescence, death and other themes in language of remarkable power, precision and beauty. Glossary of archaic terms. 80pp. 5³⁄₁₆ × 8¼. 26686-9 Pa. $1.00

BODIES IN A BOOKSHOP, R. T. Campbell. Challenging mystery of blackmail and murder with ingenious plot and superbly drawn characters. In the best tradition of British suspense fiction. 192pp. 5⅜ × 8½. 24720-1 Pa. $5.95

THE WIT AND HUMOR OF OSCAR WILDE, Alvin Redman (ed.). More than 1,000 ripostes, paradoxes, wisecracks: Work is the curse of the drinking classes; I can resist everything except temptation; etc. 258pp. 5⅜ × 8½. 20602-5 Pa. $5.95

SHAKESPEARE LEXICON AND QUOTATION DICTIONARY, Alexander Schmidt. Full definitions, locations, shades of meaning in every word in plays and poems. More than 50,000 exact quotations. 1,485pp. 6½ × 9¼. 2-vol. set.
Vol. 1: 22726-X Pa. $15.95
Vol. 2: 22727-8 Pa. $15.95

SELECTED POEMS, Emily Dickinson. Over 100 best-known, best-loved poems by one of America's foremost poets, reprinted from authoritative early editions. No comparable edition at this price. Index of first lines. 64pp. 5³⁄₁₆ × 8¼.
26466-1 Pa. $1.00

CELEBRATED CASES OF JUDGE DEE (DEE GOONG AN), translated by Robert van Gulik. Authentic 18th-century Chinese detective novel; Dee and associates solve three interlocked cases. Led to van Gulik's own stories with same characters. Extensive introduction. 9 illustrations. 237pp. 5⅜ × 8½.
23337-5 Pa. $6.95

THE MALLEUS MALEFICARUM OF KRAMER AND SPRENGER, translated by Montague Summers. Full text of most important witchhunter's "bible," used by both Catholics and Protestants. 278pp. 6⅝ × 10. 22802-9 Pa. $10.95

SPANISH STORIES/CUENTOS ESPAÑOLES: A Dual-Language Book, Angel Flores (ed.). Unique format offers 13 great stories in Spanish by Cervantes, Borges, others. Faithful English translations on facing pages. 352pp. 5⅜ × 8½.
25399-6 Pa. $8.95

THE CHICAGO WORLD'S FAIR OF 1893: A Photographic Record, Stanley Appelbaum (ed.). 128 rare photos show 200 buildings, Beaux-Arts architecture, Midway, original Ferris Wheel, Edison's kinetoscope, more. Architectural emphasis; full text. 116pp. 8¼ × 11. 23990-X Pa. $9.95

OLD QUEENS, N.Y., IN EARLY PHOTOGRAPHS, Vincent F. Seyfried and William Asadorian. Over 160 rare photographs of Maspeth, Jamaica, Jackson Heights, and other areas. Vintage views of DeWitt Clinton mansion, 1939 World's Fair and more. Captions. 192pp. 8⅞ × 11. 26358-4 Pa. $12.95

CAPTURED BY THE INDIANS: 15 Firsthand Accounts, 1750–1870, Frederick Drimmer. Astounding true historical accounts of grisly torture, bloody conflicts, relentless pursuits, miraculous escapes and more, by people who lived to tell the tale. 384pp. 5⅜ × 8½. 24901-8 Pa. $8.95

THE WORLD'S GREAT SPEECHES, Lewis Copeland and Lawrence W. Lamm (eds.). Vast collection of 278 speeches of Greeks to 1970. Powerful and effective models; unique look at history. 842pp. 5⅜ × 8½. 20468-5 Pa. $13.95

THE BOOK OF THE SWORD, Sir Richard F. Burton. Great Victorian scholar/adventurer's eloquent, erudite history of the "queen of weapons"—from prehistory to early Roman Empire. Evolution and development of early swords, variations (sabre, broadsword, cutlass, scimitar, etc.), much more. 336pp. 6⅛ × 9¼. 25434-8 Pa. $8.95

AUTOBIOGRAPHY: The Story of My Experiments with Truth, Mohandas K. Gandhi. Boyhood, legal studies, purification, the growth of the Satyagraha (nonviolent protest) movement. Critical, inspiring work of the man responsible for the freedom of India. 480pp. 5⅜ × 8½. (USO)   24593-4 Pa. $7.95

CELTIC MYTHS AND LEGENDS, T. W. Rolleston. Masterful retelling of Irish and Welsh stories and tales. Cuchulain, King Arthur, Deirdre, the Grail, many more. First paperback edition. 58 full-page illustrations. 512pp. 5⅜ × 8½.
26507-2 Pa. $9.95

THE PRINCIPLES OF PSYCHOLOGY, William James. Famous long course complete, unabridged. Stream of thought, time perception, memory, experimental methods; great work decades ahead of its time. 94 figures. 1,391pp. 5⅜ × 8½. 2-vol. set.
Vol. I: 20381-6 Pa. $12.95
Vol. II: 20382-4 Pa. $12.95

THE WORLD AS WILL AND REPRESENTATION, Arthur Schopenhauer. Definitive English translation of Schopenhauer's life work, correcting more than 1,000 errors, omissions in earlier translations. Translated by E. F. J. Payne. Total of 1,269pp. 5⅜ × 8½. 2-vol. set.   Vol. 1: 21761-2 Pa. $11.95
Vol. 2: 21762-0 Pa. $11.95

MAGIC AND MYSTERY IN TIBET, Madame Alexandra David-Neel. Experiences among lamas, magicians, sages, sorcerers, Bonpa wizards. A true psychic discovery. 32 illustrations. 321pp. 5⅜ × 8½. (USO)   22682-4 Pa. $8.95

THE EGYPTIAN BOOK OF THE DEAD, E. A. Wallis Budge. Complete reproduction of Ani's papyrus, finest ever found. Full hieroglyphic text, interlinear transliteration, word-for-word translation, smooth translation. 533pp. 6½ × 9¼.
21866-X Pa. $9.95

MATHEMATICS FOR THE NONMATHEMATICIAN, Morris Kline. Detailed, college-level treatment of mathematics in cultural and historical context, with numerous exercises. Recommended Reading Lists. Tables. Numerous figures. 641pp. 5⅜ × 8½.   24823-2 Pa. $11.95

THEORY OF WING SECTIONS: Including a Summary of Airfoil Data, Ira H. Abbott and A. E. von Doenhoff. Concise compilation of subsonic aerodynamic characteristics of NACA wing sections, plus description of theory. 350pp. of tables. 693pp. 5⅜ × 8½.   60586-8 Pa. $13.95

THE RIME OF THE ANCIENT MARINER, Gustave Doré, S. T. Coleridge. Doré's finest work; 34 plates capture moods, subtleties of poem. Flawless full-size reproductions printed on facing pages with authoritative text of poem. "Beautiful. Simply beautiful."—*Publisher's Weekly.* 77pp. 9¼ × 12.   22305-1 Pa. $5.95

NORTH AMERICAN INDIAN DESIGNS FOR ARTISTS AND CRAFTS-PEOPLE, Eva Wilson. Over 360 authentic copyright-free designs adapted from Navajo blankets, Hopi pottery, Sioux buffalo hides, more. Geometrics, symbolic figures, plant and animal motifs, etc. 128pp. 8⅜ × 11. (EUK)   25341-4 Pa. $7.95

SCULPTURE: Principles and Practice, Louis Slobodkin. Step-by-step approach to clay, plaster, metals, stone; classical and modern. 253 drawings, photos. 255pp. 8⅜ × 11.   22960-2 Pa. $10.95

CATALOG OF DOVER BOOKS

THE INFLUENCE OF SEA POWER UPON HISTORY, 1660–1783, A. T. Mahan. Influential classic of naval history and tactics still used as text in war colleges. First paperback edition. 4 maps. 24 battle plans. 640pp. 5⅜ × 8½.
25509-3 Pa. $12.95

THE STORY OF THE TITANIC AS TOLD BY ITS SURVIVORS, Jack Winocour (ed.). What it was really like. Panic, despair, shocking inefficiency, and a little heroism. More thrilling than any fictional account. 26 illustrations. 320pp. 5⅜ × 8½.
20610-6 Pa. $7.95

FAIRY AND FOLK TALES OF THE IRISH PEASANTRY, William Butler Yeats (ed.). Treasury of 64 tales from the twilight world of Celtic myth and legend: "The Soul Cages," "The Kildare Pooka," "King O'Toole and his Goose," many more. Introduction and Notes by W. B. Yeats. 352pp. 5⅜ × 8½.
26941-8 Pa. $8.95

BUDDHIST MAHAYANA TEXTS, E. B. Cowell and Others (eds.). Superb, accurate translations of basic documents in Mahayana Buddhism, highly important in history of religions. The Buddha-karita of Asvaghosha, Larger Sukhavativyuha, more. 448pp. 5⅜ × 8½. ,
25552-2 Pa. $9.95

ONE TWO THREE . . . INFINITY: Facts and Speculations of Science, George Gamow. Great physicist's fascinating, readable overview of contemporary science: number theory, relativity, fourth dimension, entropy, genes, atomic structure, much more. 128 illustrations. Index. 352pp. 5⅜ × 8½.
25664-2 Pa. $8.95

ENGINEERING IN HISTORY, Richard Shelton Kirby, et al. Broad, nontechnical survey of history's major technological advances: birth of Greek science, industrial revolution, electricity and applied science, 20th-century automation, much more. 181 illustrations. ". . . excellent . . ."—Isis. Bibliography. vii + 530pp. 5⅜ × 8¼.
26412-2 Pa. $14.95